Sturla Henriksen navigates the profound interplay between the vast, life-sustaining ocean and human civilization, unveiling its dual role as both a nurturing force and a formidable challenge. Sturla delves into the ocean's impact on history, climate, and geopolitics while blending scientific rigor with personal insights. His book, *The Ocean*, is at once a thematic journey as it explores how the ocean has shaped societies, driven global power dynamics, and a major influencer of our climate and future. Through this book, Sturla offers an ultimate reflection on how we can harmonize with this vital force to ensure a sustainable future.

Dr Kilparti Ramakrishna, Director of Marine Policy Center and Senior Advisor to the President on Ocean and Climate Policy, Woods Hole Oceanographic Institution, and a lead author of the Fifth Assessment by the Intergovernmental Panel on Climate Change (IPCC)

With broad pen strokes Sturla Henriksen shares his love and insights for our oceans. Sprinkled with enlightening detail he walks us through the crucial role oceans play in many of mankind's key challenges, but also how it provides us with solutions and hope. Sturla has a unique background as a macro economist, for many years he was a very appreciated CEO of the Norwegian Shipowner's Association and lately he is the UN special advisor for Oceans. Understanding the mechanisms of the market economy, shipping industry and current political environment, he does not shy away from contextualizing the issues at hand from an existential and moral perspective, which makes this book so relevant and important.

Carl-Johan Hagman, President and CEO, NYK Group Europe

One cannot draft, negotiate, interpret, or implement law and policy in a vacuum. It is imperative to understand the context. If you have ever wondered, 'Why is the Ocean…' from scientific, historical, geopolitical, societal, legal, economic, and cultural perspectives, this is the book you want to read. Henriksen also reveals the unprecedented degradation of the Ocean and explains why a healthy Ocean is the solution to many of our problems. This book is not just informative; it's an immersive experience, as if the author is sitting next to us, answering our questions with fascinating facts, personal experiences, and insights.

Hiroko Muraki Gottlieb, Representative for the Ocean and Head of Delegation to the High Seas Treaty negotiations, International Council of Environmental Law, and Associate, Department for Organismic and Evolutionary Biology, Harvard University

A prominent corporate leader and a UN expert in ocean affairs, Sturla Henriksen has been an insider or a direct witness of the many dramatic stories that he so elegantly describes in his book. Ocean health, governance and economics, painful relations with homo sapiens – the author presents these complicated issues to the reader in an informed and emotional manner. Gradually but systematically, individual stories coalesce to a grand vision, not only about the ocean, but also about us, humans – our destiny, moral, and future. Through Sturla Henriksen's literary talent the ocean teaches us that we need to be more humane and considerably less selfish to secure our common future.

Vladimir Ryabinin, Executive Secretary of the Intergovernmental Oceanographic Commission of UNESCO (IOC/UNESCO) and Assistant Director-General of UNESCO 2015–2024

This book touches our blue planet with a cherishing pen, guiding the path toward international cooperation in good ocean governance, even as the oceans hold a troubled past, where disputes have left a trail of sorrow. Earth remains the only planet humanity can inhabit so far, and this book illustrates approaches to protect our blue planet, helping to build confidence in a sustainable future.
> Dr Yang Jian, Senior Fellow and former Executive Vice-Chairman, Shanghai Institutes for International Studies (SIIS)

Unlocking the mysteries, the history, and current opportunities the ocean offers in one publication demands experienced and insightful eyes. Sturla is an inspiring leader who understands the past, provides a sharp analysis of the present, and has a bold vision for the future. This book offers a profound exploration of our delicate relationship with the ocean and what more we can expect from it. Take a deep breath and dive in!
> Ignace Beguin Billecocq, Ocean and Coastal Zones Lead, Climate Champions, United Nations Framework Convention on Climate Change (UNFCCC)

Sturla Henriksen has written an incredibly thoughtful look at the changing ocean and our changing relationship to it in the 21st Century. He brings a practitioner's voice of experience, a policy maker's critical eye, and a humanist's sense of self to reflections on the new ocean security landscape. His discussion of evolving maritime security threats – and the danger they pose to global peace and stability – should be assigned reading for politicians of all stripes.
> Whitley Saumweber, Director, Stephenson Ocean Security Project, Center for Strategic and International Studies (CSIS), and President Barack Obama's Associate Director for Ocean and Coastal Policy in the White House Council on Environmental Quality

I know Sturla from his role as ocean advisor to the UN. I have always regarded him as a kind, gentle man, full of energy and passion, and I have admired his drive and ability to make a difference. In this 400 page love letter to the ocean, it is easy to understand where his energy comes from. Full of excitements, fears, joys, fascination, frustrations, wonder, awe, darkness, hope, tears, and facts. This book will help you see what Sturla sees. It will give you space to feel, and it will provide you with new knowledge and the necessary overview effect. Importantly, it will give you energy to roll up your sleeves and join him, as it urges you to engage in the hard work required to make this world a better place.
> Henrik Österblom, Science Director at the Stockholm Resilience Centre and Professor at Stockholm University and the Royal Swedish Academy of Sciences

Sturla Henriksen is familiar with the ocean through his extensive career and as a UN expert, but most of all, as an ocean lover. What an experience! So, writing a book on the ocean is unsurprising for another ocean lover like me. What is surprising, though, is his decision to follow a totally inclusive and holistic approach. What an ambition! Our planet is blue, and the ocean is immense! As my Greek ancestors used to say, Ocean was a Titan, the son of the marriage between the Earth and the sky, controlling the game between them! The Ocean still controls the game on our planet, and the book proves it. Sturla discusses the whole blue economy and its sectors, the new aspects of sustainability, the green dilemmas, and the geopolitical challenges in the brave new world. And I am satisfied that his eyes look towards the future. After all, the 21st century will be the century of the oceans! In a nutshell, the book gives a complex picture of the relationship between the sea and humanity through the ages. Nothing less than that!

Maria Damanaki, Global Managing Director, Oceans, for The Nature Conservancy, and European Commissioner for Maritime Affairs and Fisheries 2010–2014

The dangers of viewing the world through a western lens are clearly set out in the book and provide a powerful reminder to those of us who seek the abolition of modern slavery. It is vital that we understand the context of colonisation and exploitation – and our profound contribution to these injustices – if we are to end the cruelty of human trafficking and forced labour.

Dame Sara Thornton, Professor of Practice in Modern Slavery Policy, University of Nottingham, and Independent Anti-Slavery Commissioner for the United Kingdom 2019–2022

Sturla Henriksen's book nails it by pointing to the pressing dilemma we face: balancing the need to exploit the ocean's resources for economic growth and green technology with the imperative to protect fragile marine ecosystems. By diving into this complex issue, Henriksen kickstarts the crucial discussion needed to find sustainable solutions, urging us to rethink how we manage and value our greatest shared resource – the ocean. A must-read for anyone interested in regulation, governance, environment, and climate, this book sets the stage for meaningful exchange on how we ensure a sustainable use of our oceans.

Andreas Nordseth, Director-General, Danish Maritime Authority, and Chair, European Maritime Safety Agency 2017–2023

This book is a masterful review of the environmental, geopolitical and economic forces shaped by the world's oceans – past, present and future. Henriksen's view is both monumental and personal, capturing these broad trends while providing an insider's view of the meetings and settings in which global ocean governance is contested and forged. This highly readable and engaging volume will be of interest to scholars, policymakers, and concerned citizens seeking to understand the politics of our liquid planet and wishing to make a contribution towards protecting it.

Elana Wilson Rowe, Research Professor, Norwegian Institute of International Affairs (NUPI), and Associate Professor at the Norwegian University of the Life Sciences

Henriksen succinctly describes the historic role and impact of the ocean economy as we know it today. The ambitious and wide ranging work helps us understand that, in spite of the existential challenges we face, we are still able to materially address climate change through achievable actions in the ocean economy. The book shows us, for example, how technology ready solutions such as the transition to new ship design operating models and new fuels can materially reduce the world's carbon emissions.

Thomas Thune Andersen, Chair of Lloyd's Register Group (LR), and Chair of Orsted 2014–2024

Sturla shows why preserving the oceans is so important to the future of our planet – and how this effort will be fundamentally determined by human cooperation. Preserving our natural environment and ensuring a just transition for workers clearly need to go hand in hand.
> *Stephen Cotton, General Secretary, International Transport Workers' Federation (ITF)*

A thought-provoking, well-written and deeply considered book which goes to the heart of our oceans, blue economy and climate change. It does not just highlight the issues but proposes fundamental solutions to mitigate the rapidly changing negative environmental damage being wrought. A must-read.
> *Guy Platten, Secretary General, International Chamber of Shipping (ICS)*

I read Sturla's book with interest and found how it wove personal accounts and storytelling with facts very compelling. The prose is very easily digestible so I think the book will accomplish its goal of improving ocean literacy! Compellingly, the book is highlighting the food-energy-water nexus and how sustainability needs to consider many aspects beyond decarbonisation. Indeed, water treatment remains the single largest use of energy and energy conversion the single largest use of water. This book consistently points out the crux of conservation and preservation, both with regard to water and energy.
> *Lynn Loo, Professor, Princeton University, USA, and CEO, Global Center for Maritime Decarbonization, Singapore*

The geopolitics of the oceans and its resources will define relationships between countries and communities in the 21st century – as it has throughout history. Sturla Henriksen's book *The Ocean* should be required reading for all who wants to be effective policymakers, diplomats, military commanders, natural resource managers, and civil society professionals today and in the decades to come.
> *Johan Bergenas, Senior Vice President Oceans, World Wildlife Fund (WWF) US*

The ocean connects us, it feeds us and it promises both opportunities and risks. Sturla Henriksen gives us important perspectives on the global commons. Read, reflect and digest these perspectives in his great book.
> *General Eirik Kristoffersen, Chief of Defense, Norway*

A book as big as the ocean. Each chapter is full of ideas, humor and poetry, and shaped by a lifetime of thinking and caring about the ocean. It is a beautiful and poetic book filled with big ideas and drawn from a lifetime of lived experience. The book is also a powerful medicine in a world of distractions and escapism. Sturla's enthusiasm for the ocean, for rolling up his sleeves and getting to work, is contagious. A potent reminder that more is possible.
> *Robert Blasiak, Associate Professor, Stockholm Resilience Center*

In addition to an extensive and quite unique knowledge related to the ocean, Sturla demonstrates in this book some amazing skills for storytelling. This reads like a novel with a lot of excitement. A must read for anyone who wants to understand why we should finally care about our Ocean, the core engine for life on our planet!
> *Vincent Doumeizel, author of 'The Seaweed Revolution – How Seaweed Has Shaped Our Past and Can Save Our Future'*

Immerse yourself in the captivating world of maritime affairs with *The Ocean* by renowned maritime expert Sturla Henriksen. This compelling book delves into the longstanding significance of oceans in global power dynamics and explores how contemporary geopolitical shifts are redefining the roles oceans and maritime issues play. Henriksen offers readers a rich and profound analysis of emerging challenges and opportunities in ocean governance, providing valuable insights into the future. *The Ocean* is essential reading for policymakers, scholars, and anyone interested in the evolving landscape of international relations. This book is a timely and enlightening guide to understanding the critical role oceans will play in shaping our world.

Ulf Sverdrup, professor at the Norwegian Business School (BI), and Director, Norwegian Institute of International Affairs (NUPI) 2011–2023

This captivating and very readable book offers a treasure trove of information in describing the critical importance of the world's oceans. Sturla Henriksen has for many years been an eloquent and strong voice in respect of the oceans, and his knowledge and professionalism is exceeded only by his boundless passion for the subject. He tries, and succeeds, in writing what is an even-handed analysis of a massive and highly complex subject. I can highly recommend this excellent book.

Esben Poulsson, Immediate past Chairman, The International Chamber of Shipping (ICS)

This eminent book is neither a thriller nor an action story, but when you start reading it cannot be put down. It is well written, flows easily and the anecdotes and topics really get under your skin. Sturla Henriksen touches on a wide range of interesting topics and in some he takes us really deep down, posing extremely challenging and often unwelcome questions. Questions to which the answers we find will shape our future. It is a remarkable and timely book, and I wish a book like this had been in the curriculum of my schools. I would then have been much more knowledgeable and better informed to have views on all the difficult game changing questions and solutions ahead. But I do take comfort in the fact that it will be available for our young generations. A great gift book.

Thor Jørgen Guttormsen, President of the Norwegian Shipowners' Association 2010–2012, and former CEO of Höegh Autoliners and Höegh LNG

I often wonder about the vast oceans – the sight, power and fury of crashing waves, the calm at daybreak or dusk, and life beneath offering a myriad surprises. A gift for the ages. And yet, too often we have taken this beauty for granted resulting in the consequences we see and feel today. There is only one Earth. Humanity must fully reflect on the incredible privilege and responsibility to enjoy this beauty that surrounds us. All this can disappear… what was handed to us will not be what we hand over to the generations that follow.

The Ocean is a wonderful series of interesting vignettes that Sturla Henriksen has woven into a compelling narrative to make us pause, think and consider. A must-read for all ages if we are to confront the realities of the darker side of individual and collective aspirations gone wrong.

I have been moved…

Gerardo A. Borromeo, Chief Executive Officer, PTC Holdings Corporation

Sturla Henriksen brings unparalleled experience and exceptional insight to bear in this holistic story of how the ocean has formed our world and will shape our destiny. The book draws the lines between the ocean as biosphere and arena for economic activities as well as describing the ocean's role in today's geopolitical rivalry. The analysis of the ocean's role in the current geopolitical rivalry intuitively connects with data available from multiple sources rarely put in context. Compelling read!
> *Svein Ringbakken, CEO, The Norwegian Shipowners' Mutual War Risks Insurance Association (DNK)*

Sturla Henriksen has produced a remarkable study of the state of our oceans. Partly personal reflection, partly academic analysis, the book speaks to the heart as well as the brain. Recommended reading!
> *Alf Håkon Hoel, Professor, UiT The Arctic University of Norway*

As group CEO of the Wilhelmsen group, a global maritime industry group, and a representative of a family that has lived with and off the ocean for more than 160 years, I appreciate the insights and perspectives shared in this book. It highlights our deep connection to the ocean, and the importance and vast opportunities the ocean represents when managed responsibly for future generations.
> *Thomas Wilhelmsen, Group CEO and fifth-generation shareholder of the Wilhelmsen Group*

The Oceans are our common responsibility and taking the maritime industry safely into the next century requires collective action. Fortunately, we see many strong players both on industry and policy side acting and shaping the future of shipping. The coming years will be crucial to meet key targets towards zero in 2050 and a book like this is a valued reminder of exactly what is at stake.
> *Bo Cerup-Simonsen, CEO, Mærsk Mc-Kinney Møller Center for Zero Carbon Shipping*

The book gives a unique insight in the wonders of the Ocean and why we should care about it.
> *Jan-Gunnar Winther, Pro-rector research and development, UiT The Arctic University of Norway, and Director, the Norwegian Polar Institute 2005–2017*

The ocean is the lifeblood of our planet. An immense treasure of resources and life, an arena for international trade and cooperation – and a battleground for geopolitical strife. Sturla Henriksen's book provides an excellent overview of the ocean's profound influence on our past and present, highlighting also how the maritime domain holds the key to some of our most pressing challenges ahead. A must-read for anyone interested in a sustainable ocean economy.
> *Rolf Thore Roppestad, CEO, Gard*

It has taken far too long for today's world to realize that without the ocean there would be no life on earth. We originate from the ocean, we depend on it and global leaders will have to urgently create a new sustainable alliance with the ocean. There could be no better guide for this journey than Sturla Henriksen.
> *Lise Kingo, Independent Board Director, and CEO and Executive Director of the UN Global Compact 2015–2020*

TO CHRISTINE, CECILIE, SILJA, MATHILDE, THEODOR AND
FERDINAND, WITH LOVE, AND IN HOPES THAT YOUR
GENERATION WILL TAKE RESPONSIBILITY
WHERE MY OWN HAS FAILED

THE OCEAN

HOW IT HAS FORMED OUR WORLD – AND WILL SHAPE OUR DESTINY

Sturla Henriksen

*Translated
by Diane Oatley*

HERO, AN IMPRINT OF LEGEND TIMES GROUP LTD
51 Gower Street
London WC1E 6HJ
United Kingdom
www.hero-press.com

This edition first published by Hero in 2025

© Sturla Henriksen, 2025
Translation © Diane Oatley, 2025

The right of the above author to be identified as the author of this work has been asserted in accordance with the Copyright, Designs and Patents Act 1988. British Library Cataloguing in Publication Data available.

Printed by Akcent Media, 5 The Quay, St Ives, Cambs, PE27 5AR

ISBN: 978-1-91716-396-5

All the pictures in this volume are reprinted with permission or presumed to be in the public domain. Every effort has been made to ascertain and acknowledge their copyright status, but any error or oversight will be rectified in subsequent printings.

All rights reserved. No part of this publication may be reproduced, stored in or introduced into a retrieval system, or transmitted, in any form or by any means (electronic, mechanical, photocopying, recording or otherwise), without the prior written permission of the publisher. This book is sold subject to the condition that it shall not be resold, lent, hired out or otherwise circulated without the express prior consent of the publisher.

Contents

PROLOGUE		17
1. ALAN'S TEARS		25
2. POWERFUL, DARK AND MYSTERIOUS		31
Jaws		31
Vast, Deep and Solitary		33
A Glass Bubble		35
The Eccentric Liquid		36
Thicker than Blood		39
Simple Arithmetic		40
A Few Fun Facts		42
Hertz 52		42
Happy Not to Be Born a Flounder		44
Only a Fish?		45
Dark Oxygen		47
The World's Worst Sex		49
Bell Jars and Bicycle Chains		51
Horse Latitudes		53
Maelstroms		54
A Sea within the Ocean		55
Bird Dung Wars		56
3. POWER AT SEA		58
The West and the Rest		58
The Story's Omissions		60
Origo's Power of Definition		63
Prone to Ornamentation, Insipidness and Brutality		65
The Sailing Eunuch		67
Fateful Decisions		68
Ciao!		69
Benches and Banks		71
Few, Dirty and Base		72
Banished to the Shadows		74
An Affable Playboy		75
The Chess Queen		76
Ecological Reunion		77
Pandemic 1.0		78
Forced Labour		79
Lost Cargo		80

Treasure, Silver and Silk	82
Cash, Not Colonies	82
Mare Liberum	84
A Stock Exchange under the Stars	85
Edward's Little Café	85
The King's Daughters	87
Ten Pounds for a City – and 200,000 for a Country	88
Drunk and Disorderly	89
Seward's Folly	91
World War Zero	92
Pirates of the Caribbean	93
A Paris Agreement	94
Tea Party	96
Hannah	97
A Gesture of Gratitude	98
Waves of Liberation	98
Cash for Colonies	99
A Classic Army Commander	100
Equality, Fraternity and Liberty	101

4. **URBANISATION, COASTALISATION AND GLOBALISATION** — 102

iPhone	102
Waterways	103
Steaming Ahead	104
The Climate Fuse Is Lit	105
A Global Shortcut	106
Time Chaos	107
Whitehouse Troubles	108
Free Ports and New Faces	109
Mass Slaughter	110
Colonial Indoctrination	111
Opium and Humiliation	112
Insults and Degradations	114
Black Ships	115
Heart of Darkness	116
United States…	118
…but Not for All	120
Our Own Backyard	121
A Great White Fleet	122
A Strategic Gift	123
The Art of Rowing under Water…	123
…and Landing on Ships	124
A Fuel Gamble	125
The Stopping Power of Water	126

	Patient Zero	128
	Insult to Injury	129
	Doomed to Fail	130
	A Maritime PIN Code	130
	The Tragedy of the Middle East	131
	The Baron and the Shah	133
	Seven Sisters – and a Frenchman	135
	Recession and Despondency	136
	An Obliterated Belief	137
	Bitter Homecoming	139
	Brimming with Confidence	140
5.	WAVES OF GROWTH AND PROSPERITY	142
	Imagine	142
	A Few United Nations	143
	A Stubborn Norwegian	144
	A Striking Paradox	145
	The Magic Capitalist Formula	146
	A Hyper-efficient Logistics System	147
	Revolution in a Box	148
	One for All	149
	Rootless, Disloyal and Invisible	150
	MAD, NUTS and NATO	152
	Doomsday Machines in the Depths	153
	Colonial Collapse	156
	Immigration and Assimilation	159
	A Poverty Trap	160
	The Curse of the Landlocked	161
	Tiger Economies	162
	Cadillacs and Coca-Cola	163
	Our Son of a Bitch	165
	Rust Belts	166
	The Actor and the Iron Lady	168
	Eurosclerosis	169
	A Westerly Wind	170
	Ping-Pong Diplomacy	171
	Little Boss with a Capital B	173
	Eggs in Two Baskets	173
	A Growth Miracle	174
6.	A NEW GLOBAL TRIANGLE	176
	How to Square a Circle	176
	Chinese Containers	177
	A Global Triangle	178
	Confirmation Bias	179

	Chain Reactions	180
	Bin Bags and Cardboard Boxes	181
	Inward and Backward	182
	Role Reversal	182
	The Rich and the Rest	184
	Shock and Disbelief	185
	Quaranta Giorni	186
	Panic Hoarding	187
	Vaccine Battle	188
	A Warped Narrative	189
	A Broader Palette	190
	A Perfect Storm	191
7.	THE OCEAN STRIKES BACK	192
	Boeing 737	192
	Wetter and Wilder	193
	A Costly Contribution	194
	Little Boy and Girl	194
	Atlantis 2.0	196
	The Doomsday Glacier	197
	Archimedes' Principle	199
	Ocean Rainforests	200
	Gender Discrimination	201
	A Carbonate Competition	202
	Respiratory Difficulties	203
	Three Tons in Three Hours	204
	The Smaller, the Worse	206
	A Painful Experience	207
	Wave Movement	208
	Invasive Species	209
	Coastal Cultivation	209
8.	BLUE GROWTH FOR A GREEN FUTURE	212
	Colourful Contrast	212
	Day Zero	213
	Water from Water	216
	Land-Based Waste	217
	Blue Fields	218
	A Conflict Commodity	219
	Caged and Seasick	220
	Highest Price for the Poorest	222
	Food for Thought	223
	Hardy Creatures	225
	Watts from Waves…	225
	…and Winds	226

Sun, Salt and Strategic Monopolies	228
Back to the Future	230
Green Corridors	233
Cruise Control	234
Cobalt Crisis	237
The Allure of the Deep Sea	239
Tiptoeing Across the Ocean Floor	241
Green Dilemmas	242

9. **THE NEW BATTLE OVER THE OCEAN** — 245

Beach Holiday	245
An Intensified Maritime Dynamic	246
Whisky War	250
Dry Land and Wet Rights	251
A Unique Island State	254
90,000 Tons of Diplomacy	255
Never Again!	256
A Great Wall in Reverse	259
A Sealed Fate?	260
A Deliberate Uncertainty	262
An Honest Mistake	263
Monroe in Mandarin	265
The Tyranny of Distance	266
A Gold-Decorated Rolls Royce	268
Malacca Dilemma	271
A Strategic Loop	272
Dystopian Hellscape	273
New Silk Roads	274
A String of Pearls	276
Strategic Speculations	277
Chronic Migraine	278
Ever Given	280
Hazardous Crossings	282
Strategic Footholds	282
Enfant Terrible	283
Blue Homeland	284
A Greek Tragedy	286
The Ice Silk Road	287
A New Ocean	289
Pariah State	290
Strategic Claustrophobia	291
Black Smoke	293
Monster Torpedo	294
Headbanger	295
No Man's Land	296

Cold Calculations	297
Little Boy Dreams	298
An Oversized Embassy	299
Reggae and Rockets	300
Out of Sight and Mind	302
Russian Jacuzzi	304
Cable Maps	305
Weapons of Mass Disturbance	307
Dark Fleet	308
Water Is Coming	310
10. HOPE ON THE HORIZON?	**312**
Generation R	312
A Silent Miracle Stalling	313
The Luxury Trap	314
Less Is More?	315
Heaven and Earth	317
Bitter Aftertaste	318
One for the Team	320
It Takes Three to Tango	322
Laziness, Cowardice and Greed	324
Emissions and Inflation	326
Missing the Mark	327
It Takes a Village	328
Poor Master, Good Servant	330
Breaking Bread	331
Dirty Business	332
Legal Risks	334
A Sea-Blue Signature	335
The Permanent Five	337
Law of the Jungle	338
Group Meetings	339
Fewer Hands Make Lighter Work	340
The G Clef	342
Anything I Can Do?	343
EPILOGUE	345
ACKNOWLEDGEMENTS	347
ABOUT THE AUTHOR	351
NOTES	354

'No ocean, no life. No ocean, no us.'
Sylvia Earle, American oceanographer

PROLOGUE

My first encounter with the ocean was not a pleasant experience. I must have been three or four years old and the memory that has stayed with me is of gulping down bitter saltwater and thrashing my limbs as I sank into the cold, dark depths. Seemingly out of nowhere, firm hands latched onto my shoulders and pulled me to the surface. My father had jumped in, fully clothed, when I slipped off the narrow, wet wooden dock where he and my uncle were cleaning fish and chatting.

The only injury I sustained was a good scare, but with time that fear was replaced by fascination with the powerful and mysterious blue world. Many years later, I can look back on hundreds of hours below the surface as a scuba diver, and on days and nights underwater in tropical and cold regions of the world. But my respect for the ocean remains unchanged. I can still sense its Janus face, how its shimmering surfaces can reflect a cloudless sky – and conceal the darkest of depths. When you enter its kingdom, it is as if the ocean itself decides whether or not it will also allow you to depart.

Through my professional career as a maritime executive, and later as special advisor to the United Nations, I have also come to better understand the ocean's existentially ambidextrous nature: how it can be both a generous ally and a powerful adversary in dealing with the most important challenges faced by humanity today.

The ocean is considered the origin of all forms of life, the planet's great mother and the human race's most important public commons. It is the earth's largest biosphere, the space that can sustain life. The ocean is in a constant state of movement, replete with sound, colour, life forms and social interaction. It determines the weather, the climate, natural diversity and our living conditions. Without the ocean, there would be no life on earth.

The ocean runs like an oscillating blue line through the history of the human race, into the present day and our future. Its vast surfaces have created barriers between countries and bridges between continents. The ocean has shaped societies, given birth to superpowers and destroyed civilisations.

THE OCEAN

It has laid the foundation for growth and prosperity, global power and national humiliation. It has taken lives, shattered dreams and inspired hope. It has forged the identities of nations, the mentalities of populations and the individual's understanding of the world. It has provided the underlying conditions for ideology, policy, culture and religion.

The ocean has formed the basis for how we act and interact across national borders, how we organise our societies and how we live our lives. It has lifted billions of people out of poverty and landed millions in despair. It provides food and livelihoods, and its trade lanes and natural resources constitute the backbone of the global economy.

The ocean is also inextricably interwoven with all the important challenges we are facing in our contemporary world. Its powerful currents carry riches and resources, and pollution and problems, from coast to coast. The ocean helps slow global warming by absorbing large quantities of heat and greenhouse gases from the atmosphere. But at the same time, this process destroys marine ecosystems and causes sea-level rise, which harms communities in coastal regions all over the world. Evaporation from the ocean surface further fuels the extreme weather events afflicting every part of our planet with increasing intensity. The mutually reinforcing interaction between the warming atmosphere and the warming ocean is the most important driver of climate change and the source of its most dangerous consequences. Simultaneously, the ocean itself holds important keys to breaking out of this vicious cycle. Nature-based marine solutions and maritime industries alone can contribute a full third of the cuts in emissions needed to reach the goals of the Paris Agreement.[1]

In this way, the ocean is where the crises of our time meet the opportunities of the future. It represents one of our most important concerns – and greatest hope.

It is only through a holistic approach, by understanding the connections and the whole, over time and across issues, that we can truly improve our ocean literacy and comprehend the ocean's significance for our history, the present day and the future. This approach defines the focus and ambition of the book. It is a book about the relationship between the ocean and the human race. It is about the ocean as a condition, a cause and a force of civilisation, how it has formed our world and will shape our destiny.

Adding a personal flavour to the discussion, I am drawing on my own experiences over the years, blending the top-down perspective of my professional maritime career, the bottom-up view of a passionate scuba diver, and my reflections as a concerned citizen and father.

The book is organised thematically. Chapter 1 invites you on a journey into outer space to gain an overview and a glimpse of 'the big picture' through the astronauts' perspective of our beautiful, shimmering blue planet. In Chapter 2 we travel in the opposite direction, into the ocean space, where we explore the ocean as a physical, chemical and biological sphere. We will look at the underwater wildlife and landscape. We will see how bodies of water, ocean currents and marine ecosystems influence the climate, weather and nature on land. Chapters 3 to 6 address how the ocean has shaped human lives, societies and world history, with a particular focus on events from the past five hundred years. We will explore how power at sea made it possible for Western countries to colonise large parts of the world, and the ocean's role as a domain and premise provider for trade, economic growth and the global distribution of labour, income, wealth and prosperity. We will discuss how the ocean has shaped international relations, the balance of power between states and the UN-led, rules-based world order of today. Chapter 7 addresses the impact of global warming, pollution and litter on marine life, and how global sea-level rise is threatening human life, homes, livelihoods, crops and infrastructure on shore. We will look at the ocean-climate nexus, the mutually reinforcing interaction between the climate and the ocean. We will address some of nature's tipping points, and how a warmer ocean leads to more intense heatwaves, drought, flooding and extreme weather events. Chapter 8 explores the vital role of the ocean in the green transition, how it can contribute to a better future for a growing world population and why the blue economy is expected to grow more rapidly than the global mainland economy in the decades ahead. Chapter 9 explores the ocean as a cause of, and domain for, the increasingly exacerbated international tensions of today. We will look at how rivalries between major powers are played out at sea, and why many countries are expanding their military fleets and reinforcing their coastal defences. We will also look at how the tense international situation of today is challenging the authority of the United Nations and the global regulatory system governing activities at sea. Chapter 10, the final chapter, gathers these many threads, asking, 'So what and what now? Is there hope on the horizon?' What must be done to halt the destructive, self-perpetuating dynamic between a warmer atmosphere and a warmer ocean? What will be required in the way of international cooperation, of policies and regulations, of political leaders and business executives – and of citizens like you and me? How are we to restore the health and productivity of the ocean, and tap the vast potential of the blue economy, to the benefit of present-day and future generations?

Each chapter can be read individually and in isolation, depending on the reader's prerequisites and interests. While the first chapters primarily describe underwater life, the marine ecosystems, biology, chemistry, physics and other factual information from the natural sciences, the subsequent parts of the book are more normative, analytical and coloured by my own assessments. This pertains, for example, to what I have chosen to include, omit and explain about history and the ocean's contribution to the development of civilisation and the current era. The influence of the ocean on the history of humankind since the dawn of time has been thoroughly explored in monumental works such as *The Sea and Civilization* by Lincoln Paine and *The Boundless Sea* by David Abulafia. I have instead chosen to limit my focus to the past five hundred years to illustrate how the ability to exercise power on, from and across the ocean was critical to the emergence of the Western colonial powers, how this laid the foundation for the organisation of today's global order and economic activity, and how this part of history continues to impact people's lives, the aspirations and anxieties of countries, and the relationship between nation states.

My opinions are also made clear when I discuss the motives and forces that are undermining trust and intensifying competition between the major powers in our times, and how these find expression at sea. My views are no less evident in the final, concluding chapter where I offer my thoughts on policy, leadership and responsibilities in addressing the serious challenges and vast opportunities offered by the ocean.

The figures and statistics included in the book are intended to provide perspective on the scale of challenges and developments, and to calibrate opportunities, solutions and alternatives. The latter are important in terms of addressing the often little-discussed, but vitally important trade-offs and dilemmas inherent to the green transition. I have also devoted some paragraphs to explain the main provisions of the United Nations Convention on the Law of the Sea (UNCLOS), often referred to as the Constitution of the Ocean. While not attempting to provide an exhaustive account of this legal framework governing two thirds of the surface of our planet, I think it is nevertheless important to understand the basics of how it is defining the rights, responsibilities and obligations in humanity's largest public commons, and how this body of rules and regulations constitutes one of the finest and most important examples of global cooperation.

Key parts of the book are focused on the social sphere, politics, economics and international relations, topics that many readers may not immediately associate with the ocean. This is because we cannot understand the ocean's significance for our history, present day and future if we only speak about what takes place under water. Neither can we understand what affects the ocean and underwater life without understanding what takes place on land.

PROLOGUE

Then, for the sake of good order, I do not believe that everything that happens in the world could or should be explained with a 'blue lens'.

In the course of my work on this book, I have received invaluable feedback and comments from prominent international experts, great colleagues, friends and family. I am deeply grateful for how they have all generously and confidently shared their perspectives, views and insights on various topics. Without their encouragement, patience and feedback, this book would not have seen the day of light. They will all be thanked in a separate chapter of the book. Any errors, omissions and misunderstandings are, however, wholly and fully my responsibility.

The views presented here are my own and do not reflect the official policy or standpoints of the United Nations.

THE OCEAN

*'When I first looked back at the Earth,
standing on the Moon, I cried.'*
Alan Bartlett Shepard Jr,
American astronaut (1923–98)

1.
ALAN'S TEARS

When Alan Shepard Jr planted his feet on the moon one day in February 1971, he was wholly unprepared for the intensity of his emotional response. After many years of hard work and determination, prevailing in numerous selections of potential candidates for this space trip, he had finally fulfilled a lifelong dream. But this was not the cause of his reaction. It was the sight of our beautiful, blue planet against the backdrop of the universe's endless, ice-cold darkness that summoned his tears. A few moments later, as the seasoned astronaut had regained his composure, he pulled out the golf club he had brought along, a modified Wilson 6-iron. He placed the little white ball on the ground and, after a few fumbled attempts, finally hit the ball with a solid thwack. Although he went down in history as the fifth man on the moon, he was the first and so far the only one to play golf there.

The experience on the moon was a watershed moment in Alan Shepard's life, and the fragile beauty of our shimmering planet has made a similar lasting impression on many of the approximately 600 people who have thus far made the trip into outer space. The American astronaut James Irwin describes his impressions from the *Apollo 15* voyage as follows:

> That beautiful, warm living object looked so fragile, so delicate, that if you touched it with a finger, it would crumble and fall apart. Seeing this has to change a man, has to make a man appreciate the creation of God and the love of God.[1]

Irwin would later refer to this space mission as his 'Christian rebirth'.[2] Regardless of faith, to the surprise of many astronauts, travel through outer space often becomes a powerful inner journey, a meeting with oneself in a weightless state of existential wonder. Many speak of sudden, compelling insights about the meaning of life. Others describe an intense, overwhelming sense of their own mortality and insignificance in meeting with the infinitude of outer space. And they all speak of how perspectives shift and awareness expands apace with the growing distance from the Earth.

The phenomenon is so widespread that psychologists have coined a phrase for it: the Overview Effect. The clear and immediate experience of connection to the Earth as a whole, and to all of humanity.

From outer space, astronauts can admire the Earth's unique cosmic signature, produced by the reflection of sunlight off the surface of the ocean. They can see how the ocean covers most of the planet, that it is so vast that if we placed all the continents of the earth side by side, like the Pangaea of geological time, the entire land mass would fit into the Pacific Ocean – and still be surrounded by water. They can sense how the depths of the ocean make up the planet's largest biosphere. They are able to comprehend that the ocean is the home of some of the world's strongest, most numerous and tenacious creatures, of the largest animals on earth, and of organisms so tiny that they are invisible to the naked eye.

Viewed from space, it also becomes clear why most of humankind's history and development has been written in shades of blue. For prehistoric coastal tribes, it was often simpler and safer to gather food from the ocean than to venture inland in search of berries and fruit or to hunt wild animals. The first primitive rafts and canoes were built by courageous individuals who ventured out to sea to fish, explore nearby islands or cross narrow straits and fjords. Through often tragic and hard-won experience, they grew more adept at building canoes and boats, better at sailing and more skilled at navigation. They ventured farther away from land, and soon the boldest of our predecessors migrated from the cradle of civilisation in Africa to journey across vast ocean areas in search of the unknown beyond the horizon. With time they learned to build even better boats and larger ships, harness the winds and currents, understand the ocean's whims and cultivate the art of navigation. The ocean was then no longer an insurmountable and perilous barrier between tribes, peoples, empires and continents. From this point on, the ocean became a bridge for friendship, cooperation, inspiration and trade – and for subjugation, rivalries and wars.

From the windows of a space capsule, astronauts cannot see state borders, political tensions, ideological differences or social distinctions. They cannot see people falling in love, forging friendships or celebrating successes, and neither can they see unfulfilled hopes, broken promises or shattered expectations. But they can discern the sharp contrast between the universe's lifeless darkness and the shimmering fields of green and blue that make up the planet Earth. They can intuitively grasp the fragile balance and clear limitations of the atmosphere's closed system. They can immediately recognise the human race's common destiny and the deeper meaning of words such as responsibility, fairness, equality and sustainability. From outer space it becomes obvious that it is not possible to compensate for the careless destruction of nature and over-consumption of our planet's resources by taking from 'someone else' or 'something else'. We can only take from one another – or from those who will come after us. It becomes evident that it is only through collaboration and peaceful co-existence that we can take care of our planet, solve our common challenges and provide good lives for more people and future generations.

For the astronauts it also becomes intuitively evident that the human race cannot take care of itself without taking care of the ocean – and that we cannot take care of the ocean without taking care of one another.

While civilisations evolve on land, the ocean is, and remains, the human race's largest public commons. Yet the ocean has been little explored, it is poorly governed and weakly managed. We know more about the surface of the planets around us than the seabed of our own, and ten times more people have made the trip into outer space than into the deepest ocean regions.

But there are too few who care and even fewer who have any appreciable knowledge about how important the ocean is – for our lives, our society and our future. We use and abuse the ocean. We overheat, overfish, pollute, litter, contaminate and acidify this vast blue space. We destroy biological resources, life forms and marine ecosystems that are of vital significance to the earth's natural balance, our current civilisation and our children's futures.

The ocean provides food, energy and transport routes. It provides livelihoods and recreation, and the basis for global economic growth and prosperity. It drives the major weather systems and performs ecosystem services of critical importance to all life on earth. The ocean has absorbed virtually all the surplus heat and a large portion of the greenhouse gases produced by human activity since the Industrial Revolution. The upper three metres of the ocean alone contain as much thermal energy as the entire atmosphere, and without the ocean's services, the average global air temperature would be more than thirty degrees higher than it is today. But these immensely important contributions come at a price. The temperature of the ocean itself is rising, which contributes to the destruction of marine ecosystems and underwater life. The temperature increase also causes the global sea level to rise, posing an imminent threat to lives, livelihoods and coastal communities all over the world. Moreover, the higher temperature causes increased evaporation from the ocean surface, further compounding global warming and feeding the extreme weather events that are ravaging the earth with heightening intensity, and for longer periods of time. The ocean-climate nexus, the interaction between the warmer ocean and the warmer atmosphere, constitutes the most important driver and dangerous force of global warming.

The impacts of the global warming, environmental degradation and loss of biodiversity are fast becoming visible and tangible, and they are emerging even sooner and with greater force than most scientists have predicted. All over the world crops are being destroyed and the production and distribution of food is disrupted by more extreme weather events, drought and flooding. The rising sea level is threatening millions of people, such as in the delta regions of Southeast Asia. The capital of Indonesia has been relocated because salty sea-water has penetrated groundwater reserves and rendered the ground conditions unstable. In the Pacific island state of Kiribati, authorities are planning to move the entire population to the neighbouring country of Fiji. In the Amazon, the rainforests and riverbeds are drying up, and in Mexico, monkeys are fainting from heat stroke. In Africa, the Middle East, India and South America, temperatures on some days reach such record-level highs that the conditions are incompatible with normal life. I happened to be in Rio de Janeiro when the city broke its own record for high temperatures in March 2024. A thermal sensation of 62.3°C was the result of temperatures exceeding 40°C in the shade, combined with unusually high humidity. This was of course not a problem for those of us fortunate enough to be staying at a hotel with air-conditioning and clean running water. I can only imagine how it must have been for the more than one million less privileged residents who suffered beneath scorching-hot tin roofs in the favelas on the hillsides surrounding the city.

As usual, those who have the least are the hardest hit: the poor populations of the Global South. But also in more affluent countries the consequences are impossible to miss. In New York, Toronto and Sydney, people have recently had to wear face masks to protect themselves from the thick smoke caused by forest fires. In Athens, Madrid and Paris, thousands have died due to heatwaves, and in Romania, Poland and Germany flooding and landslides have caused human fatalities and swept away houses, bridges and cars. In the arid desert climate of Dubai, the city and airport were flooded for the first time when a full year's precipitation fell in the course of a single day in April 2024. In the Netherlands, New Orleans and Tokyo, taller and stronger barriers are being erected to keep the ocean out.

In recent years, shipping traffic through the Panama Canal has been reduced by almost one third for extended periods of time due to drought;

there is not sufficient water to lift the ships through the locks. In the Atlantic Ocean, the surface water temperatures have reached record highs, and never before has the sea-ice extent in the Arctic and Antarctic zones been so low.[3]

The United Nations Intergovernmental Panel on Climate Change (IPCC) and the world's climate experts have been warning us about all of this for many years, and now it is happening. The development and consequences are unfolding at a faster pace than previously imagined. The future has already arrived, *c'est déjà demain*.

And once again, Oceanus, the powerful, primal river encircling the entire earth in Greek mythology, will have critical roles to play when the next chapters of the history of the human race are to be written. As John Kerry, the former US Secretary of State and Special Presidential Envoy for Climate has put it, 'You cannot protect the oceans without solving climate change and you can't solve climate change without protecting the oceans.'[4]

At the same time, the ocean holds important keys to ensuring sustainable economic growth, increased prosperity and good lives for contemporary and coming generations. Nature-based marine solutions and maritime industries can help reduce emissions by one third of what is required to meet the goals of the Paris Agreement.[5] The ocean surface, water columns and seabed contain untapped opportunities for the production of food, electricity, medicine, transport and perhaps minerals, in a more sustainable manner than on land. Most of this production can be achieved through processes that have lower air emissions, reduced environmental footprints, shorter supply lines and mitigated risks of disruptions caused by drought, flooding or extreme weather events. Therefore, the 'blue economy' is expected to grow faster than the land-based global economy in the decades ahead.

In this way, the ocean is appearing on both sides of the existential equation: it can contribute to a better future – and has the power to destroy the world.

We already have more than enough knowledge, technology, money and resources to solve this equation. We have everything we need to unleash the ocean's vast potential to the benefit of humankind and the natural world. It is fully possible: if we do it right and if we do it together.

But time is not on our side, and now the world is moving in the opposite direction. Collaboration within the United Nations is crumbling and the rules-based world order is under siege. Major powers are turning inwards and against one another. Trust between countries is withering, and protectionism, populism and divisive nationalism are on the rise. Ominous echoes of 'Make Great Again!' can be heard on all continents

In this situation of growing international tensions, the ocean constitutes, as always, an important driver and arena. All of the most powerful states, and many of the less powerful, are upscaling their naval activities and capabilities. They want to protect their coastal regions, secure trade routes, assert their rights in the ocean and strengthen their influence in the world. When the resources on land fall short, the fight over the resources found in the marine environment is intensified. Elliot Cohen, professor at Johns Hopkins University in Baltimore, writes that 'we risk a competition for the seabed and rights of transit as fierce as the European competition for Africa in the nineteenth century. In such an environment, the security of free transit, upon which the global economy rests, will be in danger.'[6]

But it is not solely military activity at sea that reflects the growing tensions and unrest of our world. Every year, hundreds of thousands of desperate refugees fleeing war, poverty, persecution and destitution attempt to enter Europe by crossing the Mediterranean in overcrowded rubber rafts and dilapidated boats. For thousands of men, women and children, the dream of a better future ends with death in the ocean's dark depths. So many refugees have drowned here that Tunisian and Greek fishermen now dread pulling in their nets for fear of finding human remains.[7]

The ocean is thus inextricably interwoven with the greatest challenges, most important trends and dramatic paradoxes of our day. It is both a cause and effect, a means and an end, in the evolution of human civilisation.

Perhaps it would have been easier to understand these connections if we could see our own planet from the perspective of the astronauts. Perhaps the 'overview effect' is comparable to viewing a mosaic wall: we must take a step back in order to comprehend the whole. If we stand too close, we can see each of the tiny pieces, but not what is most important: the image they create together.

When we adjust our perspective, we can also better understand the scale and proportions: although the ocean covers most of the surface of our planet, it constitutes only a small fraction of the earth's total volume. 'Relative to the whole planet, the depth of the ocean is thinner than the skin of an apple', write Jan Zalasiewicz and Mark Williams.[8] If we had collected all the water in the ocean in a balloon, it would have made up only one tenth of a per cent of the earth's total mass. It is this tiny fraction that is the origin and basis for all life on earth. It is this tiny fraction that has shaped our history and contemporary world, and which will determine our fate and our future.

If that recognition starts to sink in, perhaps more of us will shed tears over our beautiful, shimmering blue planet – without needing to travel to the moon to play golf.

> 'The sea, once it casts its spell, holds
> one in its net of wonder forever.'
> Jacques-Yves Cousteau,
> French oceanographer (1910–1997)

2.
POWERFUL, DARK AND MYSTERIOUS

How the ocean is a prerequisite for life on earth, about the ocean as a physical, biological and chemical sphere, about the subsea wildlife and landscape, and how bodies of water, ocean currents and marine ecosystems influence the climate, weather and nature on land.

Jaws

It was already late in the evening by the time the speedboat zipped us across the water off the coast of Key West. An instructor colleague and I were going on a night dive around the southernmost point of Florida, along the edge of the continental shelf where the ocean floor suddenly plummets, disappearing into the depths of the Gulf of Mexico. It is here, in the strait between Cuba and Florida, that the life-giving Gulf Stream begins.

One hour later we arrived. Night had fallen. On the echo sounder screen, we could clearly see the sharp edge of the shelf. Directly beneath us the depth was around thirty metres, though just a short distance away it plunged downward for several hundred metres. My pulse rate accelerated and my senses quickened into alert as we switched on our headlamps and activated our green glow-sticks, before somersaulting backwards over the edge of the boat into the pitch-black water. Navigating under water can be difficult even during the daytime, and at night the challenges are compounded. We also knew that there were sharks in the area. Earlier that day we had run into a few full-grown specimens when we dove down to the USS *Spiegel Grow*, the 150-metre decommissioned naval ship that is one of the world's largest artificial reefs.

THE OCEAN

It was not long, however, before our somewhat reserved approach gave way to joy at the sight of the fantastic world that appeared before us. The nutrient-rich water that gushes up from the depths of the Gulf of Mexico attracts teeming life in infinite shapes, colours and sizes. We crept up on fish no larger than the palms of our hands, who had literally donned pyjamas before turning in for the night. These fish weave themselves into small nets to protect them from voracious neighbours. They then fall asleep, inclining gently towards corals and stones and bobbing softly on the ebb and flow of the water. When the beam of the headlamp hit them, their eyes fluttered open like drowsy children tucked into their sleeping bags.

The powerful beams from our headlamps produced dramatic shadows and iridescence, and the soundscape was dominated by the air bubbles being released from our diving regulators. It was the ocean's nocturnal stage set. Everything except the small cones of flickering light was submerged in utter darkness, which is much more profound than any darkness on land. If you switch off the headlamp, you can barely make out the fingers on your hand, even if you place your glove against the glass of the diving mask. The darkness is so pervasive and dense that, if you suffer even slightly from claustrophobia, the conditions are ideal for the onset of a panic attack.

In the pitch dark of the ocean depths neither can you know whether any of the night hunters large and small are directly behind you, an arm's length above you, or just below you. This provides ample fodder for an active imagination, which can be anxiety-inducing if you should, just to be on the safe side, start scoping out the water around you. It is therefore a good idea to stay focused on the right things when night diving. This will help keep your imagination in check, increase your enjoyment of a fantastic experience and, above all, ensure that you find your way back to the boat.

Our night dive transpired according to plan, and my experienced colleague stayed to my right throughout the entire dive as agreed. Nonetheless, towards the end I could not escape the uncanny sensation that something was moving directly to my left. I postponed confirming this until we began ascending along the boat's anchor chain. When I finally turned and looked, the beam of my torch shone directly into the patchwork-patterned abdomen of a huge sea turtle. A peaceful, curious giant had been seeking company on its lonesome migration through the dark depths.

Safely back in the boat, my friend and I agreed that it had been a beautiful dive and a wonderful experience. But then, we would never know whether we had been inspected and dismissed by nocturnal hunters as potential midnight snacks.

Because, according to the Florida Museum of Natural History, these waters have long been *'the world capital of shark attacks'*...[1]

2. POWERFUL, DARK AND MYSTERIOUS

Vast, Deep and Solitary

The ocean covers more than two thirds of the earth's surface and constitutes close to ninety per cent of the planet's biosphere.[2] The majority of the ocean is found in the Southern Hemisphere. There the total land area makes up only a modest one fifth of the earth's surface. The rest is ocean, for as far as the eye can see – and beyond.

In common vernacular, we speak of 'the seven seas', and the origin of this expression can be traced to written sources from over a thousand years back in time. But since the ocean regions are linked and flow into each other, there are many who prefer to say that we actually have only one ocean. The Ocean.

Historically, we have not always had clear-cut definitions of the various regions. Most marine scientists will, however, agree on the four major oceanic divisions: the Pacific Ocean, the Atlantic Ocean, the Indian Ocean and the Arctic Ocean. As regions, they are all quite different. The Arctic Ocean is the coldest, smallest and most shallow, with depths that are only one quarter of the depths of the three others. The Indian Ocean is the youngest in a geological sense. It has the fewest islands and the narrowest continental shelf. Just off the coast of India, Pakistan, Iran and Bangladesh, the floor plummets abruptly downward. The Atlantic Ocean is the second largest, with a surface area equivalent to approximately one half of the Pacific Ocean.

The Pacific Ocean is the largest and deepest of them all. It contains the same amount of water as all the other ocean regions combined plus all the water found in the earth's glaciers, rivers and lakes. It is also here that we find Point Nemo, named after the mysterious captain in Jules Verne's epic tale about a submarine voyage around the world. It is called 'the loneliest place on earth' and is nothing more than a pinpoint on the map. If you drop anchor here, there is little chance that you will be disturbed. The closest island is almost 3,000 kilometres away, and neither can you expect to find much activity there, because Ducie Island is merely a modest, uninhabited atoll in the Pitcairn island group.

Then there are bodies of water that are called seas rather than oceans. Some of these seas make up parts of a larger ocean region, such as the Sea of Japan, the Barents Sea, the North Sea, the South China Sea and the Caribbean Sea. There are also some seas that are wholly or partially surrounded by land, but they contain levels of salt concentration that set them apart from large inland lakes, such as the Mediterranean Sea, the Dead Sea, the Black Sea, the Caspian Sea and the Sea of Azov. The latter is the shallowest body of water that can claim the status of a 'sea'. The average depth is only seven metres, and nowhere is the sea deeper than fourteen metres.

THE OCEAN

The average depth of the ocean is 3,800 metres, and if we levelled out the ocean floor and all the land masses on earth with a gigantic steamroller, the entire earth's surface would be almost three kilometres under water.[3] The deepest point in the ocean is Challenger Deep in the Mariana Trench, east of the Philippines, where the distance from the surface to the seabed is 10,971 metres.

Oddly enough, only a year passed between the time the first humans visited the deepest point on earth and the day the first human being was launched into outer space. In January 1960, Swiss Jacques Piccard and American Don Walsh made the descent to Challenger Deep. By April of the following year, Russian Yuri Gagarin was launched into space and became the first man to orbit the earth. Since then, many more people have travelled into space than into the ocean depths. We appear to be more fascinated by gazing upward at the sky than downward into the blue depths.

Although by far most of the seabed has not been studied in detail, we do have a good deal of broad-stroke knowledge.

At school we learn that Mount Everest is the highest mountain in the world. That is only correct if we include some important provisos, because it actually depends on what one measures. It is true that the top of Mount Everest is the world's highest point, but the highest mountain on earth is Mauna Kea, located in the ocean just off the coast of Hawaii. The distance from the underwater base of the mountain to the top is 10,000 metres, which is about twice the height of Mount Everest measured from the base of the Himalaya mountain plateau. Mauna Kea means 'the white mountain', and it is named after the snow on the peak of the mountain looming majestically above the ocean surface. Located far from the disturbance of human civilisation's pollutant emissions and artificial light, this mountain top is a favourite site for astronomers, who come from all over the world to observe outer space from the observatories there.

At school we also learn that Angel Falls is the world's tallest waterfall. The remote location in the jungle of Venezuela makes visiting the falls a bit difficult in practical terms, so I was overjoyed when some years ago I was invited to view them from a small aircraft with the Norwegian minister of trade, who was in the country on official business. It was a powerful sight when we flew past the 979-metre waterfall, which is also considered one of the most beautiful in the world.

But the guide's enthusiastic presentation of Angel Falls was unfortunately not wholly accurate. The world's tallest and largest waterfall is actually the Denmark Strait Cataract, which is located under water in the ocean between Iceland and Greenland. There the falls plunge to a depth of more than three kilometres, and the volume of the water flow is almost 2,000 times greater than Niagara Falls at peak flow.[4] The vertical movement of the water is created by the interaction of cold water running through the strait from the north with warmer, less dense water coming from the south.

A Glass Bubble

Although the ocean covers most of the earth's surface, it is like a gossamer membrane around the planet. It is therefore not immediately evident why the water does not simply evaporate and disappear into outer space or seep into the cracks in the earth. The explanation is that the atmosphere holds the water on earth inside a kind of 'glass bubble'. At the atmospheric levels between ten and fifty kilometres above the ground, it is so cold that when water vapour reaches these altitudes it is cooled into liquid and falls back to earth. That is why the total amount of the earth's moisture in the form of vapour and liquid remains relatively constant. However, some scientists are now beginning to worry about the longer-term consequences of global warming for this hydrological cycle, the continuous circulation of water between the earth and the atmosphere.

Since evaporation from the ocean surface is the main source of the hydrological cycle, some might wonder why the rain is not salty. The answer is quite simple: when seawater evaporates, the salt and other substances are left behind because they do not evaporate at these temperatures (at least at any measurable rate). In this way, evaporation acts as natural distiller.

Many of us might think that it is the seabed that forms the underlying barrier for the ocean. That is only partially correct, because a lot of water does leak into faults and porous mountain formations. All the same, the ocean is not drained, because just as much water rises from the inside of the earth. This occurs when sulphur-laden water vapour is spewed out of natural 'chimneys', such as those along the Northern Atlantic Ridge. The gases that erupt out of volcanoes around the world are also predominantly made up of water vapour at temperatures of several hundred degrees.

This tells us that there must be a lot of water inside the earth. Scientists estimate that the deep geological structures may contain at least as much water as the ocean (some estimates suggest as much as twenty-five times more!). But most of the water inside the earth is hydrous rock, not groundwater that can be pumped up or siphoned out. Some years ago, a fragment of a diamond was found in Brazil that had survived the 500-kilometre trip from deep inside the earth and it proved to contain slightly more than one per cent water.[5]

But even though there may be far more water inside the earth than in the ocean, there is still a lot on or above the earth's surface. A minuscule .001 per cent is in the form of water vapour in the air, one modest per cent is in rivers and lakes, two per cent is contained in glaciers and polar regions – what we call the cryosphere – while the remaining ninety-seven percent is found in the ocean. The ocean is made up of so much water that, if we were to distribute it equally amongst all the people living on the earth today, each of us could have filled 50,000 Olympic-size swimming pools.

The hydrological cycle in this glass bubble can serve as a reminder that everything else on this planet is also part of a closed system. Barring energy from the sun, we neither receive nor give anything outside of this system in the way of air, water, soil or other natural resources. If we consume more natural resources than we have, we must take from each other – or from those who will come after us (as long as we don't have the option of harvesting from other planets).

The Eccentric Liquid

Since the ocean has existential significance for all life on earth, it is surprising that we have so little knowledge about the vast ocean depths. More than seventy per cent of the ocean floor remains unmapped.[6] It is estimated that at least two thirds of the species on the ocean floor have still not been discovered.[7] And, according to the US National Oceanic and Atmospheric Administration (NOAA), 'scientists estimate that ninety-one percent of the species in the ocean have yet to be classified'.[8]

That said, I will hasten to add that we do, for sure, have vast and increasing amounts of scientific knowledge about the ocean and life below water. Moreover, research is being ramped up, and new scientific programmes are being initiated. The United Nations has declared the years 2021–2030 to be the Decade of Ocean Science. The Intergovernmental Oceanographic Commission of UNESCO (IOC/UNESCO), headquartered in Paris, is doing a formidable job in heading the research activities under the Decade of Ocean Science, and in promoting cooperation between its 150 member states in marine sciences to improve knowledge and management of the ocean, coasts and marine resources.[9] There is excellent research being done by many institutions, for example the Woods Hole Oceanographic Institution, Xiamen University and the Stockholm Resilience Center, to name but a few.

But much remains to be discovered and understood about the dark, abyssal depths. Despite all the knowledge we have acquired, the ocean is still a huge grab bag: if we dip our hands in it deeply enough, we never know what we will find.

2. POWERFUL, DARK AND MYSTERIOUS

But it is perhaps even more surprising that we still know so little about water itself, the most widespread chemical compound on earth. Water is not merely the liquid substance that makes an ocean an ocean, a lake a lake and a river a river. Water is considered a precondition and necessity for all forms of life: all known life originated in water and all known life will die without water. It is unnecessary to explain water's life-giving characteristics to the more than two billion people still without access to clean freshwater today, or to the four billion who experience water scarcity for at least one month every year.[10] And, for those of us fortunate enough to have access to safe drinking water from a kitchen tap or to enjoy the refreshing pleasure of the cooling drops of a summer cloudburst, water is such a natural part of daily life that we seldom bother to give any real thought to what it *is* and what it *does*.

But even for those who dedicate their careers to researching this substance, water continues to withhold many of its secrets. According to Felix Franks, the nestor of international expertise in this field, 'water is probably the most studied and least understood' of all known fluids.[11] In explaining the unique characteristics and anomalies of water, Philip Ball elaborates that 'water is not necessarily unique, or the most extreme example, in displaying any of these anomalies, but their accumulation in a single substance makes it stand out as a decidedly eccentric representative of the liquid state'.[12]

Due to its unique characteristics, water distinguishes itself from most other substances on earth. A number of these characteristics are of critical importance to the role the ocean plays in climate change and global warming. Some of these characteristics are a bit odd, such as the fact that water is a powerful solvent and extremely corrosive. This is why you can use water to remove grease from a frying pan and jam from a shirt. This is also why bolts and iron rods will rust in water.

Another fascinating characteristic can be observed when water is transformed into snowflakes. All snowflakes have a symmetric, six-armed structure and every snowflake is unique. The billions of snowflakes that descend from the sky on a winter day are an example of the great wealth of variation found on our planet. The Chinese identified the snowflake's unique combination of symmetry and individual variation long before the Common Era. In Europe it was not until the microscope was introduced in the early seventeenth century that Western scientists were able to make the same observations.

Water also has a number of characteristics that differentiate it from other substances here on earth. It can assume solid form (ice), liquid form (fluid) and gas (water vapour). In most parts of the world, the transition points between these forms constitute the reference for the temperature scale that we use in our daily lives, whether we are speaking about the weather, oven settings or bathwater. The Swedish astronomer Anders Celsius introduced this reference in the mid-eighteenth century. He proposed defining the freezing point for fresh water (the transition from water to ice) as '100 degrees' and water's boiling point (the transition from water to vapour) as 'zero degrees'. One of his countrymen, the botanist Carl von Linné, who was of a more practical mindset and literally more down to earth, later suggested reversing this into the centigrade scale of today: zero degrees for the freezing point and 100 degrees for the boiling point.

As water expands when it freezes, the ice will be lighter than the water. Without this special feature, the ice would have formed from the bottom up, not from the surface down. Then you could enjoy skating over ice on sunny winter days without any worries about falling through, as everything would be frozen to the bottom. But then, in return, all the life in the sea below you would also be frozen in ice.

Another important characteristic is water's exceptional heat capacity. Water conducts heat twenty-five times faster than air does, which is why it is important for people to stay dry when in a cold environment. More energy is required to heat water than to heat air, and water has a greater capacity to preserve heat than almost any other solid or liquid substance. The large amount of water in the ocean means the ocean has a heat capacity that is more than a thousand times greater than that of the atmosphere. This is why the ocean plays such a critical role in the curbing of global warming.[13] Since the Industrial Revolution, the ocean has absorbed almost all the surplus heat created by human activity. The ocean also absorbs approximately a quarter of the human-made carbon dioxide found in the atmosphere. Without the ocean's heat regulating effect and its absorption of carbon dioxide, it is estimated that the global average air temperature would have been 36°C higher than today.[14] This explains the vitally important role of the ocean in our efforts to meet the goals of the Paris Agreement.

Due to its large heat capacity, it is also the ocean that balances out the extreme differences in temperatures around us. As we move upward through the atmosphere, the temperature quickly becomes colder before again growing warmer. In the upper part of the atmosphere, at an altitude of around 400 kilometres, the sunbeam temperatures can be in the vicinity of a scorching 1,500°C. That is hot enough to melt iron. If we were to move in the opposite direction – from the earth's crust and inward – it also becomes steadily warmer as we progress towards the core. Ninety-nine percent of the earth has a temperature of more than 1,000°C.[15] A mere 50 to 100 kilometres beneath our feet, a distance corresponding with a couple of marathons, we find the same scorching 1,500°C as in the outer atmospheric zones. When we penetrate all the way into the earth's core, the temperature is more than 5,000°C, equal to the heat on the surface of the sun.

To preserve the basis for life here on earth, the ocean – this veil of moisture around the earth's surface that is 'thinner than the skin of an apple' – must ensure that the atmosphere can balance out the thousands of degrees of heat both directly above and directly below us.

Deviations of just a few tenths of a degree in the average atmospheric temperatures can have catastrophic consequences, as millions of people worldwide are now discovering.

Thicker than Blood

The salt in the ocean plays a vital role in the creation of ocean currents, the main drivers of the global weather systems. Ocean salt has formed over the course of millions of years and comes from land run-off and the erosion of volcanic stone on the ocean floor. The level of salt concentration in seawater varies according to location and depth. Because salt is denser than water, the deeper we descend, the greater the salt content. And likewise, since cold water is denser than warm water, the deeper we descend the colder the water temperature. This means that the water in the ocean depths is saltier, colder and denser than the water near the surface. These differences in density contribute to driving the major ocean currents and enable the water strata at different depths to move in different directions.

The content of salt in the ocean was a topic addressed when US President John F. Kennedy held a speech during a dinner for the crews participating in America's Cup on 14 September 1962:[16]

> [A]ll of us have, in our veins the exact same percentage of salt in our blood that exists in the ocean, and, therefore, we have salt in our blood, in our sweat, in our tears. We are tied to the ocean. And when we go back to the sea, whether it is to sail or to watch it we are going back from whence we came.

It is a beautiful and seductive thought, but unfortunately incorrect. The average salinity of seawater is well above three per cent and can vary considerably, according to depth and region. Human blood, on the other hand, is a stable 0.9 per cent salt, and even the most minor deviations can cause organ failure and death.

The US president would have been more in the ballpark had he said that the plasma in our blood has approximately the same concentration of salt as the ocean. This would have been closer to the truth, though of course not nearly as poetic.

Simple Arithmetic

We need not submerge our heads particularly far beneath the water surface before we notice how the weight of water increases the pressure in our ears. Most people who have tried swimming under water have experienced this. I burst one of my eardrums while descending during a dive, because I failed to equalise the difference between the pressure on the outside and inside my ear quickly enough. This caused a sharp, sudden pain when my eardrum burst and cold seawater flooded into my inner ear canal. It was quite unpleasant and can also be dangerous for divers because so much of our balance and sense of direction is decided by the inner ears. Pain, dizziness and nausea seldom provide a good basis for enjoyable and safe experiences under water.

Roughly calculated, the pressure increase equals one atmosphere of absolute pressure for every ten metres we descend in the ocean. Since by definition the pressure is one atmosphere at the surface of the ocean, at ten metres it is two atmospheres, at twenty metres it is three, and so on. This presents challenges for human beings as we move through underwater depths, but it also says a lot about the fantastic traits the creatures inhabiting these subsea regions have developed.

If you do scuba-diving, using breathing equipment, unlike free-divers you can't simply descend or rise to the surface as quickly as you like. This is because the water pressure compresses the air you breathe in the lungs and in the blood. The further you descend, the greater the pressure from the surroundings and the more the air is compressed. You can descend quite quickly, but when it's time to rise to the surface again you must be careful, because then the air in the lungs and the blood will expand as the surrounding pressure decreases. If you rise too quickly to the surface, air bubbles can form in your blood, which can lead to blood clots in the brain or the heart. Then you are in serious danger and only an immediate and extended stay in a pressure chamber will save your life.

If the bubbles are smaller, they will manifest either as pain or a prickly, itchy sensation all over your body. During a test dive in a pressure chamber that almost went sideways, I experienced how hugely unpleasant this can be. I developed what divers call 'the bends', a term that arose during the construction of the Brooklyn Bridge in New York in the late nineteenth century. Underwater breathing equipment had not yet been invented at this time. When the time came to dig the foundation for the two stone towers of the bridge in the riverbed, the labourers were sent underwater inside

2. POWERFUL, DARK AND MYSTERIOUS

large wooden crates called caissons, which were filled with air. To keep the water out, the air pressure in these caissons had to be at least as high as the pressure in the surrounding water. The caissons thereby functioned like a pressure chamber, so, when the labourers returned to the normal pressure on the surface, the air bubbles in their blood expanded immediately. They reacted by crouching or 'bending' over in pain.

Water pressure does not only affect the body physically and mechanically. When divers approach depths of around fifty metres, the environmental pressure changes the characteristics of the air. When descending to depths exceeding fifty metres, divers must use special gas mixtures to avoid causing serious injury to the central nervous system. But also before reaching such depths, the changes in the air quality can affect divers' judgements and cognitive abilities. If a diver descends rapidly to thirty metres or so, the nitrogen in the air supply will dissolve and flood into the muscles and the bloodstream. Eventually, as the nitrogen reaches the brain, the effect can be the same as that of laughing gas or intoxication. People become lethargic, risible and careless, which can put both one's own safety and that of fellow divers at risk. There are stories of individuals who have swum eagerly downward into the depths, offering air from their own regulator to astonished fish. As a diver rises towards the surface, the symptoms usually disappear right away. So, although a nitrogen rush can have extremely dangerous consequences, it is on the other hand a high without a hangover.

Even intelligent human beings can find it difficult to solve simple tasks under such conditions. Jens Stoltenberg, the former Secretary General of NATO and former Prime Minister of Norway, is an amicable and intelligent person. I can confirm this, because we have known each other since we studied together at university, and when we started working, we shared a small office. When he was prime minister, he called one day and asked if I could give him and his son Aksel a course to upgrade their basic diving certificates. Before long he would be attending an international environmental conference in the Maldives and he wanted to use the opportunity to explore the world-famous diving paradise.

Evenings spent on theory were followed by days of practical exercises under water. It was mid-winter, with sub-zero temperatures on land and ice floes on the Oslo Fjord, so the conditions were harsher than what he could expect to encounter in the Maldives. For one of the exercises we did a rapid thirty-metre descent, and there on the ocean floor he was supposed to solve a simple arithmetic problem that I wrote on a small underwater writing slate: $(3 \times 5) - 9 = ?$ I handed him the slate and marker, indicating that he should write in the answer.

It became quickly evident that even for a prime minister with a master's degree in general economics this simple calculation was far too complicated. After a number of failed attempts we therefore returned to the surface to discuss what we could learn from the exercise.

We quickly agreed that the solution did not lie in further studies in arithmetic.

A Few Fun Facts

Although the ocean covers 'only' two thirds of the earth's surface, it constitutes almost ninety per cent of the biosphere, the space on our planet that is able to sustain life.[17] This is because the ocean is so deep and every part of the ocean contains life: fish, crabs, scallops, jellyfish, and seaweed, of course, but a single litre of seawater can contain from several thousand to many millions of microscopic, one-celled microalgae.

While the ocean represents most of the biosphere, most of the biomass on Earth is found on land. Measured in gigatons of carbon, the denominator most commonly used by scientists, the biomass is more than one hundred times greater on land than in the ocean. The by far largest biomass of all is made up by terrestrial plants. Also, the biodiversity is much greater on land. Out of the earth's close to nine million living species, only one quarter are found in the ocean.

Although much smaller relative to the total biomass, marine creatures wholly dominate animal life. Fish are the true rulers of the animal kingdom, having a combined biomass ten times greater than the earth's human population and one hundred times greater than all the wild animals on land.[18] If we also include other marine animals, such as lobsters, krill, crabs, mussels, oysters, and other shellfish, a full eighty-five per cent of the animal kingdom's biomass is found in the ocean.[19]

Hertz 52

The diversity of life forms found in the ocean is staggering. In the ocean depths we find, for example, the planet's oldest, known living vertebrate, the Greenland shark, a species which can live for several hundred years.

Here we also find the blue whale, weighing 200 tons and measuring more than thirty metres in length, which makes it the largest known animal that has ever lived on Earth. Its tongue weighs as much as a medium-sized elephant, and its heart is the size of a passenger car. Its brain, however, is not much to write home about: it weighs even less than Felix, our family's playful little dog.

2. POWERFUL, DARK AND MYSTERIOUS

Given its tiny brain, the whale is more interested in consuming calories than counting them, and that may be a good thing, because it can wolf down a half million calories in a single meal. And, even though they are abnormally large, scientists believe that as a species blue whales will be even larger in the future – assuming they are able to find sufficient nourishment in the ocean.

These mighty creatures are equipped with extremely sensitive auditory abilities that enable them to communicate across distances of several thousand kilometres.[20] This produces conditions for long-distance love affairs, also of the unrequited variety. Since the late 1980s, scientists have been tracking a whale who was discovered by the US Navy while carrying out foreign submarine surveillance. The whale was named Hertz 52, because apparently it has a congenital defect that causes it to communicate at a higher frequency than other whales.[21] Unfortunately, this frequency is approximately the same as that emitted by the propellers of many of the large cargo ships crossing the ocean. This has left Hertz 52 not only with few romantic partners but also an exhausting and frustrating life. Many times he or she – scientists are not sure which – has embarked on long journeys in hopes of meeting his or her beloved, only to be disappointed and confused when the huge ship docks and switches off its engine.

Hertz 52 must be one of the world's loneliest whales.

But while some marine creatures experience extreme loneliness, in the ocean we also find many of the earth's largest communities and most numerous species. There are more copepods in the ocean than insects on land, and even though these teeny-tiny rascals are only a few millimetres long, they sometimes convene in such huge numbers that they are visible in satellite images taken from outer space. Half their body weight is pure fat, so these tiny creatures are a vital food source for marine life. Except for large predators, like the great white shark, almost all fish eat copepods, and they are even the favourite dish of the gigantic blue whale. The earth's largest creature is wholly dependent upon the earth's tiniest and most numerous, and they are both found in the ocean.

The copepod is otherwise also one of the world's strongest and fastest animals in proportion to body size. There is probably no other living creature on Earth that can compete with the force of the copepod's exertions as it moves at a thousand body-lengths per second to escape from predators.[22] For the human body, this would correspond with six times the speed of a passenger plane, through a medium that has a density more than 800 times greater than air. In meeting with the copepod, even Usain Bolt, the fastest sprinter of all time, would have no choice but to untie his track spikes and humbly bow his head in respect.

But there are others in the depths with explosive bodily force. The legs of the tiny mantis shrimp are constructed like springs that enable it to break through the shells of far larger prey. The force of these tiny legs is equivalent to that of a bullet fired from a saloon rifle.[23] Scientists have recently learned that their legs have an external layer of a ceramic substance, and an inner layer of a plastic-like material, which allows them to hit hard without injuring themselves upon impact. This has inspired researchers to fabricate tougher and more impact-resistant synthetic lightweight materials similar to those found in the mantis shrimp, which can be used for example to 'reduce the fuel consumption of airplanes and space shuttles, increase the energy efficiency of wind turbines, or increase the safety of cars and body armor', as explained in one scientific study.[24]

Happy Not to Be Born a Flounder

If in your youth you found it embarrassing when a spot appeared on your nose from time to time, you should be happy you weren't born a flounder. This creature starts its life looking like most other fishes, swimming vertically and with one eye on either side of its head. But when it reaches puberty, a sudden change occurs: one of its eyes moves to the other side. For some flounder, both eyes end up on the right side of the head, for others on the left. The process is astounding and usually takes less than a week, and sometimes only a day. Subsequent to this quick facelift, the flounder starts swimming sideways and with its body laid out horizontally, the source of the flounder's nickname, the flatfish.

It is also in the ocean that we find some of the animal world's most fascinating physiological characteristics. The history of the cephalopod, of which hundreds of variants are found, from tiny cuttlefish to giant squid measuring up to thirteen metres in length, extends 300 million years back in time.[25] That means that the cephalopod was found in the ocean long before the dinosaurs began to wander the earth. This evolutionary history has produced the modern-day octopus, which has three hearts, eight arms and eleven brains. Two of the hearts control the blood-flow to the gills, which it uses to breathe, while the third ensures blood supply to the arms. The octopus has a 'central brain' in its head, one brain in each of its eight arms, and each of its two eyes has a unique control system that can also be considered a kind of brain.

The complex interaction between these neurological functions surpasses most of what we otherwise find in the animal kingdom. Each of these individuals is a veritable wandering team seminar, because a consensus between the different brains is crucial. No wonder, then, that the octopus is considered the most intelligent of all molluscs. In the course of its brief lifespan – most octopuses live for only two or three years – an octopus can master relatively advanced tasks, such as opening beer bottles and jam jars, using tools and finding its way through mazes.[26] Because their lives are so short, they have also developed a special ability to reap the benefits of others' experiences. As for most of us, learning from the mistakes and successes of others is less risky and saves time.

Large brain capacity is, however, neither a prerequisite for nor a consequence of a lengthy evolutionary history. An example of this would be the coelacanth, which has a 400-million-year history, making it among the eldest species in the ocean. The two-metre-long fish has a brain that constitutes only a little more than 1/100 of the head's volume. Scientists long believed that it was extinct, but in 2021 a specimen was found in the Indian Ocean. The coelacanth can live for one hundred years and most of its life unfolds at a leisurely pace. The females don't reach sexual maturity until they are fifty years old and their pregnancies last for all of five years.[27]

Maybe, in its infinite wisdom, nature has determined that it's better not to have brain capacity for excessive brooding if you have to swim around pregnant for years, all alone in cold, pitch-dark depths of several hundred metres.

Only a Fish?

In addition to their individual characteristics, the social interactions of many marine animals are quite sophisticated. It is impossible not to be fascinated by the methods killer whales use to take care of their young and hunt for food. The young are cared for through a system resembling that used by kindergarten teachers when they take a group of children out for a walk in the neighbourhood. The young swim close together while the adult females follow closely behind, beneath and beside them, to protect them from danger. Similarly, the adult males hunt in groups when searching for food for the pod. They swim in precisely coordinated formations to compress large schools of fish, before hitting the schools with their tails with such force that many of the fish pass out and die. Or the killer whales swim side by side and flap their tails to create waves that knock seals off ice floes and into the water. Then fathers, mothers and their hungry children can help themselves to the delicacies.

The large schools of fish that constitute the killer whales' prey, for their own part, also have relatively advanced systems of collaboration, most of which are connected to the hypersensitive organ called the lateral line that runs through the sides of the bodies of all fish. This organ detects movements in the water, whether caused by prey or potential dangers. This is also the sensor the fish use to coordinate the movements of large schools. They function according to the principle of 'safety in numbers', as large flocks of antelope do when crossing crocodile-infested rivers in Africa: swimming in large groups makes it easier to confuse attackers, lessening the probability of individuals falling into the clutches of predators.

We also know that dolphins communicate using a complex set of whistling sounds, which differ from one dolphin to the next.[28] Each dolphin has its own unique acoustic brand, in the same way that humans have unique sets of fingerprints. Herring, on the other hand, utilise a somewhat less (or perhaps more?) advanced communication and coordination system. They fart! If half the school suddenly passes gas, they all quite simply swim away from the odour (perhaps not so strange…).[29] This is the herring's form of democracy, a practical form of majority rule ensuring survival of the group.

But the herring must contend with the air expelled by predators. Scientists have studied how the humpback whales in the Chatham Strait on the west coast of Alaska create air traps to catch large schools of herring. Every autumn, the whales migrate here to fatten up in a four-month feeding frenzy, before swimming south again on a six-month fast, followed by a mating season. Measuring fifteen metres in length and weighing thirty tons, the humpback whale must consume a large amount of food in order to enjoy the mating period after a journey of up to 8,000 kilometres. Every day during these hectic autumn months, the whale consumes around one ton of food, and herring are the dish of choice on the menu. The whales are for the most part lone wolves, but for this hunt they work in pods of six to ten whales.

The hunt starts when the whales slip underneath the herring school, usually found at fifty- to sixty-metre depths. Once below, one of the whales will swim in a large circle several times while releasing a steady stream of air. As the bubbles rise to the surface, they create a cylinder-shaped wall that agitates and confuses the herring. At the same time, the other whales emit intense acoustic signals, which frighten the herring into swimming even closer together. These high-frequency sounds can reach up to 140 decibels, which is the noise level of a passenger plane during take-off. Once the compact school of herring is trapped inside this vertical air cylinder, the whales spring into action. They attack from below with their giant, gaping jaws and chase the terrified herring upwards through the cylinder until they reach the surface, where the feast can be devoured.

2. POWERFUL, DARK AND MYSTERIOUS

Fish live in an element that is alien for most human beings. It is difficult for us to communicate with them, since for the most part they have no facial expressions, telling gestures or spoken language we can understand, and neither do they have a soft, furry coat that invites cuddling. For this reason we are inclined to show less consideration and empathy for them than we do for dogs, cats or seal pups. But fish are put together like all other vertebrates: they have both a spine and a skull. Scientists in the field of ichthyology – the branch of zoology that deals with fishes – are quite certain that these species can experience fear, joy, pain, curiosity and well-being. We know, for example, that dolphins will seek out the sting of the puffer fish, so they can enjoy the rush produced by its potent venom. Other studies have shown that fish swimming in the glass tanks of restaurants can experience stress when they witness their companions being removed, dismembered and served for dinner.

So maybe our use of the word 'fish' erases and obscures important elements of marine life's intrinsic value, meaning, and diversity. Perhaps we should instead speak of them as 'animals' and 'social beings', as we do for those around us on land.

Words and terminology carry weight, here as otherwise in life.

Dark Oxygen

The ocean is not a uniform sphere, either at different depths nor in different places on the planet. Because the pressure, temperature, light, salinity, ecosystems and marine life can be so greatly divergent from place to place, scientists have divided the different layers of the water column into what they call pelagic zones, named after the Greek word *pelagikos* meaning 'of the sea'.

The uppermost of these five zones, where there is the least amount of pressure and the greatest amount of light and heat, extends downward for 200 metres. This is the epipelagic zone, located 'above the ocean'. Here we find the majority of the marine life with which we are familiar, such as sardines, herring and tuna fish, crayfish and crabs, eels, jellyfish, coral, kelp and seaweed.

In the epipelagic zone, there is enough light for photosynthesis to occur, the process that transforms sunlight, carbon dioxide and water into oxygen and chemical energy. As we have seen previously, the surface water of the ocean absorbs approximately one quarter of the human-made carbon dioxide found in the atmosphere. Carbon dioxide is used in the process of photosynthesis, which means literally 'to put together using light'. With the (important) exception of the marine ecosystems to be found around hydrothermal vents on the deep seafloor, almost all other forms of life on Earth depend on photosynthesis. Oxygen is a key component of all of the body's most important molecules. It enables us to convert food into energy and is critical to human survival, growth and reproduction. This process is therefore often referred to as one of the world's most important chemical processes. On land it is the green plants – trees, flowers, the grass and bushes – that carry out photosynthesis. In the ocean, it is plankton, and in particular a group of single-cell algae called diatomes, which performs this vital and life-sustaining function. It is estimated that about one fifth of global carbon fixation and even more so of the oxygen production can be ascribed to these particular diatom species.[30]

Until recently, it was considered a well-documented and uncontested fact that photosynthesis is the only natural process that produces oxygen.

However, in the summer of 2024 a new scientific study showed that production of oxygen could also take place in the pitch dark at the abyssal seafloor several thousand metres below the surface, where no rays of sunshine would ever reach.[31] The study hypothesises that voltage generated from polymetallic nodules clustered on the sea floor is sufficient to provide for electrolysis, splitting the water's H_2O molecule into hydrogen and oxygen. If this finding of 'dark oxygen' is verified by other studies, it would represent one of the most spectacular and important breakthroughs in modern ocean science. It would also, not least, serve as yet another reminder of how much is still unknown about life and ecosystems in the abyssal depths. The research for the report was carried out in the Pacific Clarion-Clipperton Zone, the most important area for research and exploration of seabed minerals, which we will be addressing in Chapter 8.

Here it may be fitting to clarify a widespread misconception. Ocean advocates often speak of how the ocean produces half of the earth's oxygen, claiming that 'with every other inhalation, we can send thoughts of gratitude to the ocean'. Such enthusiastic statements appear in the speeches of, among others, France's President Emanuel Macron, and we also find them in black and white in reports from both the Commission of the European Union (EU) and the United Nations Intergovernmental Panel on Climate Change (IPCC).[32] The claim sounds important and compelling, and is certainly expressed with the best of intentions, but this does not make it accurate.

It would be more accurate to say that either 'all' or 'none' of the oxygen on Earth is produced by the ocean. The former would refer to the 'Great Oxygenation Event', which happened about two billion years ago and converted Earth's atmosphere from one with very little oxygen to one with basically current levels. It's still not entirely clear how this happened, but it was likely driven by blue-green algae which are primarily found in the ocean. In this way, it could be argued that 'all' of the breaths we take are thanks to the ocean.

The latter claim – that none of the oxygen we breathe comes from the ocean – is based on the fact that the production of oxygen on land and by the ocean, respectively, is almost equal. Each are responsible for almost exactly half. But since the life in the ocean consumes all the oxygen that is produced there, nothing remains to contribute to life on land. In practical terms, the ocean today contributes zero oxygen to the air we breathe.

In fact, the supply of oxygen for the ocean itself is diminishing. We will later see how this creates critical challenges for both marine ecosystems and life on land.

The World's Worst Sex

When we continue to descend in the ocean towards 1,000-metre depths, there is virtually no sign of sunlight. For this reason, not much photosynthesis takes place here, and neither is there as much life as closer to the surface. We are now in the mesopelagic zone, in 'the middle of the ocean', and it is in these murky, shadow-filled surroundings that we find the legendary, gigantic squids, vividly described in Herman Melville's novel *Moby Dick*. Despite their formidable sting and tentacles as long as buses, they are the preferred dish of the large sperm whales, who will often take a trip down here in search of a proper meal. Life and death battles often ensue, in the course of which many of the sperm whales perish or sustain permanent injuries. But the traffic moves both ways, because it often comes to pass that some of the creatures who live down here will go hunting for the neighbours found in the storey above. This will usually happen after sunset, when this zone is also cloaked in darkness. This traffic makes the mesopelagic zone the area of the ocean where the greatest vertical movements of biomass take place.

When we descend deeper than 1,000 metres, all sunlight is gone. From this point on, we move through the bathypelagic and abyssopelagic zones before reaching at around 6,000 metres the hadopelagic zone, 'the underworld' of the Greek god Hades. This is, in actual fact, the true underworld. In these abyssal depths, the pitch darkness, icy temperatures, high salinity and high pressure make for an extremely unpleasant environment. Since there is no light, neither is there any photosynthesis that can create organic life, so there is little on the menu to tempt those with discerning palates. In the ocean's basement, the marine life forms have basically only two choices when it comes to feeding themselves: they can eat one another or that which comes floating downward from their neighbours in the storeys above. The latter is called 'marine snow' and mainly consists of droppings, dead remains or leftovers. So, when a dead whale, shark or huge octopus floats slowly down to the bottom, the deep-sea creatures gather for a feast.

Some of the creatures able to survive down here could have been taken from a cabinet of horrors. In the ocean depths, as elsewhere on Earth, evolution has fostered the traits best suited to ensure survival. The deeper one goes, the harsher the environment and the more specialised, bizarre and fascinating the inhabitants. At these depths, nature's creativity has been granted free rein. Here we find the most incredible shapes and colours, luminous sea spiders, swimming cucumbers, anorectic squid and transparent fish. Since there is so little food to be found, these creatures must be diligent about saving their strength and making the most of the few calories at their disposal. They therefore seldom swim at all and, when they do, they swim slowly. They spend the majority of their time drifting lazily on ocean currents, trying to lure in their prey by waving their luminous arms and long tentacles.

The deep-sea angler fish is one of the many strange creatures down here. It lives its ascetic life in the total darkness of cold, salty water, often at depths of 4,000–6,000 metres. The origin of the name stems from how the female attracts her prey – using something resembling a lightbulb on the end of a long rod growing out of her head. The female has a short body, a huge head and a terrifying big mouth, studded with masses of long, razor-sharp teeth. The male is equipped with some of the animal kingdom's largest nostrils, which he needs to sniff out the few females swimming around in the vicinity. For ages, scientists were not even aware of the male's existence, and, when they finally made the male's acquaintance, they also understood that gender equality has not come far in the underworld. On the other hand, you will have to search far and wide to find a stronger love bond, because these couples literally become one. Once the male has sniffed out his intended, he chews into the flesh of her sizeable stomach. This triggers hormonal processes in both. After a while, he connects to her bloodstream. In this way he can take nourishment from her and she can take sperm from him. Once he is attached to her, he no longer needs his mouth, eyes or fins, so these body parts shrivel up, until eventually he is reduced to a small appendage on her body. No wonder then that a Norwegian news channel awarded a rare video clip of their mating ritual with the dubious honour of 'the world's worst sex'.[33]

It took scientists a long time to disclose the nature of the angler fish's romantic affairs, and, when all is said and done, we still know little about life and amorous liaisons in the vast ocean depths.

2. POWERFUL, DARK AND MYSTERIOUS

Bell Jars and Bicycle Chains

Although we have limited knowledge about life and the conditions in the ocean depths, we do know a great deal about how the ocean's physical and chemical characteristics affect the weather and climate on land. We have the movements of these enormous bodies of water to thank, whether we are enjoying a peaceful moment in the warm summer sunshine, listening to the autumn wind whistling through the treetops or gliding away on skis through the drifting snow. It is the warm water of major ocean currents that produces the differences between the pleasant, temperate climate of Northern Europe, Canada, Japan, the Korean peninsula and the north-western regions of China, and the extreme, Arctic conditions we find in Siberia and Alaska, even though all these regions are located at approximately the same northern latitudes.

There are different forces propelling the current systems at different levels of the ocean and in different parts of the planet. This was not always evident to scientists. Early on in the history of ocean science, people were puzzled by the water level of the Mediterranean. There are lots of big rivers that feed into the Mediterranean, and the current at the Straits of Gibraltar flows into the Mediterranean as well. So, with all this water going into the Mediterranean, why wasn't the water level continuously rising, like in a bell jar? The same issue was found in the confined Baltic Sea. There were lots of theories, including that there were subterranean tunnels deep beneath the seabed linking the Mediterranean and Baltic with the Atlantic, and enabling water to flow out again. It wasn't until much later that ocean scientists found a way to measure the direction of deeper water currents, and found that even though water was flowing into these seas at the surface level, deeper down it was flowing out again.

Now we also know that from the surface to around 400-metre depths, which on average constitutes the upper one tenth of the water column, the speed and pattern of the water's movement is mainly determined by surface winds and the topography of the land masses. The winds in the North Pacific Ocean are the most important force behind the Kuroshio Current, which transports warm water from the South China Sea upward past Japan, and from there, down along the west coast of Alaska and Canada. The surface winds in the southern Pacific Ocean propel the Humboldt Current's transport of cold, nutritious water from Antarctica up along the coast of Chile and Peru before it veers out into the Pacific Ocean to continue its journey towards the equator.

It is mainly the powerful surface winds, the 'westerlies', blowing across the North Atlantic Ocean that propel the Gulf Stream, the biggest and strongest of all the global-scale systems of currents. The Gulf Stream's discharge is ten times greater than all the rivers of the world combined.[34] This intercontinental conveyor belt transports water from the Gulf of Mexico across the Atlantic to Northern Europe and then up along the coast of Norway, where it splits into two branches: one turns to the north-east and brings warm water into the Barents Sea, while the other curves in the opposite direction south of Greenland, taking with it cold water on the trip home along the east coast of the USA.

As the Gulf Stream approaches Europe it converges with a deeper current system that scientists call the Atlantic Meridional Overturning Circulation (AMOC). People often confuse these different current systems. While the Gulf Stream is driven primarily by surface winds, AMOC is propelled by what scientists call thermohaline circulation, the interaction between the warmer and less dense water on the surface and the colder and denser water in the ocean depths. AMOC extends all the way from the southernmost to the northernmost regions of the Atlantic Ocean. In Chapter 7 we will see how one of the threats of global warming and the melting of ice is a potential weakening of the AMOC, and the dramatic consequences this can have for the climate and living conditions in a number of regions, including Europe and Africa.

The AMOC contributes to the exchange of nutrients and the equalising of temperature differences by sending cold water from the Arctic and Antarctic towards the equator, and warm water from the equator back towards the two poles. These opposing currents avoid collision because the warm, lighter water flowing towards the poles from the equator stays on the surface, while the cold, heavier water flowing back from the poles stays closer to the ocean floor. The stratification between the warm and cold water is reinforced by the fact that the deep ocean water masses flowing from the two poles are also heavier due to its higher salinity. This is because water with the lowest salt content will freeze first and be left behind as ice in polar regions, while the saltiest and heaviest water is sent back towards the equator. In this way, the thermohaline circulation ('thermos' meaning temperature, and 'haline' meaning salty) produces a pattern in the northbound and southbound ocean currents that resembles the rotating movement of a bicycle chain as we pedal away. The AMOC in this way also plays an important part in drawing the carbon dioxide and other greenhouse gases absorbed from the atmosphere downward into the ocean depths. A corresponding thermohaline circulation system is found in the Pacific Ocean, but due to topographical factors, among other things, this system is far weaker than the AMOC.[35]

2. POWERFUL, DARK AND MYSTERIOUS

Horse Latitudes

In the early nineteenth century, the French mathematician Gaspard-Gustave Coriolis secured himself a place in the history books as the first to explain the forces at work when the earth rotates on its axis.[36] Since the earth makes a full rotation every twenty-four hours, the velocity must be higher at the equator than at the poles. While the speed at the equator is 1,600 km/h, at the north and south poles it is zero. On the basis of this, Coriolis was able to explain why the major ocean current systems rotate in clockwise patterns in the Northern Hemisphere, and in anticlockwise patterns in the southern.

Less than one hundred years later, in the beginning of the twentieth century, the Swedish oceanographer Vagn Walfried Ekman launched his 'motion theory', complementing Coriolis's theories and contributing to our more profound understanding of the driving forces behind the major current systems and the biochemical properties of the different parts of the ocean. Ekman explained how the direction of the currents in the upper one hundred metres or so of the water column is mainly determined by the friction forces caused by the interaction between the winds and the ocean surface. This is why the westerlies are the main drivers of the Gulf Stream. However, the drag generated by the winds is gradually subsiding in the deeper parts of the water column, making these strata move more slowly than the water closer to the surface. Ekman's theory shows that at lower speed, these deeper water strata are then also more influenced by the Coriolis effect, making them deviate in a ninety-degree angle relative to the direction of the surface currents. This deviation occurs clockwise in the Northern Hemisphere and anticlockwise in the Southern Hemisphere. The deviations in the different strata of the water column creates upwelling and downwelling, called Ekman 'suction' and 'pumping' respectively. In this way, Ekman's and Coriolis's theories together explain the upwelling of nutrient-rich water in the eastern parts of the Atlantic and Pacific Ocean, and the abundance of fish stocks occurring in areas such as along the coasts of West Africa, Chile and Peru.

The Coriolis effect also explains why the large ocean currents rotate in a stable pattern of gigantic circles around five so-called 'gyres': two in the Pacific Ocean, two in the Atlantic Ocean and one in the Indian Ocean. These gyres basically function like gigantic cogs as the ocean currents meet, latch into and derive force from one another. This causes the exchange of water and marine organisms, pollution and waste between the different ocean regions. They are all interwoven in a cohesive and integrated oceanographic system, which is why some people prefer to say that there is only 'one ocean' on Earth. If we were to embark an ocean expedition by simply allowing ourselves to drift along on the major current systems, we would eventually be transported around the entire planet. But such an expedition would require great patience, since it would take about one thousand years. In order to return to our point of departure today, we would have had to commence the voyage in the Viking Era, so this is no trip for restless souls.

It is because of the Coriolis effect that hurricanes and tropical storms cannot cross the equator. The rotating weather systems produce the extreme winds and, when these try to cross over to the other hemisphere, they collide with weather systems rotating in the opposite direction. It is also the Coriolis effect that lies behind the tenacious myth that water spins clockwise when we pull out the plug in a bathtub in the Northern Hemisphere and spins in the opposite direction in Australia. This is an amusing notion, but nothing more. The volume of water in your bathtub is far from sufficient to reproduce the earth's powerful centrifugal forces, similar to how the coffee in your cup does not imitate the ebb and flow of the tides caused by the gravitational forces of the moon and the sun.

Coriolis's theories explain why there is no rotation of significance where the large wind systems and ocean currents meet along the equatorial belt. Because of this, a relatively high-pressure system of blue skies, scorching sun and little wind often prevails in these regions. During the Age of Sail this could create serious problems. To this day, seafarers still use the expression 'horse latitudes', stemming from the time when Europeans brought their horses to the colonies in America. When the wind died down and the ships drifted for days beneath a sweltering sun, the crew would sometimes throw the horses overboard when they began running out of drinking water.

Maelstroms

Although the ocean currents usually have greater force than velocity, in some places they do move at high speeds. The most rapid of all are the maelstroms, some of the best-known examples of which are found in the Naruto Strait in the north-eastern part of Japan, the Chacao Channel in Chile and off the coast of Maine, USA, where we find the maelstrom Old Sow. Two of the strongest ones are found in northern Norway, where the maximum velocity of the Saltstraumen tidal current can reach up to forty km/h. The second is the Moskenesstraumen whirlpool, where Captain Nemo and his *Nautilus* submarine were forced to surrender to the forces of nature in Jules Verne's epic tale.

2. POWERFUL, DARK AND MYSTERIOUS

Maelstroms are created when the tides are pressed back and forth through narrow straits. The moon and the sun each play a part here because they control the movements of the tides. In 1687, Isaac Newton was the first to understand and explain these connections. Although the moon is far away and only one quarter of the size of the earth, its gravitational pull is strong enough to affect bodies of water on our planet. The same holds true for the sun, which is much larger, but located so far away that its gravitational pull on the earth is less than half that of the moon's. Combined they create the daily and monthly rhythms of the tides. The tide is highest in those areas of the earth facing *towards* or *away* from these heavenly bodies. The tide is at its highest point when a so-called 'spring tide' occurs. This occurs once every fortnight, when the moon and sun are aligned and their gravitational pulls coincide.

While the diurnal tidal variations can be almost imperceptible in some places, the water levels in other locations can vary by several metres. The variations are the greatest in the Bay of Fundy in northern Canada. Here the difference between high and low tide can exceed sixteen metres, equal to the height of a four-storey apartment building.

A Sea within the Ocean

The major ocean currents have created a proper backwater in the North Atlantic, a 'sea within the ocean'. This is the Sargasso Sea, the world's only identified sea without land boundaries. Within the large region of this sea, the water currents from the several systems cancel each other out and the water for the most part remains calm within a column extending 7,000 metres below the surface. The Sargasso Sea is also considered a gyre, although it is not included in 'the big five'. The water here is clearer and saltier than elsewhere in the ocean, and the colour is often a beautiful blue. Since days of yore, mariners have known about the special seaweed found floating in the water here called sargassum. The seaweed attracts an abundant and diverse marine life, and this is the birthplace of the American and European eels.

In the wonderful *The Book of Eels*, Swedish author Patrik Svensson describes how the tiny, transparent larva of the European yellow eels are hatched in the lazily floating seaweed and subsequently carried away on the Gulf Stream for up to three years before reaching the coast of Europe.[37] At this point, the eels have grown large enough to migrate up rivers, across marshes and through ditches to adapt to a life in fresh water. Once it has found its new home, the eel lives its lonely life in the same place for up to fifty years, until one day it decides to return to its birthplace. When it makes the journey home, it is for the first and final time in its life, and on this trip many eels will swim almost 10,000 kilometres through open seas.

We still know little about how long it takes the eels to complete this journey, but it is perhaps six months or more. How the eels navigate also remains a mystery, although it is likely that their sense of smell plays a critical role. As far as we know, there are no other marine animals equipped with more advanced olfactory abilities. Eels can detect the scent of one millionth of a drop of water diluted in an Olympic-size swimming pool.[38]

We do know however that these eels can swim up to fifty kilometres a day and at depths of up to a thousand metres on the trip back to their birthplace. This is the last voyage they will make because, once they have reproduced in the brown seaweed of the Sargasso Sea, the eel's solitary life comes to an end.

The long and lonely life of an eel has a rough beginning – and an exhausting finale.

Bird Dung Wars

The teeming life in and around the seaweed in the Sargasso Sea is wholly different from the ecosystems of the five large gyres. The latter are instead veritable ecological backwaters, where few marine plants or animals can thrive. This is actually quite odd, since it is here that the ocean currents meet, so one might expect the gyres to be rich with nutrients and full of life. That is however not the case, since the ocean circulation of these large gyres generates a downwelling of water, whereas it is the upwelling of nutrient-rich water that creates an abundance of marine life. In these nutrient-poor gyres there is therefore little life, and only the most robust species can survive in this environment.

The situation is the complete opposite in the waters teeming with fish to the west of the American and African continents. Nutrient-rich water wells up from the vast ocean depths into the Benguela Current off the coast of Namibia, the Canary Current off the coast of Senegal, the California Current off the west coast of the USA and the Humboldt Current off the coast of Chile and Peru. All of these regions are located in the tropical or temperate zones, where there is enough light and heat for plankton to reproduce abundantly throughout the entire year. This produces the basis of life for large fish stocks, and although these regions constitute only two per cent of the world's ocean regions, almost twenty per cent of all wild fish catch is harvested here.[39]

2. POWERFUL, DARK AND MYSTERIOUS

The Humboldt Current, named after the German naturalist Alexander Humboldt, is particularly rich in nutrients since it hosts plankton and other organisms from the large Antarctic depths, and it is also supplied with nutrients that have flowed off land regions along the west coast of South America. In this life-giving current, one finds large schools of herring, anchovies and mackerel. This attracts, in turn, huge flocks of seagulls, who feast on the fish.

When Alexander Humboldt was here on his expeditions in the early nineteenth century, he wrote in astonishment about the amazing quantities of both fish and birds, but the notes from his diaries indicate that he was above all fascinated with what the birds left behind: large swaths of the coast were covered with layers up to twenty metres thick of droppings from cormorants and shags, gannets and pelicans. Humboldt, after whom not only an ocean current was named, but also a well-known German university, and an 'ocean' on the moon, quickly understood that the plentiful supply of this guano, more commonly known as bird dung, was every bit as interesting as the fish and birdlife.

The Incas had at this time been fertilising their fields with guano for centuries. Their understanding of the value of guano was undoubtedly an important factor behind the rise of the Inca Empire – one of the greatest and most prosperous civilisations in history. Guano contains large quantities of nitrogen, which increases soil productivity. Guano played such an important role for the Incas that there were strict rules regulating its distribution among the population and the theft of more than one's allocated quota was punishable by death.

In our times, industrially manufactured nitrogen is the main ingredient of artificial fertilisers used all over the world, but in Humboldt's day, the farmers in both Europe and the USA were struggling to coax sufficient crops from the soil. He therefore brought samples of guano back to Europe. This was the beginning of a large and extremely profitable commercial enterprise, as many ships loaded down with bird droppings were soon en route to European and American ports. The commercialisation of the valuable guano immediately triggered conflicts and, before long, war broke out between Spain and Peru. Later there were also wars between Peru, Bolivia and Chile. The last 'guano war' lasted for five years and cost 18,000 people their lives.[40]

The story of the savage guano wars provides an illustrative example of the significant and multifaceted interaction between the ocean and life on land. In the following chapters we will take a closer look at how people's use of the ocean and its resources have shaped the world of today – how it has produced superpowers, subjugated nations and contributed to waves of growth and prosperity, desperation and poverty.

> 'Whosoever commands the sea, commands the trade of the world. Whosoever commands the trade of the world, commands the resources of the world – and consequently, the world itself.'
> Sir Walter Raleigh, British explorer (1552–1618)

3.
POWER AT SEA

How the ocean has shaped history and our understanding of the world, how power at sea made it possible for Western countries to colonise large parts of the world, and how cross-ocean trading led to the emergence of large multinational companies and modern banks, insurance systems and stock markets.

The West and the Rest

Even for those of us seated towards the back of the large conference room, it was easy to see that the Norwegian minister of foreign affairs, who would later become Norway's prime minister, was in his element. Jonas Gahr Støre had the assembly of international business leaders in the palm of his hand – and he knew it. After an historical breakthrough in the Norwegian-Russian negotiations on the Barents Sea border dispute, he was a shining star in the diplomatic heavens. Demonstrating impressive knowledge and offering persuasive analyses, he outlined the major challenges of the current geopolitical situation. The conclusion of his global *tour d'horizon* was rewarded by enthusiastic applause and nods of agreement.

The contrast to the next speaker could not have been more pronounced. The well-known professor from one of the most prestigious universities in the USA had been born and raised in Singapore. After adjusting the microphone and clearing his throat with a gentle cough, he began by thanking the foreign minister for 'a uniquely interesting and thought-provoking talk based on terms and perspectives that eighty percent of the world population do not share'.[1] What followed was a lecture I believe the majority of those in attendance would not be forgetting any time soon.

3. POWER AT SEA

Many of us from the transatlantic cultural complex, the 'Western world', become sceptical when we are confronted with alternative world views and versions of history. In our narratives, the brief and Eurocentric history of the West captures the most fundamental events of the history of the world. Of particular significance are our perceptions of the events that transpired starting from the late fifteenth century – the Europeans' bold expeditions across the ocean, 'discoveries', missionaries, conquests, and the colonisation of Africa, America, Asia and Oceania. In Western schools we learn about the Renaissance, the Reformation, the Age of Enlightenment and the French Revolution. We learn how, in the second half of the eighteenth century, history took two important steps across the ocean to the west. One step moved across the English Channel, and into the modern world through the Industrial Revolution. The other step crossed the North Atlantic to the USA, a self-proclaimed exceptional nation state, conceived on ideas, born in freedom and built through the courage, determination and talent of European immigrants.

The next chapter of this story features the West's unique creativity, scientific breakthroughs, intrepid businessmen and vertical economic growth, followed by an enthusiastic account of our democratic ideals, 'universal values' and market capitalism, the combination of which provided the moral superiority, economic power and military might that would win two world wars, found the UN-led rules-based world order and triumphantly bring an end to the Cold War. All of this in the course of the twentieth century, what the historian Eric Hobsbawm calls 'the short century'.[2]

'This is the end of history,' Francis Fukuyama declared in his book by the same name, after the Berlin Wall had been hacked to pieces by cheering crowds on a cold November evening in 1989.[3] Now all of humanity had been lifted out of the ditch and placed on a well-lit, one-way street headed for a global community defined by tolerance, individual freedom, liberal democracy, market economic principles and humanistic thought. Other ideological, philosophical and religious ideas about the organisation of society, such a socialism, communism, fascism, Nazism and theocracy, had been banished once and for all to history's rubbish heap of oppressive practices and contemptuous disrespect for human life.

In this sense, the West's predominant narrative and understanding of history provide all the requisite and satisfactory explanations for the emergence of the modern world. What has taken place in the rest of the world is devalued to sideline events, background and context. In the Western world, we can thereby forget or omit information about critical historical events that occurred outside of our own part of the globe and repress the fact that many of the human race's greatest and most important advances occurred long before the West gained dominance. The American journalist Fareed Zakaria explains that 'In many countries outside the Western world, there is pent up frustration with having had to accept an entirely Western or American narrative of world history – one in which they either are miscast or remain bit players'.[4]

In many parts of the world, such as Africa, important historical events have been passed down orally from one generation to the next. These oral accounts often carried little weight in the face of the prevailing written narratives of Western colonial powers. As the world-famous South African singer Miriam Makeba put it in an interview, 'Until the White man comes to any place, nothing lives. It's only when he comes and says "poof", I've discovered you, now you exist.'[5] It is as if Africa had no history until it was described and committed to paper by a White man.

The Story's Omissions

Many people in the Western world have little knowledge and even less awareness of the fact that our view of the world and history are shaped predominantly by events dating a few centuries back in time, during a time period when Europeans set out across the ocean on plundering missions to conquer land, subjugate populations and shape the world in their own image.

I have personally fallen into this trap several times, because I was uninformed or lacked awareness. I will therefore not let the Norwegian foreign minister take his walk of shame alone. I still cringe with embarrassment over the memory of an after-dinner speech I gave in Shanghai as a young shipping executive. Eager to illustrate the long, historic ties between our two nations, I mentioned not only that Norway had, in 1949, been one of the first countries to establish diplomatic ties with the People's Republic of China; I also spoke about the Norwegian ship pilots who had been stationed on the Yangtze River (Chang Jiang) in the early years of the twentieth century. I should have omitted mention of the latter, although the host upheld Asian etiquette and merely smiled and nodded politely so neither of us would lose face. The period I referred to was one of the most traumatic in China's history, an absolute low point in the country's Century of Humiliation. The Norwegian ship pilots were working there solely because China was in the hands of Western powers. In retrospect, a shameful experience.

When I was in school, history was one of subjects that interested me the most, but there were glaring and fundamental gaps in my knowledge about China. I must of course assume responsibility for that. I nonetheless think that part of the explanation also lies in the fact that both the curriculum

3. POWER AT SEA

and version of history taught in both Norwegian and other Western school systems is shaped and disseminated through the lens of the dominant perspective of the West.

In Great Britain, for example, the historical violations are omitted from the curriculum, according to the Indian-British journalist and author Sathnam Sanghera, who writes that, although seventy to eighty per cent of British schoolchildren have learned about the Battle of Hastings in 1066, about the Tudors and the Great Fire of London in the seventeenth century, less than forty per cent have learned about the transatlantic slave trade, and less than ten per cent about Britain's colonisation of Africa.[6] Author Charlotte Mendelson, who is a graduate of Oxford, states that from the time of her early childhood until the age of twenty-one her schooling was one of the best that Great Britain has to offer. In spite of this, she was 'taught nothing about slavery or colonialism. Nothing. Ever.' Even former prime minister Tony Blair admits in his autobiography that, when the British handed Hong Kong over to China in 1997 and his Chinese host commented that this provided an opportunity to put the past behind them, he'd had 'at the time, only a fairly dim and sketchy understanding of what that past was'.[7]

Also, few British and French schoolchildren have knowledge about the injustices committed when their respective countries recruited millions of soldiers from the colonies during the First and Second World Wars. The colonial soldiers in the British and French forces fought for a freedom they themselves were denied. They had the right to die, but not to vote. During the Peace Day celebratory march in London in 1919, none of the soldiers from the British colonies were allowed to participate. They were sent home post-haste before the celebration took place.[8] Also, when the end of Second World War was on the horizon and the Germans were chased out of France, hundreds of thousands of foreign soldiers from the Free French Forces were sent back to the colonies. Only white French soldiers would be allowed to march in the parades celebrating peace on the Champs-Élysées.

In US and European schools, pupils do not learn about the numerous rich cultures of Africa. They aren't taught about the art, culture and advanced political and social structures that reigned for centuries in large empires such as Ghana, Mali and Songhai, until the continent was invaded, plundered and violated by European marauders. How many students from the institutes of philosophy at the Sorbonne, Cambridge or Princeton are able to name a single prominent African philosopher? How many recognise the name Ibn Khaldun, a historian of Tunisian descent, and can explain his non-religious philosophical masterpiece *Muqaddimah* ('Introduction'), written at the end of the fourteenth century?[9] How many of these students can explain not only the widespread salience of his ideas in the Berber and Arabic regions of the African continent, but also their influence on statesmen and scholars of the Ottoman Empire throughout the sixteenth and seventeenth centuries?

During my schooling, we learned about the French Revolution, but we were never taught about another revolution that took place at this time, in Haiti, specifically, in 1791, when African enslaved people revolted against France's atrocities and oppression. After ten years of fighting, it was enslaved Africans who founded the first free republic on the American continent, based on the ideals of the Enlightenment and the French Revolution's slogan, *liberté, égalité et fraternité*. Napoleon sent in 65,000 troops to crush the uprising but gave up in the end and relinquished control over the island after only two years. As we will see later in this chapter, the Haiti Revolution most probably gave rise to Napoleon's decision in 1803 to sell all of France's North American conquests, doubling the size of the young American nation state in one fell swoop. The Haiti Revolution inspired a number of slave uprisings in the USA and was the source of important impulses leading to the US Civil War and the subsequent abolishment of slavery. The slave uprising in Haiti is a defining event in the history of the world but was wholly absent from the curriculum when I attended school – and I suspect it remains so in most places to this day.

It is often said that history is written by the victors, and it is the perspectives of the victors that dominate the West's version of history and understanding of the world. But that history does not begin with the fights for individual freedom and universal values, as many Westerners would like to believe. It was primarily through naval military force, modern technology and organisation, and ruthless brutality that European countries, and later the USA, were able to cross the ocean and for several centuries subjugate nations and people in Asia, Africa, Oceania and America from the end of the fifteenth to the first part of the twentieth century. This was how Western powers were able to colonise India and plunder the 5,000-year-old Chinese Empire. Naval superiority was a precondition enabling the West to project its power all the way to the other side of the globe and bend other countries and societies to its will. This has since been long forgotten by the West – but not by the Rest.

When discussing the importance of power at sea as a means of conquering and colonising other parts of the world, some scholars may interpret my approach in the tradition of Alfred Thayer Mahan, the US naval officer who in 1890 published the *Influence of Sea Power upon History*, the first of twenty books and twenty-three essays on this broad subject, and whose works have had significant influence both in his home country and in other parts of the world.[10] My intention is, however, not to subscribe to the theories of either Mahan or other prominent scholars. I am simply presenting the role of power at sea in line with what I believe is conventional reasoning and scholarship based on empirical approaches.

I am, also, well aware that there have been influential thinkers in modern history who have advocated other perspectives on the conditions for global domination. Among the most prominent in the early twentieth century were British political geographer Halford Mackinder and American political scientist Nicholas Spykman, who both asserted that it was the Eurasian continent, due to its advantageous geopolitical location, which possessed the most favourable conditions for the development of military and industrial powers. Their opinions differed, however, on whether it was the 'heartland', the central part of Eurasia as asserted by Mackinder, or the 'rimland' encompassing states like the UK, Japan, and China as argued by Spykman, which played the primordial and decisive role.[11] The Heartland theory proved to be very influential for the USA's geopolitical strategy during the Cold War, and both Mackinder's and Spykman's theories appear to have gained renewed interest in the discussions about China's massive Belt and Road Initiative.

Regardless of these different doctrines and academic schools of thought, the balance of international power is now shifting, and the demographic, economic and political centres of gravity are moving more towards Asia and the Global South. This opens for alternative interpretations of history and views of the world, and also changes the perspectives of and references for the global narrative. In these parts of the world, other ideologies, perceptions and narrative traditions prevail, and increased economic growth and prosperity are also strengthening national self-image and historical awareness in these regions and in other parts of the Global South.

It is becoming increasingly clear that the West's view of the world and understanding of history is 'based on terms and perspectives that eighty per cent of the world population do not share'.

Origo's Power of Definition

There is little in the Western interpretation of the history of humankind that reflects Europe's initially modest role and contribution to civilisation. The nomenclature, concepts, narratives, time periods and geographic references employed in that history speak to a robust self-image and self-esteem.

The *Pacific Ocean* was named by the Portuguese Ferdinand Magellan, who was relieved to find that the maritime route to India he discovered in 1521 passed through surprisingly calm waters. *America* was named after the Italian explorer Amerigo Vespucci, who was the first to understand that the West Indies were not in India. He named the site where he went to land

Venezuela, after his own hometown of Venice. The word *Brazil* is a derivation of the Portuguese term for the tree that was the source of an ingredient the European textile industry used in dyes in the seventeenth century. *Argentina*, from the Latin word for silver, *Argentum*, was named after the silver jewellery the natives brought as gifts for the first conquistadors who went to shore there. The *Atlantic Ocean* was named after Atlas, the Greek god condemned to carry the vault of heaven on his broad shoulders.

The term *Asia* was also first employed by the Ancient Greeks, and none of the inhabitants of the region had or have any linguistic, historical or cultural connection to the word. *Southeast Asia*, the region east of India and south of China, is a modern linguistic construct, coined in the period after the Second World War, when the Allies formed a military body that was named the Southeast Asia Command. What the West calls the *Middle East* would from the perspective of India would have been the Middle West. *Australia* derives from Latin and means 'southern'. The etymology of the word *Africa* remains in dispute, but most historians agree that it was the Romans who gave the continent its name, referring to the 'Afri' people living there.

Moreover, ground zero for historical timelines and the calendar year is the birth of Jesus Christ, the son of the Western God. The point of reference for time zones is the Royal Observatory on the outskirts of London. Greenwich Mean Time (GMT) was established when the British Empire was at the pinnacle of its power.

The European frame of reference has also defined the timelines of historical events. One example is the printing press, wholly instrumental to the development of culture, knowledge, laws, trade and administration: the West honours Johan Gutenberg for having invented the art of printing around 1440, a technology which had at this time already been in use in China for almost 900 years.

For centuries, Europeans have placed the point of the compass at the centre of their own continent and drawn the map of the world with Europe as *origo*, coordinate zero. That is why we call the sea that breaks upon Europe's southern shores the *Mediterranean*, 'the Sea in the Middle of the World'. This is how Asia has ended up on the far right of the map, the Americas on the left and Africa and Oceania at the bottom. The way the map is drawn, using so-called Mercator projections, also distorts the proportions in a way that favours the Northern Hemisphere. On the world map, it appears as if Greenland is almost as large as Africa, while in reality Africa is more than fourteen times larger. The map also creates the impression that Great Britain and Madagascar are the same size, although the island off the east coast of Africa is in fact twice the size of the former.

3. POWER AT SEA

The Eurocentric view of the world is also subtly confirmed and reinforced by the terms we use in reference to the earth's cardinal points. On maps and globes, the Northern Hemisphere is depicted as 'up and above' while the Southern Hemisphere is 'down and below'. There is nothing in the universe that confirms this directional orientation, but it glides seamlessly into the Western perception of superiority: in all charts of hierarchies and organisations, power is situated at the top. The winners of sporting competitions stand at the top of the podium and social losers are said to be found on the bottom rungs of society. As we 'succeed', we ascend on the list of rankings, on the career ladder, and office floors. If we 'fail', we are relegated to lower levels.

'The West is the only civilisation identified by a compass direction and not by the name of a particular people, religion or geographic area. The name "the west" has also given rise to the concept of "westernisation" and has promoted a misleading conflation of westernisation and modernisation,' Samuel Huntington writes.[12] This creates the illusion of a timeless and universal entity, which in turn generates associations of 'progress', 'modernity' and 'civilisation'. In the 1960s and 1970s, it was not uncommon to use the term 'Westernised' as a laudatory badge of honour in reference to emerging industrialised nations such as Japan, Taiwan, South Korea and Singapore. These are all examples of an implicit cultural bias which feeds ideologies, narratives and unconscious assumptions about the superior qualities of Western culture.

Through such assumptions, the idea of the West as the moral authority wielding the power of definition, determining how to understand, shape and organise the world, is close at hand.

Prone to Ornamentation, Insipidness and Brutality

It can therefore be fitting to point out that, throughout the better part of the history and development of civilisation, a large portion of Europe and what we today refer to as 'the West' was an insignificant, backward region.

In large parts of Europe, we continued as fishermen, gatherers and wandering hunters for several thousand years after the great ancient cultures had developed advanced and well-organised societies. The European continent was peripheral when humanity took its first and most significant strides on the path to modern civilisation. All of the most important innovations providing the foundation for civilisation – such as agriculture, animal husbandry, the division of labour and formation of political-administrative entities – occurred up to 500 years before the Common Era and outside of

Europe. China and the Fertile Crescent (the south-western part of Asia and the Eurasian regions bordering the eastern part of the Mediterranean Sea), in particular, led the great advances of civilisation, writes Jared Diamond:

> Until the proliferation of water mills after about A.D. 900, Europe west or north of the Alps contributed nothing of significance to Old World technology or civilization; it was instead a recipient of developments from the eastern Mediterranean, Fertile Crescent, and China. Even from A.D. 1000 to 1450 the flow of science and technology was predominantly into Europe from the Islamic societies stretching from India to North Africa, rather than vice versa. During those same centuries China led the world in technology, having launched itself on food production nearly as early as the Fertile Crescent did.[13]

In the eleventh century, while the Norse Viking chieftains were performing pagan winter sacrifice rituals in primitive long houses, Indian princes, Chinese emperors and Moorish caliphs were arranging cultural evenings of music, poetry readings and dance performances in lavishly ornamented palaces.

In the fourteenth century, while the European continent was still made up of more than a thousand small states and princedoms without any strong central seat of power, Mongolian, Indian, Persian and Chinese dynasties ruled the Eurasian territories between the Arctic, Pacific and Indian Oceans. At this time the power of the African Mali Empire was also at its peak. This empire reigned over large portions of West Africa and controlled a thriving trade with Europe and Asia. The capital Timbuktu was one of the largest cities of this time, well known for its beautiful mosques and a university library that contained an estimated 700,000 manuscripts.[14] When the Malian ruler Mansa Musa made his pilgrimage to Mecca in 1324, 60,000 servants and subjects accompanied him. Musa brought along money, gold and gifts, which he liberally bestowed upon those he encountered during his long journey. Afterwards the Egyptian capital Cairo reportedly suffered high inflation rates for several decades due to all the money Musa had poured into the region when his entourage passed through the city.[15]

For more than one thousand years after the fall of the Roman Empire, the world's military, political and economic seats of power were located outside of Europe. As recently as the end of the eighteenth century, Asia stood for eighty per cent of the world economy. 'In comparison, Europe was an economic dwarf,' Yuval Noah Harari writes.[16] Howard French explains that '[b]y the late medieval and early modern eras, even literacy, especially in the

Sahel and on its fringes, was not so different from medieval Europe... In the nineteenth century, the literacy rate of enslaved Muslim Africans was often higher than that of their slaveholders in the Americas.'[17] Even after the start of the Industrial Revolution, there was little about Europe that impressed the great cultural nations of Asia. According to the Swedish historian Sven-Eric Liedman, the Chinese looked down on Europeans in contempt, viewing them as 'superficial and prone to ornamentation, insipidness and brutality'.[18]

The Sailing Eunuch

While Western culture honours Christopher Columbus, Vasco da Gama and Ferdinand Magellan as some of the greatest world explorers, few outside of China have heard of Zheng He. In the early fifteenth century, while journeys across the closed Mediterranean Sea continued to play a key role in Europe's dealings with other continents, the Chinese admiral carried out seven large-scale ocean expeditions for the emperor of the Ming Dynasty.[19]

Zheng's expedition fleets included up to 300 ships and 30,000 men. The largest vessels were 120 metres long, fifty metres wide and rigged with nine tall masts. The ships were abundantly decorated, extremely robust and impressively seaworthy by the standards of the day. The resilient sails were made of thin, woven bamboo mats. The hulls were equipped with a series of waterproof bulkheads, so they wouldn't sink even if the bottom or side of the vessel were to spring a leak. 400 years would pass before the Europeans would see fit to copy the advanced Chinese construction. Today waterproof bulkheads are standard issue and included in the obligatory, international safety requirements for all large ships regulated by the United Nations International Maritime Organization (IMO).

In Zheng's company, the great Western explorers must have seemed like scruffy sparrows hopping about in the midst of the cranes' elegant dance. On their initial and most famous expeditions, both Columbus and Vasco da Gama sailed away with crews of less than one hundred on three small, poorly equipped ships. Colombus's flagship the *Santa Maria* was barely eighteen metres long and had three short masts. When Magellan crossed the Pacific, it was with less than 300 men and a fleet of five ships.

The expeditionary fleets under Zheng's command were very large, and even in our day only the naval fleets of the most powerful nations can boast a greater number of vessels. Zheng was also an imposing figure. He was well over two metres tall, and his impressive physique was probably due to the fact that as a young Muslim he was made a eunuch in the emperor's court: castration causes hormonal imbalances that can promote growth. Zheng was so impressive, bold and powerful that he is believed to have been the inspiration for the character Sinbad the Sailor in the Middle Eastern tales of *One Thousand and One Nights*.

The expeditions cast off from Nanjing, the southern capital, which with its half a million inhabitants at this time was probably the largest city in the world. Since the Chinese had already invented the compass, the ships were able to hold a steady course across vast ocean regions even in fog and darkness. The sailing voyages passed through the South China Sea and onward into the Indian Ocean and the Middle East. There they would call at the Gulf of Aden, Yemen and the Strait of Hormuz. They also travelled down the east coast of Africa to what is today Mogadishu in Somalia.

A key objective of these expeditions was to demonstrate the empire's military strength and to secure important trade and sailing routes, including through the notoriously dangerous and pirate-infested Malacca Strait. The emperor also ordered Zheng to expand the protected buffer zone of 'vassal states', the tributary states surrounding the Middle Kingdom. These states, which included Cambodia, Vietnam, Korea, the Philippines and Thailand of today, effectively retained a type of freedom and self-determination in exchange for declaring their loyalty to the Chinese emperor. Zheng's expeditions were also tasked with exploring the unknown world beyond the vassal states.

Fateful Decisions

After Zheng's seventh expedition returned in 1433, the Chinese emperor put his foot down: there would be no more ocean expeditions. Neither the ruling Ming Dynasty nor the subsequent Qing Dynasty had ambitions of expanding their large empire, and the ocean offered no promise of growth and prosperity. Neither did the Chinese see any need to develop closer ties with other parts of the world. The uncivilised 'barbarians' inhabiting the countries beyond the vassal states had little of any value to offer the Middle Kingdom.

It is likely that the high-ranking officials of the court played a part in this decision. The jealousy of several officials had grown apace with the favour and influence that Zheng and his second-in-command, also a Muslim eunuch, eventually acquired with the emperor. But the most critical factor was the dynasty's greatest source of concern at this time: the well-organised, massive and aggressive attacks of Mongolian warriors who descended from the steppes to the west. At the zenith of their power, the Mongolians reigned over the largest kingdom in the history of the world, a region stretching from the Pacific Ocean in the east to the Mediterranean in the west, from

3. POWER AT SEA

today's Iran in the south to the Arctic Zone in the north. The Mongols had already occupied large parts of the Chinese territory for more than a century in 1370, when they were driven out by forces led by Zhu Chongba, a former monk who made himself emperor of a new dynasty, the Ming.

Since the ocean was considered a natural and protective external barrier, China turned its back on the ocean and directed its gaze towards land. From this point on, three large-scale projects were given priority with an eye to reinforcing the defence system of the empire: expansion of the Great Wall of China, relocation of the capital from Nanjing in the south to Beijing in the north, and extension of the Grand Canal to secure the supply of food and goods to the new capital.

Since the empire's resources were to be channelled into defending itself from the Mongolians, a ban on building and sailing large ships was also passed. In the year 1500, the emperor decreed that anyone who tried to build ships with more than two masts would be executed. In 1525, all remaining ships with more than two masts were destroyed, and, as of 1551, sailing the ocean on such ships was considered treason. The public authorities later set fire to more than 1,000 kilometres of China's south-east coastline to render the area uninhabitable and prevent the building of large seagoing vessels.[20]

This was how the entire fleet of the powerful Empire was eradicated – by the Chinese themselves. This would later have consequences of epic proportions, because as it turned out the ocean did not constitute the natural protective barrier that they had believed it to be.

When 300 years later the Middle Kingdom was invaded, occupied and brought to its knees by Western maritime powers, the Chinese had neither naval military strength, weapons technology nor fortresses along the coast to defend themselves against invaders. These traumatic experiences and bitter historical events would settle into the bones of all subsequent generations of Chinese military strategists.

Without knowledge of these long historical developments, it is impossible to understand the extreme sensitivity of today's Chinese leadership when it comes to the country's coastal waters, and their motives for the current maritime military build-up and activity in the nearby ocean regions.

Ciao!

It was the Mediterranean Sea's bounded atrium that would be the key to Europe's first important economic, military and political expansion on the international stage. The enclosed sea, where Europe, Asia and Africa meet, became a simmering cauldron of culture, commerce and conflicts. Until the fifteenth century, the hub of Europe's development was situated along the coast of the Mediterranean Sea.

Then, as now, the coastal nations around the Mediterranean Sea had more contact and interaction with one another – through trade or as a result of conflicts and wars – than with the neighbouring states of the continental hinterland. This included, above all, the northern parts of Africa and the southern parts of Europe. For the Asian regions of the East Mediterranean, the Silk Road – the busily trafficked transport artery to China that passed through the entirety of the central Asia – was of particular importance.

During the first millennium of the Common Era, extensive trans-Mediterranean trading activity served to link Asia, Africa and Europe in increasingly close-knit economic, social and cultural networks. On the European side, it was with time the Italian city-states that took the lead, and of particular prominence were Venice, Genoa, Milan and Naples. Venice evolved into the strongest of them all, with its own fleet of several hundred ships. Through the trade of spices, silk, precious jewels and other exotic goods from Africa and Asia, the city amassed great wealth, a glimpse of which is still visible from the sightseeing gondolas paddling through the canals between the lavishly decorated palaces at Piazza San Marco. But it was not only commerce and trade that contributed to the exchange of impulses between the continents and regions around the Mediterranean Sea. The seafarers themselves – at that time as in modern times – brought with them ideas, inspiration and impressions from the places they visited.

The slave trade was particularly lucrative for the countries around the Mediterranean Sea. Men, women and children were abducted and taken from their homes on all sides of the sea to be sold as servants, sex slaves, eunuchs, galley slaves, soldiers or forced labourers for princes and landowners. In the 900s, Prague had become the European centre for human trafficking, where southern Europeans, Vikings and Muslim merchants from North Africa and the Middle East turned huge profits. At the time, for instance the court of Cordoba in Spain alone held more than 13,000 slaves. In the Muslim regions of Africa and the Middle East, light-skinned slaves from eastern and northern parts of Europe were sold for particularly handsome sums. But also the women of the nomadic Beja people, native to the region encompassing today's Sudan, Eritrea and Egypt, were in great demand. One surviving document from this time reads: '[They] have golden complexion, beautiful faces, delicate bodies and smooth skins. They make pleasant bed mates if they are taken out of their country while they are still young.'[21]

The slave trade also provided the origin of the Italian word *ciao*. Now used as a friendly, everyday greeting, the word actually derives from *schiavo*, Venetian for 'I am your slave.'

3. POWER AT SEA

Benches and Banks

The lively trade activity across the Mediterranean Sea laid the foundation for much of what we today associate with the modern-day finance industry. The large-scale and expanding activity generated a need for new and more advanced insurance, banking and investment systems. Once more, the Italian city-states took the lead.

The fourteenth century brought the birth of what we now know as the insurance industry. Up until this point, it had been common to mitigate and distribute risk by dividing up cargo between several ships, so everything would not be lost in the event of a shipwreck or hijacking. The strategy was basically to avoid putting all one's eggs in a single basket. But with time, as trade activity expanded and cargo values increased, this became a costly and inefficient method. It wasn't long before smart Italian businessmen (they were all business*men* at that time) understood that there was money to be made here: if those who owned the ships or cargo were to pay a small fee, an insurance premium, into a joint fund, this fund could be used to cover any losses. The first known contract for these purposes was signed in Pisa on 13 February 1343, and one hundred years later, the Consoli dei Mercanti was founded in Venice, a court that specialised in marine insurance cases.

Around the same time, Italian businessmen also established a system of interest-bearing debt instruments called bonds, and the goods on board ships were put up as collateral for loans. Although highly advanced money-lending systems had already been invented several thousand years before in China, India, Mesopotamia, Egypt and Greece, it was the Italian bond loans that eventually developed into what we know as modern-day banking operations. The word 'bank' itself comes from the Italian word *banco*, in reference to the benches where the first transactions were made. Medici Bank, often considered the world's first banking and lending institution, was founded in Tuscany in 1397. Ten years later, Banco di San Giorgio was established in Genova as the world's first savings bank. From this point on, people could for the first time put their savings in a bank, instead of hiding them under the mattress or beneath the floorboards.

This was how the foundation for all modern banking operations was formed, the essence of which entails generating revenue through the conversion of short-term deposits into long-term loans. Although people are free to deposit or withdraw their own money at their convenience, the total amount of deposits is usually sufficient to enable the banks to offer these funds as long-term loans at an interest rate higher than that accrued on deposits.

THE OCEAN

Few, Dirty and Base

None of the Italian city-states or the other large commercial cities around the Mediterranean at this time had any real interest in exploring the regions to the west of the narrow Strait of Gibraltar separating the African and European continents. The existing volume of the Mediterranean sea trade generated solid profits, and voyages across the endless, stormy Atlantic ocean to the west were viewed as unnecessary and dangerous. Many were also still convinced that embarking on such a voyage entailed an actual risk of falling off the edge of the earth.

It was therefore the countries with an Atlantic coastline, first and foremost Portugal and Spain, who took the lead in Europe's great colonial conquests beyond the borders of their own continent. The two countries on the Iberian peninsula already had extensive experience with sailing along the northern part of the African west coast, and they now shouldered the risk of venturing further south. In Europe, rumours had long been in circulation about the great riches allegedly found south of the Sahara. Juicy stories were also still buzzing about how the Malian King Mansa Musa had showered money and gold upon the populace during his legendary pilgrimage to Mecca. Both Portugal and Spain therefore began sending large-scale maritime expeditions south along the African coast.

The destination was Africa – and the mission: to find gold.

Although the Portuguese and Spanish seafarers could not in any sense be described as cowardly, they were filled with excitement, fear and trepidation when they set out on these long sea voyages. They had few reservations as long as the sailing routes were restricted to the more northernly regions of the African west coast. Here they were already quite familiar with the dangers, the waters and the local Berber and Moor populations. The regions further south, however, in the 'black Africa' south of the Sahara, both enchanted and terrified them. Many ships ended up turning around and sailing home again when their crews' courage faltered. According to the rumours, there was not solely gold to be found there. It was said that the coast was populated by blood-thirsty, wild savages and that there were gigantic, whirling maelstroms in the ocean off the coast that could swallow everything and everyone. These nightmarish expectations were nurtured by the name *Sahel*, derived from the Arabic word for 'coast'. For European seafarers this was not solely a metaphoric term for the Sahara Desert's southern border with a region of spacious, wild savannas and stifling tropical jungles. It also represented the cut-off point for human civilisation, the frontier of a lawless world inhabited by savages.

3. POWER AT SEA

To the seafarers' relief and surprise, as they navigated further and further south, they did not solely encounter savage tribes. They discovered a complex world of intricate social structures. They met, as anticipated, hostile peoples and aggressive bands of robbers. But they also met peaceful, friendly and inquisitive human beings. They encountered large and small clans, tribes and kingdoms. They found poor villages and large empires led by powerful rulers. In the regions that now make up Mali, Ghana, Nigeria, Congo and Angola, they found highly developed civilisations with advanced legal systems, administrative-political structures and bureaucracies. They became acquainted with societies rich in culture and history. They discovered great dynasties with unfathomable riches. They understood also to their amazement that for several centuries extensive trade had been underway between the east and west coasts of Africa.

But the surprises were mutual. After a meeting with a Portuguese captain and his crew, the king Kwamena Ansa, of the region of modern-day Ghana, did not attempt to conceal his astonishment over how civilised the foreign visitors seemed, remarking that 'the Christians who have come here until now have been very few, dirty and base'.[22]

It was the Portuguese who eventually found the African gold which had long occupied the imaginations of Europeans. The crews could barely believe their eyes when they went ashore in 1471 on the coast of what is now Ghana. There was no need to dig into the soil to gather large quantities of the precious metal: the members of the local population openly wore necklaces, bracelets and other gold jewellery in a variety of shapes and designs. The reality surpassed the Portuguese's wildest dreams. They named the location Elmina ('The Mine') because they were certain there had to be an abundant gold mine located nearby. They immediately set to work on the construction of a fortress to defend the tiny village and a large part of the surrounding region, which from that point on they referred to as the Gold Coast.

Shortly thereafter, Portuguese ships began sailing in shuttle traffic between Elmina and their homeland. The gold was transported north, while soldiers, labourers, arms, supplies and bureaucrats were sent south. On one of these ships was a young, adventurous Italian merchant named Christopher Columbus. During this period he garnered the knowledge and experience he would later put to use in his expeditions to the west, across the vast and frightening Atlantic Ocean.

THE OCEAN

Banished to the Shadows

It is remarkable how little attention this segment of history has received in Western narrative traditions. It was the expeditions to Africa – not to the Americas, China, India or Southeast Asia – that were the most important turning point in the history of Western civilisation. These were the expeditions that marked the inception of the European 'colonisation saga'. It was the African continent that Europe initially referred to as the New World, long before the term was applied to America. Africa was not, as it is often depicted, merely a bothersome hindrance on the path to Asia. 'Most intriguing is the unshakable belief that it was Europe's yearning for a maritime route to Asia, and that obsession above all, that drove the European breakout, creating what has become known as the Age of Discovery,' Howard French writes, and continues:

> Associated with this idea is a historiographical trope that… ascribes the earliest, little-discussed phases of the Age of Discovery – meaning the first few decades of the fifteenth century, when the Portuguese led in efforts to edge their way southward down the coast of West Africa, beyond the lands of the Moors and into the world of Blacks – as nothing less than a whole-cloth bid by Europeans to navigate around Africa. The continent is rendered a mere obstacle, and if trade with it is mentioned at all, it is merely as a sideshow. Typically, in this rendering, once the Cape of Good Hope is reached by Bartolomeu Dias, in 1488, Africa drastically recedes from the narrative or disappears altogether.[23]

It is as if the African continent was banished to the shadows of Western historical awareness for 400 years, from the time Columbus went ashore in America in 1492, until the European colonial powers started the Scramble for Africa in the late nineteenth century. For example, more attention is devoted to Portuguese Bartolomeu Dias as the first European to round the southern tip of the African continent in the attempt to find a maritime route to India, than to how his homeland simultaneously established diplomatic relations and traded extensively with several of the large African empires.

Little attention is also dedicated to the fact that the wealth generated by this trading activity, along with the plundering of other parts of the African continent, provided a foundation that would rapidly transform the impoverished little country of Portugal into one of the greatest colonial powers in history.

3. POWER AT SEA

An Affable Playboy

The rumours about the existence of huge quantities of gold spread like wildfire through dry grass in Europe. It wasn't long before the Portuguese were obliged to defend their most recent conquests through armed combat. In 1478, they attacked a fleet of thirty Spanish ships loaded with gold from territories the Portuguese now considered to be their own. When the cannon were rolled into position on the Gulf of Guinea that summer, it also marked the first naval battle in history between European powers outside their own waters.

In Spain, however, Queen Isabella had little interest in going to war with her Portuguese neighbours. She therefore asked the pope to intervene so they might find a more peaceful solution for the division of the 'new regions'. Pope Alexander VI was happy to comply, since Isabella was one of the Church's most important patrons and allies. The pope himself was Spanish and otherwise known to be an affable playboy, even by the prevailing double standards of the day. Alexander plunged his hands just as deeply into the necklines of his parishioners as into the Church's coffers, and boasted brazenly about his growing flock of illegitimate children.[24] His sleazy morals and lust for women, power and money are often cited as contributing factors leading to the emergence of Protestantism further north in Europe.

Between his erotic escapades, the pope was nonetheless able to dedicate some of his time to the queen's request, and in May 1494 he finished his ruminations on the matter. He issued a papal bull, *Inter Caetera*, decreeing that the world outside of Europe would be divided in two: Spain would be granted all territories to the west of a dividing line running through the tiny island group Cape Verde, while Portugal was granted everything to the east of this line. The pope nonetheless later made an exception when he gave Brazil to Portugal. He did so because Pedro Alvares Cabral, in his attempt to find a maritime route to India, like Columbus made an error in his navigations and in the year 1500 washed up on the Brazilian coast instead. The pope, however, never went to the trouble of carving out the finer details regarding the region where east meets west on the other side of the globe. Perhaps His Holiness had not yet made up his mind about whether the Earth was round or flat.

Just a few weeks later, Spain and Portugal reached an agreement on the key points of the pope's proposal, which was formalised and signed in the Spanish city of Tordesillas. From this point on, they both wholeheartedly embraced the pope's decree for division of land, also in relation to other countries.

THE OCEAN

The Chess Queen

Several decades before this, the Spaniards had conquered the Canary Islands as a bridgehead for further advances to the south. But, when the pope announced his division of the world in two, Spain turned its gaze to the west instead. There were speculations about the possibility of another maritime route to China and India by way of the vast Atlantic Ocean and fearless explorers had already attempted the voyage. But most of them had been blown back, many had lost their lives, and up to this point none of the explorers had been successful. This was first and foremost due to the fact that at this time they did not have adequate navigation devices or sufficient knowledge about the dominant ocean currents and wind systems. More than 300 years would pass before the sextant was put to use and yet another century before Gaspard-Gustave Coriolis explained why the currents and winds north and south of the equator rotate in opposite directions.

During the fifteenth century, several technical advances were made that reduced some of the risk of overseas expeditions. More sophisticated versions of the compass and the astrolabe were developed, the latter an instrument used to establish the ship's position according to the sun and the stars. This made navigation more precise and reliable, although not anywhere in the vicinity of the standard we associate with these terms today. But the perhaps most important event was the Portuguese launch of the caravel, a new type of sailing ship, the design of which applied the fruits of the experiences from sailing missions along the west coast of Africa. These ships were fast, agile and seaworthy, and the most significant innovation was that the rigging gave the ships capacity for windward sailing.

From this point on, it was, in principle, possible to sail across the mighty and ruthless Atlantic Ocean and back again – and survive. Many more adventurous souls were therefore willing to take the risk of crossing the ocean.

The initial phase of Europe's expeditions and conquests was predominantly driven by a thirst for adventure and personal aspirations of wealth, fame and recognition. Individuals, as opposed to emperors, kings, queens or local princes, took the initiative to organise and undertake these voyages. Foremost among them was Christopher Columbus. After having completed his sea voyages off the coast of West Africa, he set out on a year-long begging mission in the halls of the princes and royal families of Europe to finance his great dream of finding the maritime route to India. In 1492 he finally received a 'sí' from Queen Isabella.

The agreement signed by Columbus and Isabella stipulated that everything in the way of land, people, wealth and resources in the regions he discovered would become the private property of the Spanish monarchy. Further, the agreement stated that Columbus would be made governor of these territories, that he would spread God's word, and personally retain one tenth of any riches he brought back to Europe.[25]

The Spanish queen was such a dominant figure in her day that a few years later she brought about the most important change in the rules of chess that had been made since this Indian game arrived in Europe in the tenth century. In 1495 the rules were amended so the queen was allowed to move across an unlimited number of squares and in all directions on the chessboard. The queen thereby became the game's most powerful piece, even though knocking out the highest formal authority, the king, remained the ultimate objective. In this way, modern-day chess reflects the balance of power in Europe at the time, and the personal relationship between Queen Isabella and her husband, King Ferdinand.[26]

Ecological Reunion

The year that the agreement was signed, an impatient and energetic Columbus sailed away. Out of gratitude to his royal patron, he named the first settlement he established in the 'New World', located on a beautiful beach in what is now the Dominican Republic, *La Isabella*. In modern times, La Isabella has been until recently an overgrown memory, hidden and forgotten by everyone except historians and archaeologists. But in recent years, biologists have begun taking a serious interest in Columbus's journeys across the ocean.

While globalisation is predominantly discussed in political, economic and sociological terms, 'it is also a biological phenomenon; indeed, from a long-term perspective it may be primarily a biological phenomenon,' Charles C. Mann writes.[27] Biologists are primarily interested in an era dating several hundred million years back in time, when all the continents of the world made up the single land mass known as Pangaea. The land mass was gradually broken apart by geological processes, and the ocean filled the spaces in between. When we look at a map it is not difficult to see how the west coast of Africa and the east coast of South America, for example, fit together like pieces of a puzzle. After the land masses slid apart, they eventually became isolated, biological enclaves. On each of the newborn continents, unique ecosystems and different animal species, birds, insects and plants evolved. That is why we find kangaroos only in Australia, ostriches in Africa, llamas in South America, and tigers in Asia (with the exception of zoos, of course).

Columbus's expeditions marked the start of a process that would reunite these disparate ecosystems. The voyages across the ocean served to stitch the continents back together, and the ocean was transformed from a physical barrier into an ecological bridge. The European ships brought horses, cattle, pigs, oats, wheat and grapevines, all of which were alien species in the Americas. They also brought hardy weed types, and it is estimated that one half of the approximately 500 species of weeds found today on the American continents came from Europe.[28] On the return trips, the European ships carried cargo such as potatoes, corn, peppers and beans.

Eventually, as the trade and relations between different parts of the world evolved, the continents' respective ecosystems also became increasingly intermingled and integrated. In our time, animals, seeds and plants are imported and exported, back and forth, all over the world. Today it is the local climate and conditions, rather than a species' geographic origin, which determine which plants and animals survive and thrive in different parts of the world. That is why Kristin and I, and our twin daughters Silja and Mathilde, can enjoy tulips, thujas and tomato plants in our garden on the outskirts of Oslo.

In this way, Columbus's expeditions came to represent the start of a process which, according to Charles Mann, was 'to ecologists… arguably the most important event since the death of the dinosaurs'.[29]

Pandemic 1.0

Columbus's expeditions did not only carry soldiers, weapons, animals and plants to the 'New World'; they also brought infectious diseases. While Europeans had developed immune systems that protected them from influenza, measles and smallpox, the indigenous populations of America were not similarly equipped. When the crews carried these illnesses ashore, the indigenous population was exposed and infected on a massive scale in what was ultimately the outbreak of a deadly epidemic.

In this sense, Columbus's 'discovery' of America became the start of one of the largest human-made tragedies in history. Scholars are still trying to establish the approximate number of fatalities, but Howard W. French describes how this came to pass:

> The waves of epidemic and expiration that followed the arrival of whites became part of what has been described as a hemisphere-wide Great Dying. One recent study has suggested that this event killed as many as 56 million people, or roughly 90 per cent of the overall hemispheric population of indigenous Americans between the time of first European contact and the start of the seventeenth century. Such a number would make the deadly transmission of diseases like these by far the largest mortality event in

proportion to the global population in human history, and second only to World War II in absolute terms with regard to the number of people killed.[30]

The death toll within the South American population was so massive that it caused a labour shortage in the European colonies. This in turn provided the foundation for another tragedy of monstrous proportions: the ruthless transport of millions of slaves across the ocean from Africa.

In the Western world, monuments have been erected in honour of Columbus, whose accomplishments are still celebrated through annual holidays in his name in a number of countries, including the USA, Italy, Spain, Argentina, Chile and Mexico. Many people have nonetheless begun viewing his historical role in a more critical light. One need not be 'woke' to recognise that Columbus's legacy begs reinterpretation in light of incontestable historical facts.

As the Covid-19 pandemic has clearly demonstrated in our own times, epidemics and viruses do not recognise national, cultural or ethnic boundaries. The colonists also brought alien illnesses home with them, and a short time after Columbus's expedition returned to Lisbon, the first known case of syphilis was recorded in Europe. From there, the illness was transmitted at an alarming pace, allegedly spreading from the Iberian peninsula northward through Europe and onward to Moscow in the time it would have taken a human being to walk the same distance.

Forced Labour

Because the indigenous population was dying in droves, a labour shortage soon arose on the plantations built by the colonists in the Americas. As a result, African rulers and European merchants glimpsed an opportunity to amass great fortunes through the sale of African slaves. Just one hundred years after the Portuguese arrived at Elmina, the revenues from human trafficking are believed to have exceeded the value of the gold exported from the African Gold Coast.[31] The profitable slave trade attracted the interest of several European countries. Denmark, among others, established fortresses and commercial enterprises on the west coast of Africa, which provided a basis for the subsequent establishment of Danish colonies on the islands of St Thomas, St Jan and St Croix in the Caribbean Sea.

There were already long traditions of human trafficking on the African continent, so there was little in the local culture to prevent rulers from refurbishing the supply of slaves. The Europeans offered their African allies weapons and money, and soon there was a dramatic rise in the number of terrified human beings being chained together and led by force from the inner regions of the continent to the coastline. There they were loaded onto ships to embark on a dangerous and horrific transatlantic Ocean journey lasting several months. Men, women and children were tightly packed into the hot stifling darkness below deck, where they were manacled in place. Many of the women were pregnant, having been raped repeatedly before they were stowed on board. During the crossing, the slaves received the bare minimum of food and water – or not even that – and the stench of fear, sweat, urine and excrement must have been unbearable. When the ships reached their destination on the other side of the Atlantic, the slaves were sold, while the ships were loaded with cotton, coffee, sugar, tobacco and other goods from the colonies and sailed back to Europe. There the cargo was unloaded and the ships, on the third and final leg of the journey, returned to Africa carrying arms, ammunition and other European goods that would be used as payment to local kings and residents who then helped to capture further slaves.

This was the Atlantic 'Triangular Trade' which started in the first half of the sixteenth century and would continue for more than three hundred years.[32] The triangle was part of a larger network that also included other parts of the world, such as the slave trade between West and East Africa. The exact figures are not certain, but it is believed that from ten to twelve million enslaved Africans were shipped across the Atlantic Ocean to the colonies in the Americas.[33] The population of Africa was so depleted by the slave trade that by the mid-nineteenth century it was very likely only half the size of what it otherwise would have been.[34]

Lost Cargo

The slave trade was a heartless and heart-breaking operation. It is estimated that only four out of ten slaves survived the gruelling trip from their villages in Africa to the plantations in the Americas. Of those who survived the journey, only two out of three were still alive after four years in the service of ruthless plantation owners.[35] This, however, did not make much of an impression on those who were reaping a handsome profit from the lucrative purchase and sale of human beings. For the slave traders, the loss of human life was a tedious but unavoidable economic cost, a 'natural wastage' of the business model. For the shipping companies that owned and operated the ships, these were risky but highly profitable journeys. Several of them had agreements with the Spanish rulers, so-called *asiento de negros*, which granted them a monopoly on segments of the slave transport to Spain's colonies in South America.

3. POWER AT SEA

There are numerous stories of the inhumane treatment of enslaved Africans, but few are as unsettling and well-documented as that of the slave vessel *Zong*, owned by the Liverpool shipping company Gregson.[36] In the middle of August 1781, the *Zong* set out from Accra in today's Ghana with 442 slaves on board, twice as many as the ship normally carried. When the ship finally arrived in Jamaica after four months at sea, only 300 slaves remained. In the subsequent trial the crew claimed that the voyage took longer than planned due to navigational errors and for that reason they ran out of fresh water. On 29 November 1781, while they were still far from land, according to the crew the shortage of fresh water on board was so critical that they 'found themselves obliged to' throw fifty-four women and children overboard. Women and children were discarded first because their sales value was less than the men's. In the next few days, seventy-eight male slaves suffered the same fate. In desperation and panic, another ten slaves committed suicide by jumping overboard and into the ocean. When the ship reached Jamaica, the remaining slaves were sold on the market square for an average price of thirty-six pounds each.

So far, this gruesome account is not particularly different from many other stories about the slave trade. What makes this case unique is the trial that took place when the insurance company refused to honour the shipping company's compensation claim for the enslaved Africans 'that had to be thrown overboard'. The shipping company had, as was common at this time, insured the slave ship against 'loss of cargo'. The case was therefore solely about the formal legal technicalities related to the shipping company's compensation claim, not about the grotesque and criminal aspects of drowning 142 men, women and children in the ocean. The slaves were considered 'cargo', as opposed to individuals with intrinsic value, families, knowledge and culture.

The case was therefore handled as an insurance case, rather than a penal case. The key question before the court was whether it had been 'necessary' to throw the slaves into the ocean, since the insurance company could prove that rain had been pouring down on the day in question. It could also be documented that the ship still had almost two tons of fresh water on board when it finally reached Jamaica. The shipping company's claim was upheld by the court, but the subsequent appeal was denied.

Lord Chief Justice William Murray, who administrated the case in both courts, was later promoted and named Earl of Mansfield. The case, however, attracted so much attention and awakened so much disgust in its day that it became an important symbolic catalyst for the abolitionist movement, which played an integral role in ending the slave trade.

Treasure, Silver and Silk

While the European colonial powers were securing footholds in ever greater portions of coastal areas around the world, the emperors of China had their hands full with the reinforcement of a system of defence against the Mongolians. To finance the relocation of the capital and the massive construction projects for the Grand Canal and the Great Wall of China, the Ming Dynasty's emperor decreed in 1571 that only silver would be accepted as payment for taxes levied in the vast Chinese Empire. Until then, subjects could pay their taxes with rice, fruit, vegetables, goats, cattle, tools, pots and vats, or anything else the imperial court deemed of value. This system was, however, poorly suited for the financing of construction projects and also produced practical problems in terms of warehousing and storage. Large portions of the tax revenues were literally consumed by mould, rust and rot. From this point on, only silver was accepted as valid payment and this created a dramatic upswing in the demand for the precious metal in the Chinese population. The significance of silver in the economic history of China is reflected by the Chinese language: the Chinese character signifying money is 'silver' and the character for bank is 'silversmith'.

China's tax system overhaul coincided with the discovery of large quantities of silver in the European colonies in South America. Especially abundant were the silver mines in the Spanish colonies in Bolivia, Colombia, Peru and Mexico. The colonial powers' newly discovered wealth, combined with China's redesigned tax system, produced the basis for a worldwide silver trade, across the ocean from South America to Europe, and from there onward to China. In exchange, the European vessels brought home silk, porcelain, jewels, spices and other exotic goods from China. Although the cargo volume was modest, by the standards of the day the silver trade was extremely valuable.

This silver-based, seaborne trade between the three continents is considered by many historians to have been the starting shot for what would later be referred to as the initial phase of 'globalisation'.

Cash, Not Colonies

In Europe, more countries began acquiring overseas ambitions, and the Dutch were the first to challenge Spain and Portugal's world monopoly by papal decree. This occurred in the wake of the Reformation, which began on All Saints' Eve in 1517, when Martin Luther hammered his ninety-five theses on the church door in Wittenberg. Luther criticised and protested against the Catholic Church's endeavours to amass riches through its self-appointed role as a customs station and the gatekeeper granting entrance to the Kingdom of God. He viewed the men of the church, such as the above-mentioned Pope Alexander VI, as not only greedy and power-hungry; in his eyes, they also represented costly, disruptive and unnecessary middlemen in the relationship between God and his flock here on Earth.

There were few, however, who had the pleasure of reading Luther's ninety-five theses at this time. Only one per cent of the German-speaking population were literate. The Reformation improved the literacy rate in Europe long before introduction of the school system, creating favourable conditions for Europe's cultural, technological, political and economic development. Joseph Henrich writes:

> Embedded deep in Protestantism is the notion that individuals should develop a personal relationship with God and Jesus. To accomplish this, both men and women needed to read and interpret the sacred scriptures – the Bible – for themselves, and not rely primarily on the authority of supposed experts, priests, or institutional authorities like the Church.[37]

The Dutch adopted Calvinism, a strict and God-fearing branch of Protestantism, and refused to be ruled any longer by papal decrees and bulls of excommunication. Due to the extremely strained relationship between the Catholic church and Protestant renegades, and the Netherlands' rebellion against the colonial monopoly, the industrious Dutch soon found themselves in conflict with Spain and Portugal. The Dutch ships therefore went to great lengths to avoid the waters controlled by the powerful Spanish and Portuguese fleets. This is the origin of the legend of the *Flying Dutchman*, the ship condemned to sail the seven seas for all eternity.

To evade the Spanish and Portuguese warships, the Dutch also tried to make the old dream of finding a northern maritime route through the Arctic Ocean to China and the Asian continent come true. They searched for a *Noorwegen* ('the northern way', possibly the etymological origin of the word 'Norway') along the extended coastal territory in the north, and in 1691 Wilhelm Barents set out on his second expedition in search of the Northeast Passage. The expedition foundered on the stony, frozen shores of Novaya Zemlya, where the entire crew perished from the cold, hunger and illness. But Wilhelm's legacy has been immortalised in the name of the Barents Sea.

The Netherlands was first and foremost a trading nation; its goal was cash, not colonies. The Dutch authorities held the view that governing and defending conquered territories was a waste of time, money and energy. For that reason they left management of the colonies in the hands of private companies, in exchange for taxes and dividends paid to the monarchy. The Dutch West India Company was established in the Caribbean, Brazil and North America, and in 1626 they purchased the peninsula of Manhattan from the local indigenous population for 60 Dutch guilder, equivalent to US $24.[38] Here the company operated the bustling trade centre New Amsterdam for forty years, until the English Royal Navy sailed in, seized control of the peninsula and renamed it New York.

At the beginning of the seventeenth century, several of the Dutch companies in Africa and Southeast Asia were merged into a single mega-company, Dutch East India Company, which acquired a monopoly on the highly lucrative spice trade. Just a few decades later, it is held that the company had expanded to the point that it stood for half of Europe's trade with the rest of the world.[39] The company owned hundreds of ships, had its own fortresses, warships and armies, and ruled the coastal regions of the Dutch colonies in South Africa, Mauritius, India, Indonesia, Malaysia, Thailand and Vietnam with an iron fist.

Back home, the Dutch were also profiting handsomely from partnerships with private investors, companies and European princes who lacked the capabilities required for management of the fast-growing world trade. The Dutch had top-notch seafarers and first-class shipbuilders and could in addition offer loans, investment capital, market knowledge and business acumen.

Mare Liberum

Inspired by the growing rivalry between European naval powers, specifically pertaining to an ongoing conflict between Portugal and the Dutch East India Company, in 1609 the young Dutch jurist, humanist and theologian Hugo Grotius formulated his view on a number of fundamental principles for traffic on and use of the ocean. In *Mare Liberum* ('the Freedom of the Seas'), he advocates the idea that all human beings and countries have a nature-given, universal right to freely use the ocean for trade and fishing, among other pursuits. With time, Grotius's approach and principles were universally accepted and would later constitute the foundation for all modern-day thought about maritime law, as this is set out in the United Nations Convention on the Law of the Sea (UNCLOS).

Grotius's principles were nonetheless controversial in his day, above all because many felt that the Dutch interpreted and enforced them with an eye to promoting their own interests. The Dutch, for instance, invoked these principles as a basis for the right to fish in the waters near the coast of England. Two decades later, in 1631, the English jurist John Selden countered these principles in his response *Mare Clausum* ('the Closed Sea'), arguing that coastal states must have exclusive rights on waters along their own shores. With time, this idea gained increasing international support. There is also a long historical connection running from Selden's principles to the UNCLOS's modern regulations for coastal states' exclusive rights and obligations in their territorial waters.

A Stock Exchange under the Stars

In the sixteenth century, abundant trading activity, an extensive international network and a close-knit cluster of capital and commercial expertise made Amsterdam the most important and innovative financial hub in Europe, a position the city would retain for the next two centuries.

In 1602, the Dutch East India Company became the first company in history to issue shares. Despite the massive size of the company and its monopoly privileges, its operations were capital-intensive and risky. The company's enterprising financiers therefore hatched a revolutionary plan: they would finance new investments by dividing the company into small ownership stakes, shares, to be sold against the promise that the buyer would benefit from appreciations in value and receive a portion of future profits.

To make the system more attractive, the company simultaneously established a meeting place where investors could purchase and sell shares among themselves. This outdoor marketplace, located on Warmoestreet in Amsterdam, became the world's first stock exchange. Almost ten years would pass before the enterprise moved into a building and acquired a roof over its head. For several years it was also solely Dutch East India's own shares that were traded at the Amsterdam stock exchange, until other enterprises eventually joined.

The stock market gave companies access to large sums of money for investment and expansion and contributed to the distribution of risk across several owners.

Edward's Little Café

When trade with the European colonies overseas skyrocketed in the sixteenth and seventeenth centuries, the demand for more specialised forms of risk relief also grew. The voyages were long, the dangers numerous, and the number and size of the ships transporting valuable cargo overseas continued to grow. The insurance agreements were refined and standardised, and people with strong mathematical skills were hired to analyse risk and calculate insurance premiums. The actuary profession was hereby born.

THE OCEAN

On Tower Street in London, just a short walk from the busy quays along the Thames, lay Edward Lloyd's little café, a well-known and popular meeting place for the shipowners, captains, financiers and merchants of the day. Amidst the heavy smoke from cigars and steaming pipes, stories, gossip, rumours and the latest news from the outside world were exchanged. Business agreements and insurance documents were signed between teacups and brandy glasses.

The café was the forerunner to Lloyd's of London, today one of the world's largest marketplaces for commercial insurance. It was Lloyd's that would later insure the slave ship *Zong* and take part in the court case regarding the enslaved Africans who were thrown overboard. With time, the incoming payments of insurance premiums amounted to such large sums that Lloyd's also established its own bank. Known as Lloyd's Banking Group, with its 65,000 employees and more than 30 million customers it is today one of Great Britain's largest financial institutions.

One need not be a skilled actuary or even especially quick-witted to understand that the probability for accidents and mishaps is greater if a ship is shoddily constructed, undersized or poorly maintained. To increase profits, shipowners sought to economise on all kinds of costs, all the while assuring both the insurance companies and cargo owners that the ships were safe and of a good standard. The insurance companies for this reason did not fully trust the shipowners' own claims about the condition of their ships.

This led to the emergence of the so-called classification companies, independent third parties who were tasked with assessing and certifying the seaworthiness of commercial ships. It was Lloyd's who in 1760 founded the world's first classification company, Lloyd's Register (LR), which remains one of the largest to this day. This company was also the first to publish annual lists of the conditions of each individual ship. On the list, each ship was ranked and classified according to criteria such as how it was built and maintained, the type of cargo it could carry and whether it was suitable for inter-continental ocean transport or merely traffic in coastal waters. The focus was on the ship's technical standard. Almost one hundred years would go by from the time the industry began keeping records of the number of ships that disappeared or sank, until the classification societies and other industry stakeholders went to the trouble of keeping systematic statistics of the number of sailors who perished or were injured.

The origins of nearly all major modern classification companies, such as the Norwegian Det Norske Veritas (DNV), French Bureau Veritas, Japanese Class NK and the US Bureau of Shipping, can be traced back to this era of sailing vessels and colonial trade. For most of these companies, the share of revenues from maritime activities is less in our time than previously. Today, these companies will often be responsible for independent quality assessments and classifications of everything from trains, passenger planes and space stations, to high rises, bridges and tunnels.

It is hardly a coincidence that the Latin word *veritas* appears in the name of both the Norwegian and French companies. Management of the 'truth' has proved to be a lucrative business.

The King's Daughters

England and France were also enticed by the infinite opportunities of the big wide world. In the late sixteenth century, there were still no states that had proper navies, but the English and French commenced building warships to compete with Spain, Portugal and the Netherlands. There were few other European countries at this time who could hold their own in this competition. The Habsburg empire of Austria-Hungary had only modest maritime traditions, and the Italian city-states and German principalities would not consolidate into their respective nation states for many years to come.

In the seventeenth and eighteenth centuries, France was at the outset better positioned than England to become a colonial superpower. Author Harry Magdoff cites how France 'had the largest population and wealth, the best army while Louis XIV ruled, and, for a time in his reign, the strongest navy'.[40] The French were, however, continental introverts, notoriously focused on wars and conflicts on the European continent, and at the time had only a limited interest in pursuing grand ambitions of world power. For this reason, their initial expeditions to North and South America in the sixteenth century were poorly equipped and half-heartedly executed. It was only when the Spanish Empire began showing signs of weakness in the seventeenth century that the French upped their game. They sent their navy to establish colonies in the West Indies and North America, such as in Haiti, around the Mississippi and in Canada, and granted the Compagnie Française des Indes Orientales a monopoly on trade with France's colonial conquests in the Southeast Asian waters.

In North America, however, there was a continued shortage of French settlers to build, operate and defend the newly acquired colonies. This led the authorities at home to devise creative and untraditional solutions. Early in the seventeenth century, the powerful Cardinal Richelieu issued an ordinance stipulating that any 'natives' who converted to Catholicism would be considered 'naturalised Frenchmen' with the right to reside in the colonies or in France. Later Louis XIV sent one hundred women, in common vernacular referred to as 'the king's daughters', to the colonies to seduce residents into French reproduction. During one period they tried killing two birds with one stone by releasing prisoners in France on the condition that they would marry prostitutes and move to the colonies in North America. The idea was a kind of French 'Kinder egg': the ambition was to empty the prisons, clean up the red-light districts and populate the colonies in one fell swoop. But the entire scheme met with furrowed brows and lukewarm interest on the part of the settled, pietistic French colonists.

In the eighteenth century, France's colonial territories in North America were larger than any of the other European powers. Despite energetic efforts on the part of the French authorities to increase the population, towards the end of the century the ratio of English to French settlers in North America was almost 150 to one.[41] With time, because of this numerical superiority, the English language and population became politically, economically and culturally dominant throughout all of North America.

Ten Pounds for a City – and 200,000 for a Country

The late spring of 1588 was a red-letter period in England's historical journey towards global power and national wealth. In May of this year, more than one hundred ships from the Spanish Armada set sail for the Netherlands, where they would pick up Spanish soldiers and subsequently invade England. King Philip II was sick and tired of the arrogant, obstinate people of this tiny island kingdom in the north, who were not only Protestant, and thus in opposition to the Catholic Church, but had also begun to challenge the Spanish Empire's dominion on the other side of the Atlantic Ocean.

The English had far fewer and more poorly equipped war ships than the Spanish, and to mobilise a sufficient fleet Francis Drake, the pirate who had just been appointed admiral, had to rely heavily on so-called privateers, crew and ships paid to participate in the battle. He understood that the odds were by no means in his favour when he raised the sail and went to battle against the period's largest military power at sea. But, with a mixture of great tactical skill and equal portions of luck, the English won a stunning victory. A sudden change in the weather, followed by a violent storm, sent the Spanish fleet scattering in all directions. Individually, the Spanish ships became easier targets for the English cannon-bearing ships. When the battered remains of the proud Armada finally returned to their Spanish ports, one third of the fleet had been sunk or destroyed.

Although the Spanish Empire remained dominant for a long time, the Armada's crushing defeat was a contributing factor in Spain's gradual decline and England's emergence as the largest empire in the history of the world. The victory over Spain also provided further fuel for England's already sizeable and burgeoning self-confidence. Seduced by rumours about great riches in the East and inspired by the success of the Dutch, they founded the East India Company and sent the Royal Navy on a mission to Asia.

It would soon become clear, however, that the English start-up company was unable to compete with the already well-established and powerful Dutch East India Company in the regions of Southeast Asia. The English therefore concentrated their activities on the western parts of India instead, and here the company at last achieved a foothold when they were allowed to take Mumbai out of the hands of England's own King Charles. The king had received the city as a dowry when he married the Portuguese queen in 1661, but he soon tired of arguing with the Portuguese about who was entitled to the revenues from port duties. In 1668, he therefore formed an agreement with the East India Company, authorising it to take over the entire city in exchange for an annual lease fee of ten British pounds. He also received a loan from the company for 50,000 pounds at a six per cent interest rate, providing him with an injection of disposable income for wars, parties and other amusements.[42]

This laid the foundation for India's destiny as England's largest and most important colony.

The temptations of colonialism had major consequences also for several of the European countries that never reached the Premier League of global powers. In the early eighteenth century, Scotland made a valiant but failed attempt to jump onto the colonialism wave when it invaded the small country that today is Panama. The venture ended in a humiliating defeat and generated a national debt of 200,000 pounds. The sum perhaps does not seem outrageous, but the debt was enough of a burden that the Scottish government in the end accepted Queen Anne's proposal to cover the debt in exchange for entering into a union with England. The two countries had at this time for more than one hundred years made several failed attempts to form a union, and also lived under the same king for a period. But in the end it was money, or the lack thereof, that was the deciding factor for the Scots.

This is how Great Britain was formed in 1707 for the price of 200,000 pounds, and from there we can trace the long historical lines leading up to today's intense and agonising political discussions about Scotland's autonomy, national independence and relation to the European Union.

Drunk and Disorderly

When Peter the Great (who at a height of two metres cut quite an impressive figure) ascended to the Russian throne in 1682, obviously he could not stand by watching other European countries voraciously helping themselves to territories all over the world. Russia was already at this time the largest country in the world, but Peter had even loftier ambitions. He was also obliged to contend with Russia's eternal military headache: the lack of unrestricted, year-round access to the great ocean areas of the world. At this time, the country's only large port was Arkhangelsk, the City of the Archangel, in the far north by the windswept White Sea. Since the sea froze over in the winter there, Russia was landlocked by none other than nature itself.

The fear of being shut in has always been a source of strategic claustrophobia on the part of Russian military planners, and this 'encirclement syndrome' continues to define the underlying premises for their modern-day thinking. It is not only the ice in the ports that creates problems; hostile navies can also easily block the narrow passageways to the ocean in the south, east and west.

In 1697, Peter the Great therefore set out on an ambitious tour of Europe. Accompanying him was an entourage of 300 people known as the Great Embassy. Peter wished to enlist the support of European countries in the fight against the Muslim Ottomans who were threatening the Russian empire's southern border and creating problems for vessels sailing through the Strait of Bosporus between the Black Sea and the Mediterranean. He also wanted to learn the principles of modern shipbuilding since a powerful empire must have a powerful navy.

But it was not only military considerations that made a navy so important for Peter the Great. With a hard-hitting fleet he could wage war without being as dependent on the Russian noblemen who were responsible for recruiting the farmers to the huge army on land. The navy was thus also a key to bolstering the tsar's power and position and undercutting the interests of a powerful nobility.

Since his attempts to secure the sailing routes between the Black Sea and the Mediterranean were largely unsuccessful, Peter instead trained his gaze on the Baltic Sea. Here he attacked Sweden, who controlled Finland at this time. In 1703, he established the city named after him in the conquered region. St Petersburg replaced Moscow as the empire's capital and became the country's 'window to the west'. The port that was built here became from this point on and up until today also the most important for Russia's foreign trade operations.

Peter the Great was a brilliant strategist, but he had a complex and demanding personality. He was fond of the bottle, had an express distrust of people who didn't drink themselves senseless, and purportedly said that 'one has to be drunk every day and never go to bed sober'.[43] Like some of the group charter trips of our times, the Great Embassy's European tour was concluded with a massive booze-up. It took place in a private palace the entourage had rented in a central district of London, and on the itemised bill from the angry and extremely dissatisfied owner we can read that 'all the floors were covered in grease and ink … The curtains, quilts and bed linen were "tore in pieces". All the chairs in the house, numbering over fifty, were broken, or had disappeared, probably used to stoke the fires. Three hundred windowpanes were broken and there were "twenty fine pictures very much tore and the frames broke."'[44] Even the large, beautiful garden was completely destroyed.

Down the road, alcohol would also play a key role in determining Russia's access to the ocean, as compellingly explained by Norwegian journalist Hans-Wilhelm Steinfeld in his book about Putin.[45] Peter the Great never succeeded in holding his own against the Ottoman troops, but Catherine the Great (at this time 'the Greats' abounded) later enlisted the help of Ukrainian Prince Potemkin to oust the Turks. The Crimean peninsula thereby ended up in Russian hands in 1775, at a time when Ukraine was in a union with Russia. When the union's 300-year anniversary was celebrated in 1954, Soviet President Nikita Khrushchev, who was raised in Ukraine, was so exceedingly intoxicated that in a burst of effusive spontaneity he gave Crimea back to Ukraine as a kind of anniversary gift. At the time of the collapse of the Soviet Union in 1991, the newly elected Ukrainian President Leonid Kravchuk was therefore fully committed to signing an agreement for the return of the peninsula to Russia. But during these chaotic and fateful days under considerable duress, Russia's President Boris Yeltsin turned to the bottle as usual to calm his nerves and refused to listen to the insistent counsel of his generals and admirals. Yeltsin was so many sheets to the wind that he basically forgot about the entire business and the Crimean peninsula remained in Ukrainian hands.

A subtle historical irony lies in the fact that it was the alleged teetotaller Vladimir Putin who ordered the occupation of Crimea in 2014, the same peninsula that one of his predecessors gave away in an inebriated state and the other could have recovered with a simple signature – had he only managed to stay sober.

Seward's Folly

Peter the Great was also drawn to the opportunities in the north. In 1724 he tasked the Danish-born naval officer Vitus Bering with the mission of finding the maritime route to China along the northern coast of Russia. After having crossed the strait which today bears the sub-lieutenant's name, the industrious Dane instead sailed onward down the US west coast. This paved the way for Russia's establishment of the colony of Alaska in the late eighteenth century, with New Archangel as the capital. Here, as in other parts of the colonised world, the indigenous population was displaced, enslaved and oppressed. A few generations after the arrival of the first Russian settlers, eighty per cent of the indigenous population were dead because they lacked the immune system that would protect them from infectious diseases from 'the Old World'.

However, Alaska never received much attention from Russia. The colony had little to offer in the way of economic or strategic added value for subsequent rulers in the Kremlin, who were more interested in their wars on the European continent. In particular, the Crimean War against France, Great Britain and the Ottoman Empire drained the Russian state treasury. In the mid-nineteenth century, Tsar Alexander II therefore offered to sell Alaska to the USA. It was Secretary of State William Seward who signed the agreement when the US government put US $7.2 million on the table in 1867, equivalent to around US $120 million in today's value. Many Americans were critical of the purchase, and for a long time referred to Alaska as 'Seward's Folly'. But critical voices were silenced when large deposits of gold were found on the other side of the Canadian border and Alaska became the gateway to the famous gold rush in the Klondike of the late 1890s.

Seward's 'folly' would later be viewed as both a political stroke of genius and a strategic fluke. One can only imagine what might have transpired, had the Soviet Union stationed large military forces on the North American continent during the Cold War.

World War Zero

At the beginning of the eighteenth century, Spain had colonies in North and South America, Africa and Southeast Asia. The Spanish Empire was still the largest in the world, but the government in Madrid was substantially weakened when the death of the ailing and childless King Charles II in 1701 triggered the War of the Spanish Succession. This war would last for twelve years and become a battle over dominance in Europe and control of the colonies. 'This was world war zero,' according to writer George Choundas.[46] On one side of the conflict were Spain and France, on the other an alliance between England, the Netherlands and Austria-Hungary.

At the time of the war, the key to Spain's power and wealth lay in the silver and gold mines in Mexico, Peru and Bolivia. The maritime routes to Europe passed through the Caribbean Sea, and, in Spain's view, all the islands in this region were part of the Spanish Empire. It was Spain, after all, who had financed Columbus's 'discovery' of the West Indies, and the pope himself had decreed that this part of world belonged to Spain. Nonetheless, England managed to gain a foothold on a number of the islands, including Jamaica, the Bahamas, Bermuda and Barbados, and wasted no time installing bases from which to attack the Spanish fleet transporting valuable cargo back to Europe. The English could hereby cut off the flow of capital funding the floundering Spanish Empire.

Money also came to play a critical role in the evolution of the struggle for control between Spain and England. It wasn't until the late sixteenth century that the types of fleets and ships we now associate with navies saw the light of day. Up to this point, during battles at sea, ships would sail in close formation so the soldiers could board and fight on enemy ships. When full-rigged ships armed with cannon were introduced towards the end of this century, the nature of naval battles changed, because the ships could keep their distance and fire broadsides at one another. The ships became larger and faster, but building, outfitting and crewing them also became much more expensive.

Moreover, ship building required shipyards and supporting industries, which became some of the largest commercial enterprises of their time.

Pirates of the Caribbean

With colonies in many parts of world that required defending, and after several years of war with Spain, the English monarchy was virtually scraping the bottom of its war treasury barrel. Queen Anne was having difficulties paying the salaries of the sailors of the Royal Navy, so she offered them something else in lieu of cash: the right to rob and steal. The queen issued what were called *letters of marque*, basically a royal licence to attack, plunder and sink Spanish ships. This state-sanctioned piracy scheme was at the time both well known and popular among kings, queens and intrepid fortune hunters. For example, Francis Drake, who was awarded a knighthood for being the first Englishman to sail around the earth, went on to earn his living as a licensed pirate in the Caribbean in the employ of the English monarchy, until he was called back to lead the fight against the Spanish Armada in 1588. Throughout the seventeenth century, widespread licensed piracy also prevailed along the north-western coast of Africa and in the Mediterranean Sea.

For Queen Anne, weakening the rival Spanish Empire was her primary objective, and she therefore allowed the sailors of the Royal Navy to keep the spoils from the ships they attacked. Piracy in the Caribbean hereby flourished with the blessing of the monarchy, and war was waged by privateers. The scheme functioned according to plan: the Spanish state treasury suffered severe attrition from the ongoing war, and the constant pirate attacks throttled the transport of gold and silver from the South American colonies. The Spanish king therefore copied his royal British colleague: he issued letters of marque authorising the attack of English vessels. In this way, the Caribbean branch of the Spanish coast guard, the Guarda Costa, was funded by all the booty the sailors could get their hands on. Then as now, governments resorted to the privatisation of public enterprises when state funds were running low.

When the War of the Spanish Succession ended, England and Spain agreed that they would no longer use letters of marque to fight their battles in the Caribbean. The pirates were not only a threat to adversaries; they were also beginning to cause trouble for their patrons. When both England and Spain declared the practice illegal, the market for these self-employed pirates abruptly disappeared. Tens of thousands suddenly lost their source of income and livelihood and were left to their own devices on the islands of the Caribbean Sea. This was a guaranteed recipe for disaster. In the words of historian Ed Fox, '[t]here was particularly massive unemployment amongst a community of people who knew nothing but sailing, fighting and stealing.'[47]

In the years that followed, pirates from both sides of the conflict joined forces to form sailing bands of robbers. Several of the pirate captains, such as Blackbeard, Hornigold, Jennings and Black Sam, became renowned pirates in their lifetimes. Johnny Depp's character Jack Sparrow from *Pirates of the Caribbean* is purportedly based on the life of the shrewd and legendary Captain Blackbeard. If this is correct, here comes a small spoiler regarding Jack Sparrow's fate: he was eventually taken prisoner and executed by the English Navy, who dangled his head off the end of a sloop's bowsprit when the ship docked in Virginia on a late November day in 1718.[48]

It was not until the mid-nineteenth century that European countries also ceased employing state-sanctioned pirates in other parts of the world. The United States retained the system until 1907, when it signed the Hague Convention, the precursor to the Geneva Convention which would later form the basis of modern international humanitarian law.

A Paris Agreement

When the War of Succession finally ended in 1714, Spain lost substantial portions of its empire. It was in this context that the Rock of Gibraltar, located on the narrow strait separating Europe and Africa at the western entrance to the Mediterranean, was transferred from Spain to Great Britain. Today the tiny British stronghold off the Spanish mainland seems to be a recalcitrant territorial anomaly, a bizarre artifact from a bygone era. But a British naval base is still located on Gibraltar, serving as a strategic checkpoint on the most important trade route between Western Europe and the countries of Southeast Asia and the Middle East.

3. POWER AT SEA

In the mid-eighteenth century another large-scale war broke out on the European continent. The Seven Years' War was in reality a battle over world dominion between Great Britain and France. France formed an alliance with Russia and Austria-Hungary, while the Prussians fought on the side of the British. Great Britain also received support from its former arch enemy Spain to fight the French in the American colonies.

The Seven Years' War ended with a bitter defeat for France. When the countries made peace in 1763, the French were obliged to sign the Treaty of Paris, under which they agreed to surrender all of Canada and the regions east of the Mississippi to the British. While the Paris Agreement of our times is about saving the world, at this time it was about dividing it up.

The Portuguese had by then already long since been on the margins of the global balance of power. The tiny country had been hard-hit by the repercussions of its colonial overreach and in practical terms had been devoured by its Spanish neighbours. Spain still controlled territories larger than those of Great Britain, but they were no longer a match for the superior British power at sea. All the same, it was ultimately the old-fashioned, inflexible and weak Spanish economy which dealt the *coup de grâce* to the country's position as a leading colonial power. The Spanish colonies were exploited as a veritable resource grab, and almost all of the revenues went to the monarchy. Giant sums were spent on waging war and building beautiful palaces, but the vast wealth from the colonies generated little domestic economic development or industrial growth. According to historian Terje Tvedt, Spain also lacked a network of suitable rivers that could provide the necessary infrastructure for rapid economic growth in the soon imminent Industrial Revolution.[49] The Spanish economy crumbled, and the country fell hopelessly behind the development elsewhere in Europe.

With France defeated, Portugal out of the picture and Spain weakened, Great Britain continued to expand its territories overseas. In 1770, James Cook went ashore on the north-east coast of Australia after having first stopped off in New Zealand. Captain Cook planted the flag and claimed both territories on behalf of the British king. To begin with, Great Britain did not derive many benefits from its latest acquisition, but the monarchy wanted to consolidate its control over the Australian continent. Would-be British settlers were therefore offered free labour in the form of convicted criminals from English prisons. This was how the first settlements *down under* were formed, around Sydney, as a mixture of farming and penal colonies.

The local Aboriginal people naturally attempted to defend their territories and traditional ways of life against the colonial invaders. But their spears, slingshots and boomerangs were a poor match for the British modern weaponry. The majority of the indigenous population who resisted were imprisoned, killed or displaced. To this day, more than two hundred years later, many Aboriginals remain overrepresented in the lower and marginalised strata of Australian society, with consistently low rates of completed schooling and higher levels of unemployment and social problems.

While Australia celebrates its official national day on 26 January, there are many who instead take part in protest demonstrations and call the day Invasion Day, in commemoration of the fate of the indigenous population. In several places, Seas of Hands are set up, physical installations of hundreds or thousands of cardboard hands planted in the ground to symbolise support for the Aboriginals and other First People.

Tea Party

Although Great Britain's global power was expanding, the empire met with growing resistance from the North American colonies, where their own countrymen were rebelling against British rule. The rebels of English descent were especially upset about having to pay taxes to the British treasury without receiving any particular benefits in return. These hardened and hard-working settlers who were struggling to survive had little interest in contributing to the British war chest or the monarchy's lavish luxuries. It was also considered a provocation that they were expected to pay taxes without being represented in Parliament in London: they had all the obligations but none of the rights of ordinary Englishmen, which was the origin of the rebels' slogan 'No taxation without representation!' Eventually, in response to prolonged and escalating protests, the British government repealed all taxes on imported goods in the North American colonies – with the notable exception of tea. The tea tax hereby became not solely an irritating added expense; it also became a unifying symbol for all those who supported full independence from Great Britain.

The conflict was brought to a head when a group of rebels, disguised as Native Americans, boarded a ship docked in Boston Harbor on a December night in 1773. The rebels threw forty-five tons of bags containing tea leaves overboard and disappeared without a trace into the night. None of the rebels were killed or injured, but the government in London immediately tightened its grip, imprisoning rebels, restricting autonomy and deploying more soldiers. It is said that King George III was so furious that he vowed 'to keep the rebels harassed, anxious and poor, until the day when … discontent and disappointment were converted into penitence and remorse'.[50] The king's intractable reaction to what came to be known as the Boston Tea Party nonetheless only served to fuel further resistance and rebellion in all of the thirteen British colonies along the Atlantic coast. The episode would later be considered one of the most important events on the road to the American Revolution and the achievement of full autonomy just three years later, when the revolutionaries signed the Declaration of Independence and founded the United States of America in 1776.

In our times, the American Tea Party movement continues to derive inspiration from this incident, encouraging right-wing Republican voters to oppose higher taxes and undue intervention on the part of the federal government. On protest signs, 'TEA!' is now an acronym for 'Taxed Enough Already!' The momentum of this political, grassroots movement was one of the key forces leading to the election of Donald Trump as president of the United States in 2016.

Hannah

Subsequent to declaring its independence on 4 July 1776, the young American nation state immediately started building up its navy. The first US president, the plantation owner with virtually no maritime experience, was fully aware of the importance of a hard-hitting naval fleet. In a letter, George Washington wrote: 'It follows then as certain as that night succeeds the day, that without a decisive naval force we can do nothing definitive, and with it, everything honorable and glorious.'[51] His foremost concern was preventing the British government from sending troops and supplies to the southern parts of the country, where there were large pockets of discontent and unrest. In 1775 he had already scraped together a tiny fleet of schooners which, with a few simple tweaks, were converted into armed ships. The first of these was the *Hannah*, named after the owner's wife, which would later enter into the US Navy's history as their 'first armed vessel'.[52]

After the USA declared independence and the threat posed by the British no longer loomed as large on the horizon, the Navy's mandate was expanded to include impeding European influence on the islands and countries in the Caribbean Sea. It was also necessary to protect merchant ships from the ongoing threat of pirates in the region.

In the early nineteenth century, the US Navy was sent on its first overseas mission, a campaign targeting pirates and slave traders in the Mediterranean Sea. The fleet was also overhauled at this time, equipping the wooden hulls of sailing ships with armour plating. This was one of the most important military innovations of the day, and the ships proved effective in battle. But the fleet still comprised only a few ships and, even towards the end of the century, only 6,000 men served in the US Navy.

THE OCEAN

A Gesture of Gratitude

In 1789, the French Revolution sent shock waves through the ballrooms of the European nobility. It was a violent and brutal social revolution, born out of oppression and destitution and fuelled by rage, aggression and frustration. The people could not simply 'eat cake if they had no bread', as Queen Marie Antoinette had famously enjoined before she and her husband King Louis XVI were beheaded.

When the French Revolution began, Spain and Portugal, along with the British, exploited the situation of chaos and confusion by launching a failed attempt to invade France. This would prove to be a strategic error with far-reaching geopolitical consequences, because the French Revolution had also opened the door for the country's youngest general, Napoleon Bonaparte. He was born in Corsica in 1769, the same year that France assumed control of the island after having purchased it from the Italian city-state Genoa. Had it not been for this transaction, one of history's most famous Frenchmen would have lived his life as an Italian.

Also known as the 'little Corsican', Napoleon drove back the attacks and subsequently spearheaded France's invasion of the Netherlands in 1795. William V, Prince of Orange, fled for his life, seeking exile in London, and from there commanded his military troops at home in the battle against the French occupying forces. In a generous gesture of gratitude to his English protectors, he gave most of the Netherlands' colonies in Southeast Asia to the British king. In the historically significant 'Kew letters', named after his lodgings in London at the time of their writing, William quite simply conveyed Malacca, Sumatra, Sri Lanka and the southern parts of India to his royal and hospitable British friend. India hereby became Britain's largest and most important colony.

From this point on, Great Britain was not only stronger, but also larger, than the Spanish kingdom. The British Empire could begin its 150-year reign as history's largest and strongest global empire.

Waves of Liberation

At the start of the nineteenth century, Spain's global power was so depleted that the settlers of the Spanish colonies in South America did what the North American settlers had done a few decades before: they declared independence and liberated themselves from their European masters. Between 1808 and 1826, all Latin American countries, with the exception of Cuba and Puerto Rico, severed their ties with Spain. Portugal was also essentially powerless when Brazil followed suit and declared its independence in 1822.

As was the case in North America, it was the settlers of European descent in South America who rebelled against the European colonial authorities and took control when the countries became independent. To this day, it is the white or *creole* populations, people with European branches in their family trees, who totally dominate the worlds of politics, finance and business and constitute the upper social strata of all the South American countries. The indigenous population, and the descendants of enslaved Africans, are still heavily overrepresented in the lower ranks of South American society.

Cash for Colonies

The young Napoleon Bonaparte soon demonstrated the scale of his imperial ambitions, launching a campaign to challenge Great Britain's dominant control over the Mediterranean Sea. With a fleet of more than 200 ships, 2,000 cannon and 30,000 troops, he beat the British in a brutal naval battle off the coast of Alexandria in 1798. He thereby effectively cut off the shortest route between Great Britain and India, their most important colony. The British immediately deployed huge naval forces to launch a counterattack led by Rear Admiral Sir Horatio Nelson, who had lost an eye and an arm in previous battles. After an initial victory, the French fleet was beaten in several battles in rapid succession in the Mediterranean Sea and before long, they also lost control over Egypt. The French fleet was sunk, and Napoleon's army stranded under the scorching desert sun. The decisive victory would reverse the strategic situation in the Mediterranean and solidify Great Britain's global hegemony. The bold and innovative tactics used by Admiral Nelson in these battles would also change naval warfare in much the same way that Napoleon's novel tactics had changed war between armies on land.

His defeat in Egypt notwithstanding, Napoleon was so popular at home that he had the enthusiastic support of the people when he seized power in France through a military coup in 1799. The French economy at this time was seriously debilitated due to the revolution at home. At the same time, the wars on the European continent were expensive and required many soldiers. The costly and failed attempt to quash the large-scale uprising of the enslaved Africans in Haiti, and aid to America's fight for independence from the British, had also put a sizeable dent in the holdings of the state treasury. As the death tolls mounted and the state treasury was expeditiously drained, rumours began to circulate that France wanted to sell its American colonies and bring the soldiers home.

In 1803, the US government therefore put US $15 million on the table, in what they themselves viewed as a brazenly low offer, for the purchase of

New Orleans, Louisiana and the other French colonial territories west of the Mississippi. But Napoleon was in dire straits, and to the great surprise of the US government, he signed the agreement. In the course of a mere forty years, France thereby went from controlling more territories in North America in the mid-seventeenth century than any other European power to ceding all of its territories on the continent.

Simultaneously, through this agreement the USA doubled the size of its own land area, just twenty years after the country had gained independence. The newly purchased territory encompassed the land area of fifteen current US states.[53]

A Classic Army Commander

In a military sense, Napoleon was a classic army commander, with an emphasis on both army and commander. He was not a man of the sea, put no stock in the merits of a navy and had even less in the merits of admirals. He became even more convinced of the rightness of his convictions when the army's artillery managed to sink – from land – hostile British vessels positioned near the coast of Spain. Yet it was only by crossing the English Channel that he could make good on his plans to conquer Great Britain.

In preparation for the all-important battle, in the autumn of 1805, he mustered an invading army of more than 150,000 French troops which were to be transported across the English Channel. By then, however, the Royal Navy had become aware of Napoleon's plan and blockaded the French warships off the coast of Brest and Cadiz. Napoleon probably still thought he had a good plan, but the execution was poor and the outcome was disastrous. In the Battle of Trafalgar, the French ships were bombed to smithereens by a British fleet led by Napoleon's nemesis in life, Admiral Nelson, who did not lose a single ship in the violent battle. He did, however, unfortunately lose his life, and never had the chance to bask in the afterglow of his recently earned fame and heroic status in his homeland.

The Battle of Trafalgar is often referred to as the most important naval battle of the nineteenth century. It saved Britain from invasion, and France's period as a major naval power suffered a coup de grâce. For the next one hundred years Great Britain remained wholly dominant at sea. Until a newly united Germany began its armament in the lead-up to the First World War, the British Royal Navy was larger than the fleets of all the other European nations combined.

3. POWER AT SEA

Equality, Fraternity and Liberty

The French Revolution was inspired and fuelled by an uprising led by the intellectuals of the day. This was the Age of the Enlightenment, a movement espousing reason, rationality, tolerance and humanistic values. It was the golden era of the emergence of European philosophers such as Montesquieu, Rousseau, Diderot, Locke, Hume, Bacon and François-Marie Arouet. The latter is more commonly known as Voltaire, which is an anagram of the Latinised spelling of the author's surname, Arovet Li.

The Enlightenment was a sweeping intellectual revolution, driven by a fundamental search for truth, insight and understanding, which would inspire cultivation of the principles of modern scientific method. The systematic, knowledge-based verification of hypotheses, assertions and dogma became an effective tool in the pursuit of research, innovation and technological development, but would also prove to be an existential challenge for the feudal and religious rulers of the day.

Traditional ideas regarding the relation between the rulers and the ruled were challenged and turned upside-down. Power based on succession, religious incantations, traditions and fanciful tales was a poor match for rationality's unbiased contemplation and science's dispassionate dissections. The king's subjects no longer accepted an existence as his personal property, and were no longer willing to dedicate their lives and destinies to the kingdom and its rulers. The idea was born that the obligations of society must be mutual, and that the state's *raison d'être* must be to provide protection and services to its citizens. Also, by extension, the state's use of power and coercion must be expedient, proportional and just. Based on the premise that every human being has intrinsic value, integrity and inviolable rights, all citizens were to have the right to live a good life and follow their dreams, hopes and ambitions.

Hence, when French revolutionaries stormed the much-despised Bastille on 14 July 1789 to liberate political prisoners, they chanted the battle cry *'Liberté, Égalité, Fraternité – ou la Mort!'*: Freedom, Equality, Fraternity – or Death!

'Supply chains and connectivity, not sovereignty and borders, are the organizing principles of humanity in the 21st century.'
Parag Kannah, Indian American author

4.
URBANISATION, COASTALISATION AND GLOBALISATION

How the ocean served as a domain and premise provider for trade and economic growth in the wake of the Industrial Revolution. How attacks from the sea by Western powers threw China into its Century of Humiliation, and why the naval powers' transition from vessels propelled by steam to diesel oil engines sealed the fate of the Middle East. How attacks at and from the sea drew reluctant Americans into World Wars I and II, and how the United States emerged to become a global superpower.

iPhone

A forest of hands flew into the air when I asked how many of the students owned a smartphone. I had been invited to a secondary school in Oslo to speak about the role of the ocean in the context of climate change, the environment and economic sustainability. This led to a discussion of how countries are interconnected through worldwide trade.

Few things are more representative of our times than these addictive gadgets, and for that reason I often use my own iPhone as an example.[1] It was designed at Apple headquarters in Cupertino, California, and is made of hundreds of components that are produced by more than 200 companies in forty or so countries all over the world. The majority of these components are shipped on merchant vessels criss-crossing the ocean between countries and continents, until a factory in China assembles all the pieces to create the sleek, sophisticated iPhone. The phones are then distributed all over the world. By the time it reaches the shops, such as the one in Oslo where

4. URBANISATION, COASTALISATION AND GLOBALISATION

I had purchased mine, the total transportation footprint of each iPhone is typically more than 800,000 km, equalling the distance to the moon and back, most of which can be attributed to ships.[2]

The network of seaborne transport enabling the smartphone's global supply chain is a key characteristic of the modern world's economic organisation, the backbone of what we commonly refer to as 'globalisation'.

Waterways

While intellectual, scientific and political revolutions were sweeping the European continent, the Industrial Revolution was taking place in England. The starting shot was fired in 1764, the year of the introduction of the spinning jenny, a spinning frame that made it possible to increase the garment production output of each worker many times over.

The spinning jenny became the first of many English technical innovations that would drive awe-inspiring economic growth. Yet it was not only superior English ingenuity that lay behind the occurrence of this in England, because there were many innovative communities to be found in a host of other countries as well. It was also the infrastructure, first the rivers and canals, and then, one hundred years later, the railways, which gave England a critical advantage and head start.

Recent research findings suggest that England became the birthplace of the Industrial Revolution in part due to uniquely favourable natural conditions. While, for example, Spain suffered from lack of inland waterways, Norwegian historian Terje Tvedt explains how England's deep, calm and stable rivers initially provided two critical parameters that were instrumental to the industrial and economic development. First, the relatively stable climate and calm, steady flow of water through the country's many rivers provided reliable energy year-round to overshot waterwheels powering axles and machinery. This made large-scale industrial production possible in different regions. Economic activity thereby expanded, and it was possible to exploit a larger portion of the country's human and natural resources. Second, the widespread network of suitable rivers and human-made canals provided a particularly effective infrastructure for the year-round shipment of large quantities of goods between different parts of the country. Because of the relatively stable climate on the British Isles, the rivers do not flood, dry out or freeze over. The Industrial Revolution was based on the logic of capitalism, which calls for efficient supply chains – access to the right goods in the right amounts at the right time in designated locations – and, in this sense, few other nations in Europe or other parts of the world had better natural conditions than England. 'The fact that England was part of an island and had ocean on three sides also facilitated the national market's connection to trade routes crossing the ocean,' Tvedt writes.[3]

When the steam engine made its entrance a few decades later, economic activity was further liberated from geographic limitations: the reliance on rivers and windmills to power machinery was reduced and eventually eliminated. Instead, the factories were concentrated in areas offering the best access to labourers and raw materials and the simplest route to large markets. In the port cities on the coast, ships could both deliver large quantities of raw materials from the colonies and transport finished products from these cities to domestic and international markets.

Urbanisation, coastalisation and globalisation occurred hand in hand with the Industrial Revolution. In England, on the European continent, and subsequently in other parts of the world, people migrated from inland rural areas to the cities, most of which emerged around the bustling ports that constituted the hubs of international trade. In this way, the countries transitioned from agrarian to industrial societies.

Today there are more than 500 coastal cities of more than one million residents throughout the world, and the majority of these are located where cargo was shipped in and out following the Industrial Revolution.[4]

Steaming Ahead

In the mid-nineteenth century, Englishman Henry Bessemer discovered how to produce steel from liquid pig iron. The invention of steel made possible the creation of machines that tolerated greater pressure and higher temperatures, and beams and rails that could bear heavier loads. Shortly thereafter, Bessemer's countryman George Stephenson constructed a steam engine small enough to fit inside a locomotive, the 'iron horse', and this was the start of large-scale construction of the English and European railway network. From this point on, the transport of large quantities of goods between ports and inland regions became more time and cost-efficient. Later, Britain built railways in the colonies, such as in India and Kenya, to transport raw materials from the interior to the coast, where the cargo was loaded onto ships headed for Europe.

Before long, steam engines were also installed on ships, spurring further strong growth in maritime freight shipping. The number of ships increased, as did the cargo volume of each ship, to meet the demand of expanding economies, populations and prosperity in the growth-hungry European countries. The steamships were initially constructed with large paddles on the sides or behind. Subsequently, the propellor's revolutionary design was

4. URBANISATION, COASTALISATION AND GLOBALISATION

introduced, which reduced water resistance, thereby increasing the ship's speed and navigational agility. Hulls made of steel were introduced, and the steamships became larger, faster and less sensitive to currents, wind and waves. This significantly increased the capacity and efficiency of seagoing transport and improved the predictability of cargo delivery times. At long last, the factories in England could trust that the raw materials they had ordered from the colonies would be delivered more or less on time.

This marked the fumbling and troubled birth of the just-in-time concept, because there were still many factors that could hinder and delay the ships on their journeys. All the same, the foundation was put in place for the expansive and increasingly close-knit network of supply chains, weaving together the continents, countries and companies of the world. From this point on, companies in different regions could focus on producing what they did best and distribute their goods simply and cost-effectively to almost every corner of the earth.

This is how the supply chain's logic became the organising principle of the modern-day world economy and the genome of globalisation: 'Supply chains and connectivity, not sovereignty and borders, are the organizing principles of humanity in the 21st century,' Parag Khanna writes. He continues: 'Indeed, as globalization expands into every corner of the planet, supply chains have widened, deepened, and strengthened to such an extent that we must ask ourselves whether they represent a deeper organizing force in the world than states themselves.'[5]

The Climate Fuse Is Lit

The steam engine is, as the name suggests, powered by steam. Again, it was water that enabled civilisation to take new strides. In the steam engine, water is heated until it reaches the boiling point and becomes steam, which expands. This expansion creates pressure that is converted into a mechanical force, which in turn powers axles, wheels and machinery. Initially, charcoal fire was used to heat the water. Because charcoal is produced by burning wood in an oxygen-starved environment, large forest areas were razed to feed the hungry engines. This caused vast natural destruction and lit the fuse of the environmental crisis and global warming. When trees are chopped down, they can no longer absorb carbon dioxide from the atmosphere, and when charcoal is burned the greenhouse gases found in the wood are released in the form of emissions.

This development marked the first large-scale human-made environmental impact in history, and this time period would later become the point of reference for the United Nations Convention on Climate Change of 1992 and the international climate agreement of 2015, the latter today only referred to as the Paris Agreement. When we speak of 'limiting global warming to well below 2°C, and preferably 1.5°C', it is in reference to the rise in temperatures that has been underway since the start of the Industrial Revolution.

The environmental impact was exacerbated in the late nineteenth century when charcoal was replaced by black coal, the substance that we today refer to as simply 'coal'. Black coal, which is extracted from mines, produces more intense heat and burns longer than charcoal. This means that more heat can be extracted from every kilo of coal; consequently, the same amount of energy can be generated using less coal and smaller engines. The higher heating value made coal much more suitable for use in locomotives, steamships and factories, such as in industrial processes that converted iron ore into steel. The latter provided further momentum to the production of beams, bridges, rails, ship hulls, bolts and screws. But the downside of these benefits is that every kilo of coal releases a much larger amount of greenhouse gas emissions, since the higher heat energy is due to a greater concentration of carbon.

The second half of the nineteenth century also brought the discovery of how steam pressure can be used to produce electric power. In 1882, the unstoppable Thomas Alva Edison invented a generator that converted the power produced by a coal-fired steam engine into electricity. Mere decades later, most of the large cities in the world were surrounded by huge coal-fired power plants that supplied electricity for lights, heat, cooking and industrial production. In just one century, the human race had taken a gigantic step forward on the path towards what we generally associate with modern-day society.

But the price of progress and modernity could be discerned in the yellowish-brown smoke pouring out of the power plants, factories, homes, locomotives and steam ships. Greenhouse gas emissions rose into the sky, forming a blanket of thermal insulation around the earth, while particles of sulphur, nitrogen and soot rained down on areas in the vicinity of the coal-fired power plants. It wasn't long before the air in the cities became thick with pollution, black soot settled in layers of filth on buildings and in nature, and people fell ill and died from inhaling the toxic substances.

A Global Shortcut

The Industrial Revolution in Europe spurred further growth in the demand for goods from the colonies, and the long-held dream of a canal connecting the Mediterranean and the Red Sea finally came to fruition. In the mid-nineteenth

4. URBANISATION, COASTALISATION AND GLOBALISATION

century, France escalated the work on history's first human-made 'global shortcut'.

When the almost 200-kilometre-long Suez Canal was completed in 1869, rumours were already circulating about how one of the most well-known composers of the time, the Italian Giuseppe Verdi, had been commissioned to write an opera for the grand opening ceremony. But although Verdi composed in a rapid fury, *Aïda* was not completed until two years after the last of the celebrated guests had gone home. As a compensation prize, the premiere performance was held at the opera house in Cairo. It was an immediate success. *Aïda* is one of the best-known works in the history of opera and still receives enthusiastic ovations from audiences when performed in major theatres all over the world.

The French sculptor Frédéric-Auguste Bartholdi was equally unfortunate in his attempt to convince the Egyptian authorities to erect a gigantic statue by the canal's Mediterranean entryway. The female figure draped in robes and holding a torch high above her head was intended to be both a beacon and a symbol of how the Suez Canal would 'carry light to Asia'.[6] To Bartholdi's great disappointment, the idea hit the wall with the Egyptian authorities, but almost twenty years later the US authorities expressed an interest in the proposal. The iconic Statue of Liberty holding her torch hereby found her home at the entrance to New York Harbor.

With the opening of the Suez Canal, the sailing distance between Southeast Asia and Europe was halved, and the crews were spared the long, demanding and dangerous route around the Cape of Good Hope on the southern tip of Africa.

In the year of the opening of the Suez Canal, the USA completed its first transcontinental railway. The network of steel rails connecting the cities on the east and west coasts spawned new opportunities for collaboration and the division of labour in the young nation state. The railway also generated wholly different requirements for the synchronisation of time and labour processes.

Time Chaos

When ships could sail across the ocean and trains cross continents in far less time and much more reliably than previously, this also changed the perception of time. Vague promises such as 'in the course of the day tomorrow', 'sometime next week' or 'maybe next month' no longer passed muster in the increasingly streamlined logic of the industrial supply chains. From this point on, the calendar and the clock would have a wholly different impact on people's daily lives. Even in a backward country such as Norway, a common saying held that 'he who lives with neither calendar nor religion flings caution to the wind'. In the nineteenth century, the calendar was the most widely used book in Norway after the Bible.[7]

As a number of social sectors were obliged to acquire a more active and binding relationship to time, the question 'which time?' was inevitably raised. It was the US railway companies who first encountered the problem on a broad scale. The railway network sprawled across the entire continent and the trains ran between countless cities and stations, all of which had their own local time, based on when the sun was at its zenith in the sky. In 1870, for example, passengers travelling on the recently opened railway from San Francisco to Washington DC would have to reset their watches 200 times (!) if they were to synchronise the settings with the local time zones they passed through on the journey.[8]

It was a formula for chaos and accidents, as it was almost impossible for the companies to coordinate traffic safely. In 1883, the directors of the US railway companies therefore convened and agreed to divide the country into five different time zones, a system still in effect in the USA.[9] The next year, the director general of the Canadian railway organised an international conference for the purpose of standardising time references internationally. They quickly formed an agreement that divided the world into the twenty-four time zones of today, based on the number of hours it takes the earth to complete one rotation on its axis. There was further discussion about the geographic 'reference point' for these time zones, and in the end the choice fell on the Observatory in Greenwich Park in east London.

Of course, because did it not go without saying that the capital of the world-wide British Empire should also be the point of reference for time all over the world? Since that day, clocks all over the globe have been set according to Greenwich Mean Time, GMT.

Whitehouse Troubles

Until the introduction of the electric telegraph in the early 1840s, it was the speed of ships, trains, horses and carrier pigeons that determined how quickly news was spread. When the British Lord Nelson destroyed Napoleon's fleet in the Battle of Trafalgar in October 1805, three weeks went by before the news appeared in print in *The Times* in London. When large deposits of gold were discovered in California in January 1848, all of eight months passed before the news reached New York and another seven weeks before it found its way into print in London. When Abraham Lincoln was assassinated by one of the actors during a performance of the comedy

4. URBANISATION, COASTALISATION AND GLOBALISATION

Our American Cousin in Washington DC in April 1865, the news of his death did not reach London until two weeks later.[10]

The telegraph transmits electric signals that are either connected or interrupted by a sender. In 1842, Samuel Morse was experimenting with the transmission of signals through copper wire encased in hemp and Indian rubber that was laid across the bottom of New York Harbor. Eight years later, the first commercial cable across the floor of the English Channel was installed.[11]

After several years of work riddled with problems and setbacks, the first underwater cable between Europe and the USA was officially opened in August 1858. For the formal inauguration ceremony, Queen Victoria sent a congratulatory telegram to US President James Buchanan, who awaited its arrival with great anticipation in the White House. Quite soon, however, it became clear that the quality and speed of the transmission were sadly below par; it took almost sixteen hours to send the ninety-eight words in the telegram. A sweating and agitated chief engineer Wildman Whitehouse(!), who had originally trained as a surgeon, tried to fix the problem by increasing the voltage. This remedy proved particularly unsuccessful. The dejected Whitehouse was fired and held responsible for the short-circuiting of the cable that ensued. Another ten years passed before a new cable was installed, but, fortunately, this time it functioned.[12]

Free Ports and New Faces

While nothing can compete with cables when it comes to the transmission of large quantities of information between different parts of the world, no other means of transport can compete with ships in the long-distance transport of large quantities of goods. International shipping therefore played a critical role in the strong economic growth and burgeoning international trade fuelled by the Industrial Revolution.

As the British Empire expanded, so did the need for overseas ports that could serve as military strategic bases and commercial transit ports, while ensuring the navy and merchant fleets access to supplies, maintenance and refuelling. From the first half of the nineteenth century, such ports were established, among others, in Singapore, Hong Kong and Aden, and on several of the Caribbean islands. In order to stimulate activity in these ports, the government in London arranged particularly favourable economic conditions, and with that they created the model for what we today refer to as 'free ports' and 'economic free zones'. Later and up until today, countries all over the world have adopted this system, where special ports and fenced land areas in practice function as legal and fiscal enclaves with far more liberal rules for economic activity than in the host country in general. In order to attract international companies and investors, these places are typically exempt from customs and VAT, tax rates are extremely modest, capital can be moved in and out without special obstacles, and they often have far more relaxed working time regulations, environmental regulations, visa rules and requirements for business.

The establishment of free ports and economic free zones has in many places led to strong growth in economic activity. An example of this is Dubai, which became a free port in the 1950s and which today is the Middle East's leading centre for trade, finance and tourism. In recent times, China has also established dozens of such economic free zones, and Hainan is building what is expected to be the world's largest free port when it is completed in 2025.[13]

Although the British merchant fleet in the first half of the nineteenth century was by far the largest in the world, it was unable to keep up with the rapid development of the world economy. Great Britain's own fleet eventually became a bottleneck for the country's economic development. In 1849, Great Britain therefore found itself obliged to rescind the two-hundred-year-old Navigation Act, which up until this time had given British vessels a monopoly on trade with the colonies.

From this point on, international shipping companies were permitted to begin trading with the British colonies, and this laid the groundwork for the emergence of Greece, Germany, Denmark and Norway as some of the world's largest and most advanced shipping nations of today.

Mass Slaughter

The Industrial Revolution led to an explosive surge in the demand for oil and lubricants for machinery, pumps, pistons and axles. Many years would pass before the world began pumping crude oil out of the ground and the ocean floor, and until that time, whale oil was used. Obtained from whale blubber, meat, bones and entrails, whale oil was also used to make soap, paint, margarine, lamp oil, perfumes and cosmetics. From the late 1700s and long into the next century, whale oil was a more important commodity in the world economy than fossil fuels.

The demand for whale oil unleashed a powerful surge of hunting activity, which would come close to decimating the population of the blue whale, the majestic giant of the ocean.[14] Initially, the USA and Norway took the lead in the massive slaughter of these magnificent animals. Whale hunting underwent its own industrial revolution when Norwegian Svend Foyn invented

4. URBANISATION, COASTALISATION AND GLOBALISATION

the grenade harpoon in 1863. This harpoon contained an explosive charge that was triggered when the grenade penetrated the whale's body, causing extensive injuries, suffering and ultimately death. The following year, the newly developed grenade harpoon was installed on the world's first steam-driven whaling schooner, also designed and financed by Foyn. While previously Foyn had landed one or perhaps two whales per season, the inventor of the exploding harpoon now logged up to one hundred.[15] Foyn became a wealthy man, and affluent Norwegian, British and US shipowners were quick to exploit the new and highly lucrative opportunities found thanks to Foyn's new inventions.

Soon the swift whaling schooners were accompanied by large factory vessels, specially equipped to boil, store and ship whale oil. After more than one thousand years of whale hunting using traditional methods, the harvest of the world's largest mammal was converted from the hunt of individual whales to the slaughter of huge pods. In the twentieth century alone, more than three million whales were killed.

In biomass this represents the largest slaughter of any type of animal in history.[16]

Colonial Indoctrination

The Industrial Revolution and the growth in world trade created wholly new parameters and conditions for the relations between the colonies and the colonial powers. Up until this point, the colonies had mainly supplied gold, silver and consumer goods such as tobacco, sugar, exotic fruits, silk, porcelain, precious jewels, spices and rum. But now the colonies also became key suppliers of raw materials such as cotton, metals, timber, jute and animal hides for industry. A need also emerged for the import of more food from the colonies, since the industrialisation of Europe led to the relocation of populations from farms in rural districts to large towns and cities on the coastlines where the factories and seaports were located.

At the same time, several of the colonies had expanded and developed to such an extent that they became attractive markets in their own right, fuelling a demand for imported goods from Europe. Their economies grew so quickly that the local population had to be recruited and educated to keep the wheels turning; employing 'expat' workers from Europe was no longer sufficient. This in turn created a need for other forms of political, economic and administrative management of the colonies.

In the late eighteenth century, Great Britain therefore introduced its own laws, bureaucratic structures and school systems in the colonies, and English as the official language. The school curriculum was based on the British culture, history and world view, while little if any effort was dedicated to instilling knowledge or a sense of pride about local conditions and traditions.

The same process was implemented in all parts of the world under European rule. The local populations were force-fed the colonial powers' languages, school curriculums and administrative systems. In this way, the colonial powers did not merely steal resources; they also divested the locals of their own language, culture, history and traditions – everything defining their own identity and self-esteem.

Portuguese became the official language of Brazil and in parts of Africa, and Spanish became dominant in most of Latin America. Dutch was introduced in Surinam and Curaçao, and French in the colonies in Africa and Southeast Asia. Today, Spanish and English, together with Chinese and Hindi, are the languages spoken by the greatest number of people in the world. Spanish and English are currently the first language for more than six times as many people as the combined populations of Spain and Great Britain.

Opium and Humiliation

Prolonged wars and persistent conflicts on the European continent did not discourage the colonial powers' continued expansion into other parts of the world, and in the beginning of the nineteenth century they set their sights on China. The Middle Kingdom would be obliged to pay a high price for having turned its back on the sea and neglected the protection of its coastline since the time of the Ming Dynasty.

It was the British East India Company that initiated the chain of events leading to China's 'Century of Humiliation'. The company complained to its own government that China constantly imposed barriers to the highly lucrative sale of opium. The production and sale of opium, used to make heroin, was at this time legal in Great Britain, but prohibited in China. East India Company had large poppy plantations in India and profited handsomely on the sale of the white powder to Chinese smugglers, who carried it over the border to Hong Kong and Guangzhou. The local authorities there were increasingly concerned about the social and economic repercussions of a rapidly proliferating heroin abuse.

The government in London was a compliant master because, despite the healthy revenue stream from the colonies, it continued to struggle with national budgetary and foreign trade deficits. The British government therefore attempted to pressure China into legalising the opium trade, but

4. URBANISATION, COASTALISATION AND GLOBALISATION

their repeated requests were flatly denied. The British therefore did what they had done with such success elsewhere so many times before: they sent in the Royal Navy to impose their will by brute force. Since the Chinese Empire had been without a navy and coastal defence for three hundred years, they were wholly unprepared for the powerful British fleet. In 1839, the Royal Navy easily crushed China's hapless attempts at self-defence, and the British took control of the ports in Hong Kong and Guangzhou. This marked the start of what has later become known as the First Opium War. Three years later, the Chinese emperor had no choice but to relinquish Hong Kong into British hands and to open the country's largest ports for trade with Europe. The emperor was also pressured into signing an agreement legalising the opium traffic.

Through military force, the British Empire hereby compelled the most populous country in the world to open up to the free trade of opium and heroin. The British government became in a practical sense the largest drug dealer of all time, and almost 160 years would pass before Hong Kong was returned to China in 1997.

Wanting to further extend their trading rights in China, the British were still not satisfied. Under the pretext of a British-registered ship named *Arrow* being boarded by Chinese officials on suspicion of piracy, Britain in 1856 launched new attacks which would be the start of the Second Opium War. A year later the French, who also sought to expand their overseas markets and establish new ports of call, joined forces with their British arch enemy to attack the Chinese Empire in the war that did not end until 1860.

Once the doors to the Middle Kingdom had been broken down, the other colonial powers quickly followed their lead. A few years later, Warren Delano, the grandfather of President Franklin Delano Roosevelt, stood on the dock in Shanghai welcoming ships from the US Navy that would take part in the occupation of China. With the exception of the British, few others profited more from the opium trade during this period than Delano and a handful of other Americans, such as the wealthy Forbes family. Portions of the highly lucrative revenues from opium sales were channelled into the funding of the first railways in the USA and the founding of a number of the country's most famous and prestigious universities such as Yale, Princeton, Harvard and Columbia.[17]

Throughout the rest of the nineteenth century, military fleets from Great Britain, France, Germany, Italy, Portugal, Austria-Hungary, Japan and the USA attacked the long coast of China. The latter was wholly defenceless when its invaders occupied cities and ports, helped themselves to riches, burned down palaces, and imposed trade and banking institutions on the country, along with customs duties and Christian missionaries. The Western naval forces were so superior that the US Navy was able to patrol more than one thousand kilometres up the Yangtze River, deep into inland China.

For the Chinese population, the Century of Humiliation became a traumatic watershed era, the dark, long shadows of which extend throughout the entirety of China's modern history and into the present day.

Insults and Degradations

The bitterness over the invasions and imperialist entitlement of Western countries eventually culminated in the Boxer Uprising at the start of the twentieth century. The uprising was quashed before long by an alliance of the USA, Japan and a handful of European nations. Insult was added to injury when the occupying powers demanded that China cover the damages incurred by the uprising. The damages claim was set at one *tael* (around 50 grams of silver) per citizen. The total amount was so exorbitant that China was forced to take out loans from foreign banks such as the British HSBC and the German Deutsche Bank. The terms of the loans were also so unreasonable that the Empire's annual state revenues were allocated in their entirety to the payment of interest and instalments to the occupying powers. This continued for fifty years, until Mao's communists gained control over China's own territory.

In 1912, the external pressures and internal conflicts were so extensive that the Qing Dynasty was no longer able to preserve national unity. The millennia-old imperial dynasty of China collapsed and the country descended into a period of gruesome and traumatic civil war that lasted for more than forty years. This also made China vulnerable to invasions and occupation by Russia and Japan. In the course of one hundred years, the Middle Kingdom was hereby transformed from an economic and political superpower to a demolished and humiliated victim. The empire that had represented more than thirty per cent of the global economic value creation when the British commenced their attacks contributed merely six per cent by the time Mao assumed power.

After the communists under Mao's leadership emerged victorious from the civil war in 1949, it would take China the remainder of the twentieth century to rise from the ashes as a country. Among all the bitter experiences of the Century of Humiliation, there are three in particular that are etched into the Chinese consciousness and which also clearly influence their attitudes and mentality of today: first, a deep-running scepticism of Western ideals, political intentions and social systems. Second, a fundamental conviction that only a strong central power can keep the huge empire united. Third,

4. URBANISATION, COASTALISATION AND GLOBALISATION

and lastly, an almost manic fear of once again losing control over their own coastal waters.

Mao and his leaders were so concerned about the possibility of new attacks from the sea that during the Cultural Revolution in the 1960s and 1970s they established what they called the 'Third Front': they moved factories and important heavy industry inland, into deep valleys between tall mountains, where they were easier to defend. In total, 1.6 million labourers were tasked with the construction of some 500 large factories and 100 research centres in the barren and difficult-to-access regions of south-west China.[18] Today most of this is disused and abandoned, but this 'rust belt' deep in the interior of the huge country continues to serve as an industrial monument to China's fear of attack from the sea.

Without knowledge of these historical developments, it is impossible to understand how fundamentally these traumatic experiences have shaped modern-day China's strategic thinking, territorial demands and often aggressive conduct in its coastal waters and nearby seas.

China will never again leave itself open to an attack from the sea.

Black Ships

Japan, up to this point extremely isolated, also fell prey to the West's aggressive and expansionist expeditions across the ocean. In the summer of 1853, the US Navy, under the command of Matthew C. Perry, sailed into Tokyo Bay with his little fleet of 'black ships', all of which were paddle steamers. The USA wanted access to Japanese ports for the refuelling and refurbishment of its whaling schooners, trade ships and battleships sailing in the Pacific and Indian Oceans.

Although Perry's fleet consisted of just four ships, the Japanese were overwhelmed by the noise and smoke produced by these terrifying black vessels, which could even move without the use of sails. The Japanese were of course well acquainted with the brutal and humiliating fate of the Chinese and quickly understood that there was little room for negotiation. Perry's fleet also arrived at a time when the Shogunate feudal military rule was already faltering under the pressure of internal conflicts and social unrest.

The USA's demands forced the isolated country to fully open its ports. The trade and calling ships significantly increased foreign goods, missionaries, and new impulses. With time, this lit the spark of the Meiji Restoration of 1868, which in the course of a few short decades completely transformed Japanese society. Out of fear of suffering the same fate as China, the Japanese ousted the 650-year-long Shogunate military regime and installed an emperor in its stead, the fourteen-year-old Mutsuhito. This marked the start of a massive political, economic and social transformation that was unparalleled in its time.

The Japanese rapidly modernised their system of governance, legislation, administration, school system and armed forces. They built railways and roads and installed telegraph lines. They developed a modern banking system and founded research institutions based on the Western model. The modern Japan was born, today the fourth-largest economy in the world.

Japan adopted Western cultural trends in fields ranging from architecture to fashion.[19] The Japanese also embraced with enthusiasm the Western countries' use of superior maritime military force as a means of subjugating other countries and peoples. They soon built a large fleet of modern warships, which were quickly deployed in frequent and aggressive attacks on countries throughout eastern and southern Asia.

Japan behaved like a hungry cuckoo in the Asian nest. The country's naval strength and traditional warrior culture, cultivated for centuries under Shogunate military rule, contributed to bolstering their identity, national self-esteem – and effectiveness in combat. In 1894 Japan attacked China, taking control of the Korean peninsula. In the course of the subsequent decades and up to the end of the Second World War, Japan attacked coastal states and strategic locations in large parts of Asia. One by one, Taiwan, Hong Kong, Vietnam, Cambodia, Laos, Thailand, Malaysia, the Philippines, Singapore and Myanmar were made the targets of Japan's brutal pillages and imperialistic ambitions.

For centuries a closed and isolated country, Japan became in a short period of time every bit as aggressive and expansionist as the Western colonial powers. To this day, the traumatic memories from this period continue to leave a heavy and defining mark on the relations between Japan and several of the coastal states of Asia. This is particularly true for the relationship between Japan and South Korea.

Heart of Darkness

While Japan was busy attacking its Asian neighbours, the Western nations continued to exhibit a virtually insatiable appetite for the conquest of further countries. Every year during the period 1871–1914, new non-European territories were colonised equivalent to the size of France.[20] At the end of the First World War, European countries and the USA controlled almost ninety per cent of the world's inhabited regions.

On the vast, resource-rich African continent, the European colonial powers were fighting among themselves as much as with local peoples and tribes.

4. URBANISATION, COASTALISATION AND GLOBALISATION

The competition over Africa was fierce, above all between Great Britain and France. Both countries were trying to usurp control of the coastal regions in particular, since the trade routes providing connection to Europe were by sea. There were few Europeans bold enough to make the trip into the interior of the huge continent, and even as late as the 1870s the majority of Africa had still not been mapped. This provided fertile soil for the cultivation of myths, fears and fantasies about what might be hiding in Africa's hinterland. In the minds of Europeans, Africa still embodied the darkness, the uncivilised and the untamed.

The mysterious and mythical perceptions of what was hidden in the African interior provided inspiration for one of the most popular novels of the day. In 1891, Joseph Conrad wrote *Heart of Darkness*, based on his eight years of experience working as a river boat captain in Congo. The story of the well-educated and sophisticated Charles Marlow's journey up the Congo River and into the 'heart' of Africa serves as a metaphor for a voyage into human nature's dark, uncivilised and primitive brutality. The book would later provide the basis for one of the best-known films of the 1970s, *Apocalypse Now*, directed by Francis Ford Coppola.

When the German princes came together to establish Germany as a nation state in 1871, Otto von Bismarck became the country's first chancellor. A mere decade later, he invited the major European powers to a conference in Berlin to discuss how they might divide Africa up between them. Germany demanded what they would later describe as 'a place in the sun'.[21] All the larger European countries had colonies, and a unified Germany now considered itself a major power. Bismarck expected the gentlemen of the era to form an agreement on some joint principles for the partitioning of the uncivilised and resource-abundant African continent between the European powers.

In keeping with the spirit of the times, no African countries or representatives were offered a seat at the table – not a single African person participated in the meetings. Among the Europeans, only a small minority went to the trouble of listening to the voices of the African people. The latter were mainly viewed as 'people of nature' who didn't understand what was in their best own interests. The only thought given to the interests of Africa's indigenous population was the argument that colonisation would lead to 'commerce and Christianity,' as David Livingstone put it: Africa could become a part of the world economy as a market and supplier of raw materials and the local population would be civilised and guided to the Christian faith.

When the negotiations concluded in 1885, the new map of Africa was drawn with 'a ruler, cigars and cognac'. The borders between the European colonies were defined without consideration for the fact that the continent was inhabited by more than 3,000 ethnic groups in a patchwork of clans, tribes, religions, languages, cultures and societies, each with their own unique history and distinct features.

In direct contradiction with its intention, the General Act of Berlin also intensified the competition over Africa between European countries. The agreement contained a 'first-come-first-served' principle for regions that had not yet been allocated, and, because of this, several European countries stepped up their efforts to seize control of new territories. French and English troops consequently found themselves on the brink of war when they met in Sudan. It was only due to Britain's superior military capacity that war was avoided and France retreated in the end.

The European countries immediately introduced the nation state's political-administrative concept for the division, organisation and rule of the vast African continent. Since the imposed borders cut through landscapes without taking into account ethnic, cultural, social or historical affiliations, a need quickly emerged for the creation of specific national 'identities' to consolidate the artificial and politically motivated divisions. National symbols such as flags and national anthems were created. Passports were required for border crossings, and to distinguish those who were citizens of the state from those who were not.

New narratives, administrative systems and physical borders were created that defined the terms of inclusion and exclusion and imposed artificial distinctions between individuals and groups. Tensions, hostilities and violent conflicts resulted, the onerous impact of which can still be felt in large parts of the African continent.

United States...

The settlement of North America predominantly took place in three waves of immigration from Europe. The first settlers started arriving in the late seventeenth century, but up until the early nineteenth century, there were no more than four million European settlers in the USA and Canada. The population growth of the first part of this century almost exclusively stemmed from natural causes: the settlers had children. The second wave was much larger and arrived in the mid-nineteenth century. At this time, the revolutions of 1848 were spreading throughout Europe. These widespread and failed political uprisings were led by liberals and intellectuals who wanted to do away with the monarchy and introduce democracy. At the same time, the Great Famine struck in Ireland. As a result of these events, millions of persecuted and starving people fled across the Atlantic Ocean, the majority from northern European countries such as Ireland, Great Britain, Germany

4. URBANISATION, COASTALISATION AND GLOBALISATION

and France. In 1867, the European settlers to the north of the US border cut ties with the Old World and founded Canada as an independent nation state.

The third and largest wave occurred when the railway and steamship were introduced towards the end of the nineteenth century. The railway made it easier for people from Central and Eastern Europe to reach the seaports on the coast. Steamships could accommodate more passengers, the fares were lower and the transatlantic crossing took less time. While sailing ships would often spend six to eight weeks on the crossing, the new steamships made the trip in one week's time. From 1880 to 1930, it is estimated that some twenty-five million people made the journey, many from southern and eastern European countries such as Italy, Greece and Hungary.[22] At this time there were also many immigrants from Scandinavia who sought their fortune in the new world. In some parts of southern Norway more than forty percent of the population packed their suitcases and departed.[23]

The USA hereby became the world's first 'universal country', made up of people of different origins, ethnicities and nationalities, but with common dreams, hopes and expectations for the future. The country was built by energetic, independent and freedom-seeking individuals. Many were fleeing war, famine, unemployment, oppression or persecution. Some were opportunistic fortune hunters, but the majority were basically ordinary people in search of a better life.

Ships continued to transport millions of young, enterprising and skilled farmers, craftsmen, businessmen and intellectuals across the Atlantic Ocean to the promised land. Europe's 'brain drain' became the new country's 'brain gain'. Many had strong ideological or religious convictions, others had traumatic experiences from the authoritarian regimes of the Old World – and all had a desire for a life of freedom. Therefore, religious values, freedom of speech, the right to bear arms to protect oneself and one's family, and the ingrained scepticism of a strong central power became important parts of the foundation of the United States. It should be a 'country of and for free people', a country where everyone could follow their dreams and forge their own paths in life. Together they wanted to create the American Dream, a new country built by people from the Old World. Therefore, several of the states, areas and cities were also named after the immigrants' European homes, and often with 'New' in front, such as New York, New Hampshire, New Jersey and New England.

The young nation state and the American dream were shaped by white Europeans for white Europeans. Because not everyone was invited to share in the freedom and dreams. Not everyone was united in the United States of America.

...but Not for All

The original population of the North American colonies, the indigenous peoples or First Nations as they are called in Canada, suffered the same fate as their brothers and sisters in European colonies all over the world: they were fought, massacred, displaced and subjugated. The contradictions in culture and way of life were so great that in 1758 separate 'reserves' were established to mitigate the conflicts and 'shield' the indigenous people from the European colonists.

Even in our time, a fifth of the indigenous Indian population, one million people, still live in three hundred reservations scattered around the United States. In practice, these reserves function as poorly as the name suggests: they are isolated communities that are, for the most part, dependent on the support of the larger society.[24] All land is owned by the US government, so its citizens cannot take out loans secured by real estate to build homes, businesses or futures. Indigenous people who live outside the reserves as part of the general society also have, on average, worse living conditions and a lower standard of living than the rest of the American population.[25] They live shorter lives and have higher unemployment, lower incomes and greater social problems.

In 2019 the Democratic governor of California, Gavin Newsom, issued a formal apology to California's Native Americans for 'historical mistreatment, violence, and neglect', adding that: 'That's what it was, a genocide. No other way to describe it. And that's the way it needs to be described in the history books.'[26] From the other end of the political spectrum, former Republican senator Rick Santorum in a speech two years later asserted that '[w]e birthed a nation from nothing. I mean there was nothing here. I mean, yes we have Native Americans but candidly there isn't much Native American culture in American culture.'[27]

Still in our time, there are also regular reminders of the historical abuses against the indigenous population on the other side of the border, in Canada. In 2021, unmarked mass graves were discovered with the remains of more than 1,300 young children who had died in Catholic boarding schools, most of them as a result of illness and accidents. For over a hundred years right up until the 1990s, more than 150,000 children from the Canadian indigenous population were taken from their parents and put in Catholic boarding schools, while another 20,000 children were placed in white 'foster homes'.[28] The official Truth and Reconciliation Commission of Canada has characterised what happened as 'a cultural genocide'.[29]

The purpose was to eradicate the identity and traditions of the indigenous population, and force them to convert to the Christian faith and a 'modern', Western way of life.

4. URBANISATION, COASTALISATION AND GLOBALISATION

Our Own Backyard

The two newly formed independent nation states on the large North American continent were self-sufficient in terms of food and natural resources. The USA had abundant supplies of wheat and meat from the endless plains and pastures in the Midwest, wine and vegetables from California, cotton from the plantations in Louisiana, and gold and minerals from the mines in Colorado. The country had its own supply of coal from the large strip mines in Wyoming and Pennsylvania, steel from the factories in Cleveland and Pittsburgh, and on the flat plains of Texas the oil pumps would soon be nodding in time with the ever-accelerating pulse rate of the US economy. The USA therefore had no need to conquer other nations to secure resources, and neither did it have overseas colonial ambitions like those of the other European nations. In the 1800s, the Americans had their hands full with the task of completing colonisation of their own, vast continent.

This did not, however, prevent the USA from recognising the need to protect itself from political and military threats from nearby regions to the south. First and foremost, the USA would no longer tolerate the continued meddling of European powers in its own 'backyard'. The Caribbean region was economically significant and the only region where potential enemies had only to make a short trip across the sea to reach the American mainland. The core of the Monroe Doctrine, named after the US president who introduced it in the early nineteenth century, was designed to secure US control over this region.

The USA therefore implemented a long-term strategic campaign to gain what it viewed as wholly critical economic, diplomatic and military influence in the countries in and around the Caribbean Sea. This set the stage for the Spanish-American war, which ended with a crushing defeat for Spain in the spring of 1898. Under the peace treaty, Spain was obliged to relinquish Cuba and Puerto Rico to the USA, marking the end of Spain's 400-year-long colonial rule in South and North America. At the same time, the USA took its first steps into the world as a global superpower by obliging Spain to surrender both Guam and the Philippines as well. The Americans had now acquired both colonies and naval bases in Southeast Asia.

In the autumn of the same year, the USA also annexed Hawaii. The location of the island facilitated the US Navy's protection of the west coast, and later the Panama Canal, from the potential threat of hostile naval fleets approaching from the other side of the Pacific Ocean.

A Great White Fleet

The large ocean expanses surrounding the USA were at this time considered a natural form of protection from external enemies. The country's topography also offered uniquely advantageous conditions for the US naval forces. The long Pacific, Atlantic and Caribbean coastlines provided guarantees against the threat of blockades. US Navy planners would never be burdened by the chronic, strategic claustrophobia which the Chinese navy and, to an even greater extent, the Russian navy have had to endure at all times. Along the east coast of the USA are a number of naturally protected, ice-free deepwater ports, which to this day serve as important naval bases. One of these is in Norfolk, Virginia, home of the world's largest naval base and the headquarters of the US Navy.

There are few individuals who have influenced the terms of US military strategic thinking more than the naval officer Alfred Thayer Mahan. In the late nineteenth century, he introduced the term 'sea power', convincing the leading American politicians of the day that a strong Navy was the key to a powerful national military defence and an independent foreign policy.[30] When the USA built up its military power, it was therefore first and foremost as a naval force. The US Navy became the girder of the country's armed forces, designed to project global power and secure safe ocean trade routes. In pursuit of these goals, it would also later develop a forward-deployment system able to resist and respond to adversaries before they could approach the country's own coastlines.

When the USA pulled out all the stops in the build-up of their Navy, an intense arms race was already underway at sea. The European colonial powers had a need to protect their own heavily trafficked trade routes, Germany had begun flaunting its naval military prowess and Japan was in pursuit of domination in the Asian maritime regions. Great Britain was at this time still the world's uncontested superpower, but its empire was being challenged all over the world. The British were also fully aware that control at sea was critical to maintaining their global position. The British parliament therefore passed the Naval Defence Act, formalising what was also the reality at that time: according to the act's 'two-power standard', the size of the British fleet at any given time was to be at least as large as the two next-largest fleets combined.[31] When the act was approved in 1889, the USA was not even one of the two next largest.

But it was not long before the US fleet began to make its presence felt in earnest. In 1907, sixteen white naval vessels with weapons on display assembled at Hampton Roads in Virginia to set out on a voyage around the world. With the Great White Fleet, President Theodore Roosevelt wanted

4. URBANISATION, COASTALISATION AND GLOBALISATION

to demonstrate to the world that the USA was a superpower to be reckoned with.[32] Since the US fleet was initially so modest, the US Navy had no burdensome legacy to uphold, in terms of ships, weapons or crews. The USA built its navy as it was building the country: using the newest and most advanced technology invented since the start of the Industrial Revolution.

As a result of the rapid build-up, the US Navy was already on a par with the British Royal Navy in terms of both military might and firepower by the end of the First World War. This prompted Neville Chamberlain, who would later become Britain's prime minister, to declare in a speech in the House of Commons that the mere thought of a naval arms race with the United States at sea was in its own right a highly effective deterrent.[33]

A Strategic Gift

To reduce its dependency on trade with its former European colonial rulers and protect its trade routes, the USA aspired to wield greater control over key ocean regions. This also induced a gradual shift in the US orientation from the east towards the west and the countries in the Pacific Ocean, a 'pivot to Asia' of the day.

When the French abandoned the construction of the Panama Canal in 1904 and handed the project over to the USA, it therefore constituted a much welcome strategic gift. When the canal opened in August 1914, the maritime trade route between the resource-rich regions of Southeast Asia and the rapidly growing industrial regions in the eastern and southern USA was dramatically shortened. The canal also reduced the US Navy's sailing distance when moving warships between the American east and west coasts by more than 15,000 kilometres – equivalent to one third of the distance around the earth at the equator. This provided a wholly new strategic flexibility for the deployment of US naval forces to different parts of the world as required.

Less than thirty years later, the USA would take full advantage of this flexibility, when Japan attacked Pearl Harbour and virtually eradicated the USA Pacific Fleet. In the days and weeks following the Japanese attack, a deluge of US fleets transited the canal.

The Art of Rowing under Water...

At the turn of the twentieth century, every country with international ambitions invested substantial resources in the development of naval forces. They all expanded their arsenals of naval vessels, building larger and more advanced ships with more sophisticated weaponry. While previously warships had been outfitted with sails, wooden hulls and muzzle-loaded, cast-iron cannon, the new vessels had steam-driven engines, armoured steel hulls and breach-loading steel cannon with rifled barrels. The ships were no longer at the mercy of wind conditions; they sailed faster, were manned by smaller crews and were better protected. The cannon could be loaded more quickly and had greater range and accuracy.

The ultimate breakthrough for the bold and ambitious visions of 'underwater ships' also occurred at this time: a vessel that could navigate under water and remain hidden in the depths before launching a sudden attack on an enemy ship on the surface. The challenges involved in the construction of such underwater vessels were considerable. The hull had to be able to withstand the water pressure and a sufficient supply of fresh air for the crew and engines had to be secured. Upon meeting these requirements, the next challenge was to create instruments enabling navigation of the murky darkness of ocean depths.

More than 300 years would pass from the time the first conceptual designs were completed until submarines were deployed in combat. As far back as the late sixteenth century, the English mathematician William Bourne drafted a design proposal and fifty years later the eccentric Dutch investor Cornelius van Drebel launched the first undersea vessel in history.[34] The vessel had a cigar-shaped wooden frame covered in leather with oars sticking out on the sides through 'waterproof' leather sleeves. The sides of the vessel were flexible so they could be collapsed, allowing it to descend beneath the surface. The maximum depth was limited to four or five metres, and brave men allegedly rowed the submarine for several hundred metres just below the surface when it was tested in the brown, sewage-infested water of the River Thames in the 1620s.

It was nonetheless the Germans who were the first to develop these underwater vessels to a standard that was adequate for use in combat (after they had apparently stolen the designs from the French). By the start of the First World War, German engineers had successfully built combat-ready submarines equipped with self-propelled torpedoes. They also constructed submarines with vertical pipes for the release of horn mines, which would explode if a ship made contact with the horns sticking out of the mine.

...and Landing on Ships

The beginning of the twentieth century also saw the deployment of ships serving as airstrips at sea for military planes. Already in the latter part of the previous century, decommissioned merchant and military vessels had been converted to launch hot air and gas-filled balloons for observation purposes. From these high-flying balloons, spotters equipped with state-of-the-art binoculars could detect targets and approaching vessels which would otherwise be hidden beyond the horizon. It was not until the outbreak of the Second World War that the radar would be deployed for military and civilian purposes, so until then naval ships had to rely on manual, eyesight observations.[35] In 1910, seven years after the Wright brothers became the first to successfully fly an aeroplane, the first fixed-wing plane departed from a US Navy vessel.

However, this concept was still so nascent that during the First World War only 'balloon carriers' were used in combat. Such carriers were extensively used by both the USA, France, Great Britain, Germany and Russia, and even the navies of smaller countries, like Sweden, were able to launch balloons from their ships.

Then, in 1922, Japan commissioned *Hōshō*, the world's first purpose-built aircraft carrier. A couple of years later, the UK's Royal Navy got their first, and by the end of the 1930s, the UK, Japan and the USA had each deployed several aircraft carriers, which came to play a crucially important role in the Second World War. As we shall see later, the aircraft carriers not only changed the tactics of naval warfare; they also fundamentally changed war on land and the conditions for global power projection across the ocean.

Due to the technological developments and proliferation of naval capacities, the strategies and techniques of warfare at sea changed more during three decades in the early twentieth century than over the course of the previous three thousand years.

A Fuel Gamble

While the USA was building the Panama Canal to secure a shorter maritime route for its Navy, the British were building new engines for the Royal Navy to increase the speed and agility of their fleet. The British could no longer match the economic and industrial capacity of the USA or the emergent German military power in the construction of numerous large warships. They therefore decided instead to equip their fleet with diesel-powered internal combustion engines that would increase the speed, and thereby improve the tactical flexibility of each individual vessel. Under the direction of the headstrong and dynamic naval minister, and later prime minister, Winston Churchill, in the course of a few years the entire British naval fleet was converted from the coal-fired steam engine to the newly invented diesel-engine, running on heavy fuel oil.

It was an expensive, bold and controversial move at the time, and there was no shortage of criticism from the political opposition and Churchill's fellow party members of Parliament. The diesel technology was still in its nascent stages, the Royal Navy had a large number of vessels that required retrofitting, and tens of thousands of seafarers had to be trained and reskilled to operate the new engines. But the challenges of transitioning the British navy to another fuel type were not limited to the technical, practical and financial requirements; such a changeover had first and foremost broad-reaching strategic ramifications. The Royal Navy was Great Britain's most important tool for exercising global power and the backbone of the country's own defence. Britain had its own coal mines to fuel the steam engines of the entire Royal Navy, while at this time there were no known oil deposits on the British Isles. Without access to oil, a diesel-powered navy would be paralysed.

Nonetheless, under Churchill's steadfast leadership the entire Royal Navy transitioned from steam to diesel in the mid-1910s. To secure access to the new fuel, Great Britain simultaneously set its sights on the Middle East, where the existence of large oil reserves had already been confirmed.

In the space of a few brief years, other nations followed suit, retrofitting their navies with diesel-engines. At the same time, the use of internal combustion engines in tanks, trains, merchant vessels and automobiles virtually exploded. The entire industrialised world, including Japan, was now wholly dependent on oil to fuel their military forces and keep civil society up and running. The demand increased so rapidly that not even the oil-rich USA had sufficient supplies to cover its own consumption.

From this point on, oil was no longer merely a profitable commodity. For the first time in history, it had become a strategic resource of critical significance. This sealed the tragic and turbulent fate of the Middle East because, from this point forward, it was no longer solely the Suez Canal and the trade routes to Asia that gave the region strategic importance.

The eyes of the whole world were now focused on the oil-rich desert states of the Middle East.

The Stopping Power of Water

Each year, true to a tradition extending back to the US Civil War, George Washington's farewell address is read aloud in the US Congress on his birthday, 22 February. In the address, which was published in newspapers all over the country, he issued three urgent warnings: the first, to remember 'the immense value of your national Union'; the second, to refrain from putting 'in the place of the delegated will of the nation the will of a party' and the third, to 'steer clear of permanent alliances with any portion of the

4. URBANISATION, COASTALISATION AND GLOBALISATION

foreign world'.[36] George Washington saw no reason for the USA to involve itself in conflicts on other continents, or between colonial powers with whom the young nation state had previously severed ties by way of armed combat. He was also convinced that it would promote national unity if the USA remained neutral and avoided permanent alliances with other states.

His warnings have proved to have been in vain. The modern-day USA has shown few reservations about involving itself in conflicts in every part of the world. Given the extreme political polarisation of today, nobody would describe either Republicans or Democrats as refraining from putting 'in the place of the delegated will of the nation the will of the party', and today the 'national Union' is more characterised by unabating distrust and intense partisan contradictions.

But at the start of the twentieth century, most Americans still shared George Washington's scepticism about intervention in the affairs of foreign countries. At this time, the USA was 'a nation unsure about the role it wanted to play in the world, if any', as historian Robert Kagan writes, arguing that 'Americans had no grand international plan and no clear direction'.[37] US trade and industry were already operating in high gear, further accelerated when the First World War broke out in Europe. American factories and banks profited handsomely on the sale of goods and on loans to parties on both sides of the war. Initially, they were just as willing to do business with Germany as with France and Great Britain.

The Atlantic Ocean also provided a broad and protective buffer against the horrors of war. USA enjoyed the advantages of what political scientist John Mearsheimer would later call 'the stopping power of water'.[38]

The USA therefore assumed a distanced and neutral stance towards the war in Europe, at least initially. When German submarines began targeting civilian merchant ships and passenger vessels sailing between Europe and the USA, however, this provoked powerful reactions from the US population. The sinking of the passenger ship RMS *Lusitania*, off the southern coast of Ireland, by a German submarine on 7 May 1915, marked a critical turning point. The ship was the largest and most well-known ocean liner travelling between New York and Liverpool. The Germans were aware that the ship was also smuggling large quantities of ammunition for British and French troops in Europe. Before its departure from New York on this fateful crossing, the German ambassador to the USA had therefore issued an official notice that was printed in *The Times*, warning that the ship could be attacked. Unfortunately, nobody took the warning seriously, and the *Lusitania* was booked to capacity when torpedoes struck and sank the ship. Of the 1,198 children, women and men who perished in the ice-cold, dark waves, 124 were American citizens. 'In God's name, how could any nation calling itself civilized purpose so horrible a thing?' a shocked President Woodrow Wilson exclaimed.[39]

This tragic incident contributed to altering public opinion about both the war in Europe and the country's capacity to exercise military force abroad. In 1916, the US Congress adopted the Big Navy Act, calling for the build-up of the largest and most powerful navy in the world. This expansion of US military might at sea was first and foremost intended for defence purposes.[40] But in the winter of 1917, when Germany escalated its attacks and sank more ships transporting American civilians, it was the last straw.

On 6 April 1917, the USA declared war against Germany, and only a few months later, American troops headed across the ocean to enter the war on the European continent.

Patient Zero

The American soldiers who arrived in Europe did not only bring with them equipment, weapons and ammunition. When they went ashore, nobody knew, the soldiers included, that they were also carrying a supremely contagious, fatal virus that attacked the lungs and respiratory system. The flu-like illness was first detected in the spring of 1918 in US soldiers in training at Fort Riley, Kansas. 'Patient Zero' is believed to have been Albert Martin Gitchell, the son of an American father and Norwegian mother.[41] The twenty-eight-year-old butcher had been drafted as a cook for the army, and when he one day reported for duty coughing and sneezing and with a high fever, he was transferred to a field kitchen. There he continued to cook and serve his fellow soldiers in a feverish state, his eyes glassy, nose running and breathing increasingly debilitated.

The infection spread quickly through the camp and to other parts of the USA, and by the time the troops crossed the Atlantic Ocean many of the American soldiers were sick. Not long after they went ashore, the 'Spanish Flu' also began spreading its damp, foul-smelling carpet of death, suffering and desperation across the European continent.

Four hundred years after Columbus brought syphilis to Europe from the New World, another American-born virus ravaged Europe. This was the worst known pandemic in history, taking the lives of an estimated fifty to 100 million people worldwide in the course of the next two years.[42] More people died from the Spanish flu in a single year than from the Black Death in its entirety, and the number of flu-related fatalities exceeded those of the First and Second World Wars combined.

4. URBANISATION, COASTALISATION AND GLOBALISATION

The name of the illness has its own history: because of the war, both the US and European press were under strict censorship, so the newspapers in neutral Spain were the first to report the illness. It is the mention, not the origin, of the pandemic which would forever connect it to Spain.

Insult to Injury

The USA entered the First World War late and dragging its feet, but its military contribution was instrumental to Germany's surrender and brought the war to an end in 1918. Moreover, the USA played an important role in the subsequent peace talks, and Woodrow Wilson became the first sitting US president to make the trip across the Atlantic Ocean. A century and a half after throwing off the yoke of European colonialism, the Americans returned to Europe to broker peace.

The peace talks in Versailles formally ended the First World War – and laid the foundation for the Second. 'The Talks', under the direction of the USA, France and Great Britain, became a lesson in how the victors should *not* behave. The two countries, which a mere two decades later would become Europe's strongest military powers, were excluded and denigrated. Russia was boxed out of the talks because none of the Western Allied powers was willing to recognise Lenin's revolutionary communist Bolsheviks, who were about to take power in Russia. Germany was down for the count and no effort was spared in ensuring that it would remain so in the future. Germany alone was held responsible for the war, obliged to surrender its colonies and parts of its domestic territory, and pay astronomical sums in reparations to the Allied powers.

The political architecture that was intended to establish peace and security in Europe was constructed on a foundation of exclusion and humiliation. The consequences would prove fatal.

In the aftermath of the war, the European states took further advantage of Germany's weakness by redistributing the African colonial territories among themselves. In July 1919, the following notice was printed in a Norwegian business paper:

> A telegraph from Paris: Baron Criffier announces an agreement formed between England and Belgium on the force of which a portion of Congo and Tanganyka has been exchanged for large territories within German East Africa. Belgium hereby gains an abundant region with a population of seven million and England three million, and the connection between the Cape and Cairo is secured.

As recently as a century ago, in my grandfather's youth, the Western colonial powers were still trading countries and peoples among themselves as if these were merely real-estate transactions.

Doomed to Fail

The peace talks in Versailles were the first time the USA assumed a key role on the international stage. This was also the moment when the USA expressed, for the first time in an international context, what it viewed as its unique mission of promoting democracy and freedom and its moral responsibility to prevent oppression elsewhere in the world. Woodrow Wilson argued with great pathos for the inviolable freedom of the individual and every country's right to self-determination. His words were understood, and rightly so, as an unveiled and damning critique of the European colonial system. Although Russia was excluded from the talks, the American president received the verbal support of Lenin and his Bolsheviks. The European colonial powers suddenly found themselves being challenged by a paradoxical alliance between the liberal, capitalist USA and the revolutionary, communist Russia.

In an attempt to establish institutions and dialogue mechanisms that could preserve peace and prevent new wars, the US president proposed formation of the League of Nations. The mandate of the new organisation would be to promote disarmament and international cooperation, and to improve people's working and living conditions all over the world.

It was the first time such lofty objectives formed the basis for a comprehensive international collaboration, and in 1920 Woodward Wilson was awarded the Nobel Peace Prize for his initiative.[43] But the League of Nations had only a few member nations, and both its authority and influence were weak. The visionary ideas and moralising of the American president did not sit particularly well with Great Britain and France: the power, wealth and prosperity of these two countries still relied on the resources of their colonies. Strong isolationist forces in Congress also did their part to ensure that the USA itself never joined the League of Nations.

The ambitious initiative was doomed to fail, and the League of Nations was dissolved when the United Nations was formed in 1945.

A Maritime PIN Code

After the First World War, as a measure to reduce the risk of another large-scale war, the major powers formed an international agreement regulating the relations of power at sea. In the so-called Five Power Treaty, the USA,

Great Britain, Japan, France and Italy agreed on a type of 'PIN-code' stipulating the size of the fleet permitted for each country: 5-5-3-2-2.[44] The treaty calibrated naval capacity so that the USA and Great Britain would be both the largest and the same size, and Japan second largest, followed by France and Italy.

For Great Britain this implied an open acknowledgement that they no longer 'ruled the waves' alone. At sea, the power of the British Empire was now equal to that of its former subjects on the North American continent.

For Japan, the agreement represented an international acknowledgement of its role as the 'Far East's' leading regional power. However, the increasing scale of Japan's imperialist ambitions led to their withdrawal from the agreement in 1934. They would no longer allow themselves to be dictated by random numerical codes established by the Western powers.

The Tragedy of the Middle East

During the First World War, oil had already become a critical strategic resource. While the war was being waged from muddy trenches flanked by barbed wire in Europe, the French and the British formed a secret agreement – behind the backs of the USA, Russia and the other Allied forces – outlining how they would share the regions and oil fields of the dying Ottoman Empire between them after the war. The modern architecture was thereby created for the chronically tension-fraught and besieged Middle East. The conditions underpinning the endless wars, gruesome conflicts and human suffering that to this day continue to prevail in the region were laid down.

It is not possible to understand the Middle East of today without understanding the events that transpired in the wake of the First World War.

The British intelligence officer Mark Sykes, who before long would die from the Spanish flu, and the French diplomat Francois Georges-Picot were assigned responsibility for the secret negotiations. Both conducted themselves with the self-assured arrogance of colonial masters when they divided up the Ottoman Empire, 'the Sick Man of Europe', between England and France. The partitioning was designed with blatant disregard for historical, ethnic, sectarian or topographic distinctions, and without consideration for or input from the affected populations.

Through the Sykes-Picot Agreement of 1916, large areas of the Middle East were divided into nation states with randomly assigned borders. The delineation of these artificial political and administrative entities was defined by French and British interests and to reward Christian groups and Arab tribes who had assisted the two countries in the First World War.[45] France initially established Lebanon by allocating a territory between the Lebanon Mountains and the Mediterranean Sea to the Christian Arab Maronites. When the French subsequently gave up on the idea of a 'Greater Lebanon', Syria was partitioned off as a separated state. The French rule was oppressive and extremely unpopular. When the Syrians celebrate their national day today, they are commemorating the moment the last French soldier left the country on 17 April 1946.[46]

Iraq was pencilled in on the map above the Euphrates and Tigris Rivers. Here the British enthroned the Sunni Muslim Hashemite Faisal as king, inciting violent demonstrations and armed rebellion on the part of Kurdish and Yazidi residents, Christian Syrians and Shia Muslims. The British simultaneously assumed that they could keep Iraq on a tight leash by blocking its access to the ocean. They therefore established Kuwait as a British colony on the innermost shore of the Persian Gulf.[47] Jordan was born when the British rewarded the Hashemite Bedouins and their King Abdullah ibn Hussein with a region east of the Jordan River. Almost the entire Arabian peninsula, up to this time known as Hejaz, was given to King Ibn Saud and his nomads, who had been British allies during the war. The new state was called Saudi Arabia, and thereby became one of very few countries in the world named after a family.

At this time, the majority of the population in the territory between the Jordan River and the Mediterranean Sea was Palestinian, while the Jewish population made up less than ten per cent. Many of the Jews living in this region had fled persecution in Europe and had come here for deeply religious and historical reasons. Since the time of their displacement by the Romans in the first century of the Common Era, Jews all over the world have dreamt of returning to Zion, the historical Jerusalem and the surrounding regions. In Great Britain the political influence of the Zionist movement was growing, and in 1917 the British government issued the Balfour Declaration stating support for 'the establishment in Palestine of a national home for the Jewish people'. The League of Nations would later declare the territory a Mandatory Palestine, whereby the British Mandate of Palestine was granted. In 1922, the British made good on their promise to facilitate Jewish immigration into the region by proposing the division of Palestine into one Jewish and one Palestinian state. This met, however, with such strong resistance from the Palestinian and Arab camp that it was not until after the Second World War that such a division could be forcibly imposed. But the Palestinian-Israeli issue has never been settled, and remains today the main source of conflict in the Middle East.

Just as when the European colonial powers had divided Africa up among themselves three decades before, the artificial borders in the Middle East cut through the region's historical relations and traditions. Friends, families and allies were divided, while tribes and clans with long histories of conflict, war, feuds and vendettas were made a part of the same nation state and subjected to the same political-administrative system of governance. In the newly born nation states, tensions, contradictions and underlying conflicts between hostile factions immediately rose to the surface.

From the start, the new Middle East nation states were therefore put under the administration and 'supervision' of France and Great Britain.

The Baron and the Shah

Persia, which changed its name to Iran in 1935, was never colonised in the traditional sense by Western powers. Yet the Persian dynasties were both before and after the First World War under persistent pressure to grant concessions that served the interests of the colonial powers. In 1872, when the shah granted the German-British baron Julius de Reuter a monopoly on extraction of the country's oil, gas, coal, iron and copper resources, it was therefore difficult to ascertain whether this was due to the pressure from the West or the ruler's psyche. Reuter also received exclusive rights on the construction of roads, railways, telegraph lines and irrigation systems for a period of seventy years.[48] The shah effectively gave his country away to a foreign businessman in exchange for a portion of the revenues. This triggered such violent reactions at home that the shah was obliged to rescind the agreement after only a year. Although the baron protested vigorously, he had no choice but to surrender the rights. He then returned to his own company, the leader in telegraphy and news transmission of the day. In our times, Thomson Reuters remains one of the world's largest and best-known news agencies.

Just a few years later, the shah again fanned the flames of domestic discontent and unrest when he granted British East India Company a monopoly on the Iranian tobacco trade. For this reason, few of his own countrymen shed any tears when the shah was assassinated. The incomprehensible preferential treatment of foreign over national interests incited such widespread protest and frustration in the population that it also led to the Persian Constitutional Revolution of 1906, through which the constitution was amended to restrict foreign influence and the power of the shah.

This was, however, not sufficient to prevent the flamboyant Englishman William Knox D'Arcy, with the solid backing of the British government, from simultaneously landing a particularly lucrative deal with the new shah. D'Arcy had neither an organisation nor a company, merely a secretary who helped him manage his extensive correspondence. But, by way of glowing promises, cynical intrigues and liberal bribes, D'Arcy acquired a monopoly on the exploration, production, refinement and transport of all Persian oil – for all eternity. All oil reserves found in the Persian empire would according to this agreement fall to this man, and in exchange the shah would receive sixteen per cent of the running profits (though D'Arcy never submitted any certified accounts, so he basically paid what suited him).[49] D'Arcy's partners quickly discovered the first oil deposits, and in 1909 the enterprising Englishman founded Anglo-Persian Oil Company.

When the British Royal Navy converted from steam to diesel in the mid-1910s, oil became such an important strategic resource that the British government forcibly nationalised parts of D'Arcy's company.[50] Churchill and his government, who had actively assisted D'Arcy previously, did not want this one businessman to be in a position to exercise undue influence over the United Kingdom. The government therefore snatched up a controlling share of the company in a hostile takeover and invested further capital.

Torrents of oil were pumped out of the giant reserves. 'Fortune brought us a prize from fairyland beyond our wildest dreams,' a delighted Winston Churchill wrote.[51] The oil gave the British strategic oil reserves, cheap fuel and large sums for the national treasury. In the years following the Second World War, Great Britain amassed tax revenues from Persian oil that were three times greater than Iran's.[52] The euphoria was understandably not equally exuberant on the part of Iran. Discontent simmered at every level of the Iranian population, and in 1951, on a wave of anti-British sentiment, Mohammad Mosaddegh was elected prime minister. His most important campaign promise was that he would nationalise the country's oil industry and throw out the Anglo-Persian Oil Company. This proved to be more than the Western major powers could stomach.

Mosaddegh's promises of nationalising Iran's oil resources constituted a dangerous strategic threat to Western powers, who feared that also other oil-rich countries in the Middle East would be inspired to follow suit. The USA and Great Britain consequently intervened directly in Iran's political sphere. Acting at the behest of US President Eisenhower and British Prime Minister Churchill, the two countries' intelligence services, the CIA and MI6, conspired to depose the democratically elected Prime Minister Mosaddegh in favour of the shah, the Western sympathiser Mohammad Reza Pahlavi, who had been a figurehead until then.[53] The conspiracy succeeded in 1953, only two years after the election of Mosaddegh. With the shah now fully in power, the USA and Britain were able to lock down an agreement granting each of them forty per cent of Iran's oil reserves, while the remaining twenty per cent would go to Iran. The Anglo-Persian Oil Company then

4. URBANISATION, COASTALISATION AND GLOBALISATION

changed its name to the British Petroleum Company Ltd, known today as BP. Britain no longer even pretended that the company was operating in collaboration with Iran.

The shah's heavily UK- and US-backed, autocratic regime eventually became so unpopular that it paved the way for the Iranian Revolution in 1979. Iran was hereby established as the world's first Islamic republic, appointing the previously exiled Ayatollah Khomeini as political and religious leader for life. From now on, the deeply conservative, theocratic regime used the derogatory epithet Great Satan when referring to the USA, while the UK and Israel where both dubbed Little Satan. While the ousted shah had been a staunch supporter of Israel, the Islamic rulers actively supported anti-Israeli and pro-Palestinian militias such as Hezbollah in Lebanon, Hamas in Gaza and the Houthis in Yemen, which were all established in the 1980s and 1990s.

From here there is a direct line to today's armed conflicts in the region.

Seven Sisters – and a Frenchman

When the victors of the First World War made themselves at home in the Middle East, this spurred powerful growth for Western oil companies. In 1928, several American, British, and French oil companies struck a deal concerning the oil resources in territories that formerly comprised the Ottoman Empire within the Middle East.[54] The Red Line Agreement divided the rights to the region's petroleum resources between companies from these three countries. The name of the agreement allegedly stems from the red pen that was used to draw on the map what the parties 'seemed to recall' had been the borders of the now-defunct Ottoman Empire (apparently nobody realised that in 1916 the borders had already been meticulously defined by Sykes and Picot).[55]

Although several of the large oil companies at the time had long histories, it was through the licences in the Middle East that they truly flourished, eventually achieving such dominance that they would later be known as the 'Seven Sisters': the Standard Oil Company of New Jersey (later Exxon), the Standard Oil Company of New York (Socony, later Mobil, which eventually merged with Exxon), the Standard Oil Company of California (Socal, later renamed Chevron), the Texas Oil Company (later renamed Texaco), Gulf Oil (which later merged with Chevron), Anglo-Persian (later British Petroleum) and Royal Dutch/Shell. The French government feared coming into a situation of strategic reliance on the US and British companies and thus decided to establish what today is known as TotalEnergies, which was also granted a generous portion of the attractive rights.

The Red Line Agreement was essentially a Western natural resource grab that gutted the basis for social and economic development in the region. The vast income from the valuable oil reserves was funnelled to large Western companies and local Middle Eastern elites.

Little was left for the general populations of the countries where the oil was extracted.

Recession and Despondency

In the years following the First World War, self-confidence and optimism about the future blossomed in the still-young USA. The country had for the first time demonstrated its economic, political and military power in the international arena, and was already in the process of displacing Great Britain as the uncontested global hegemon. But the economic growth was followed by new social challenges at home. In 1919, the US temperance movement won support for its cause when the sale of alcohol was banned through the National Prohibition Act. Five years later F. Scott Fitzgerald would write *The Great Gatsby* about the decadent lives and wild parties of the rich and famous in the US period of Prohibition. The era came to an abrupt and dramatic end when the Wall Street stock market crashed in late October 1929, the ripple effects of which spread through the industrialised world. Fortunes were lost, companies went bankrupt and millions of people exited factory gates to join the ranks of the unemployed throughout the USA and Europe.

Economic setbacks and collective despondency defined the following decade. It was a period characterised by economic recession, unbridled inflation and widespread social destitution. In the USA, one quarter of the labour force was unemployed in 1933. Every third adult male in Germany was without work at this time, and the German inflation rate was so high that people trundled wheelbarrows full of banknotes to the stores to buy bread. A sense of hopelessness ran rampant, and governments were at a loss as to how to come to grips with the situation.

'The solution' became that countries turned inward – and against one another. It became everyone's fight against everyone else. Protectionism and isolationism supplanted free trade, toll barriers were imposed and new trade restrictions introduced. The import of goods was considered a threat; countries preferred to produce domestic goods to create jobs for their own populations. Exports thereby also dried up, since one country's import is another country's export.

When economic growth subsided, confidence in the political system and establishment deteriorated. Benito Mussolini's fascists took power in Italy, and in Germany the stage was set for Adolf Hitler's national socialist movement. Germany was in a profound crisis and wobbling under the weight of national humiliation and unsurmountable war debt following its defeat in 1918. Hitler promised to 'make Germany great again' by putting an end to unemployment, generating economic growth and restoring *das Vaterland*'s pride and self-confidence.

While the prelude to the First World War was so elusively complex that the outbreak of war took the majority by surprise, the Second World War was a foretold catastrophe. It was the predictable outcome of dark forces and ideologies that fostered nationalism, nostalgia, xenophobia and social polarisation.

An Obliterated Belief

When Hitler's Germany annexed Austria in the spring of 1938, there was little to tempt the Americans to intervene once more in an internal conflict on the European continent. The idea that the USA would be protected by the vast ocean expanses in the event another large-scale international war were to break out was still very much alive and well. The country was also grappling with its own economic, political and social challenges, and there was a broad consensus that this time the Europeans would have to do their own housekeeping.

Yet again it was an attack from the sea that pulled the USA into war. Christmas decorations decked the streets of Honolulu on the day in early December 1941 when the city was caught completely off guard by low-flying formations of several hundred Japanese bombers and fighter planes that swarmed in to attack US naval vessels stationed in Pearl Harbor. Japan had made no formal declaration of war, and US intelligence had failed to intercept the covert threat of the Japanese aircraft carriers and other naval vessels that had taken position off the coast of Hawaii. In a few short minutes, the idyllic base of the US Pacific Fleet was transformed into an inferno of desperate screams, gunshot volleys, explosions and burning vessels. By the time the Japanese planes returned to their mother ships, more than 2,400 American soldiers and civilians had been killed, several hundred US fighter planes had been destroyed and dozens of warships had been sunk.[56] Just days later both Germany and Italy declared war on the USA, and Hitler implemented Operation Paukenschlag ('Operation Drumbeat'), a massive campaign of submarine attacks on US and Allied ships. The German submarines came so close to the US coast that some of them surfaced at night-time to see the lights of the city of New York.

The US belief in the ocean's guarantee of protection was hereby obliterated. Pearl Harbor was the 9/11 of the time, when without warning and for the first time the USA was attacked on its own soil. On this occasion as well, Americans became suspicious of a particular ethnic group within their own population. Just months after the attack, more than 100,000 US citizens of Japanese descent were forcibly displaced into internment camps out of fear that they would collaborate with the Japanese armed forces.[57] The latent racism of American society bubbled to the surface. Oddly enough, citizens of German and Italian descent, even self-proclaimed Nazis or fascists, were not targeted in a similar manner. It has never been proved that any of the interned US citizens of Japanese descent collaborated with the enemy, but more than forty years would pass before this segment of the population received an official apology from the government. In the meantime, a mountain, a think-tank and a nuclear-powered submarine had been named after the architect behind it all, Secretary of War Henry Stimson.

When the USA entered the Second World War, the large American growth machine was reset to a war economy in record time. Men of an age fit for combat were recruited to the military and stay-at-home wives enlisted to work in the factories. Aircraft manufacturers such as Boeing, Douglas and Wright-Martin changed over from passenger planes to bombers and fighter planes. Automobile factories such as Chevrolet, Dodge and Ford pivoted from the manufacture of shiny family cars to camouflage-drab jeeps, lorries and armoured vehicles.

Simultaneously, the capacity of the US Navy was further upscaled. There was an urgent need for everything from aircraft carriers, frigates, submarines and landing craft to ships for transport and supplies. All the components of the US war effort in Europe, Africa and Asia were to be transported by sea over long and vulnerable supply routes: soldiers, weapons, aeroplanes, artillery, tanks and replenishments of ammunition, fuel, uniforms and provisions. Virtually everything and anything that could float was commissioned for combat or transport. The shipyards were repurposed and upgraded. The Liberty ships were born – large, standardised transport vessels, each with cargo capacity for close to 3,000 jeeps or sixteen million day rations of food on each trip across the ocean.[58] But the large cargo capacity came at a price. The Liberty ships were often unable to sail at more than ten knots (eighteen km/h) as they battled through the waves of the Atlantic or Pacific Ocean.

The large convoys of ships carrying vital supplies across the ocean to the Allied forces were therefore also extremely vulnerable to the relentless attacks of German or Japanese submarines lurking in the depths.

4. URBANISATION, COASTALISATION AND GLOBALISATION

Bitter Homecoming

Almost all these convoys also included civilian merchant vessels manned by ordinary seafarers who had very little, if any, training to prepare them for the terrible ordeals in store. Merchant ships in both military convoys and ordinary traffic were attacked and sunk by bomber planes and torpedoes launched by enemy submarines. Throughout the entire war and all over the world, sailors anxiously scanned the surface of the ocean, at all times on the lookout for the bubbling trail of a torpedo. Tens of thousands of seafarers perished, many more were wounded and the majority of those who survived bore the brunt of traumatic memories for the rest of their lives.

Over the ten years I headed the Norwegian Shipowners' Association, I gained a more profound understanding of the history of the Norwegian war sailors. I therefore wish to dedicate a few words to the welcome they received upon their return to Norway after the war. This does not stem in any sense from a wish to diminish the trials and sacrifices of war sailors of other nationalities. To the contrary, I believe that in most countries, homecoming sailors were 'welcomed' in a similar manner.

A significant number of the ships in the transatlantic convoys between the USA and war-ravaged Europe were Norwegian. At that time, like today, Norway had one of the largest merchant fleets in the world, and when war broke out, every single Norwegian captain on ships in international and neutral waters – with the exception of a handful of ships stuck in Sweden – refused to follow the order issued by the German authorities to return to occupied Norway. The ships instead called at nearby Allied ports and were subsequently organised under Nortraship, a newly formed shipping company. The company was administrated and run from New York and London by Norwegian shipowners, in collaboration with the Norwegian government in exile in London. With its more than 1,000 ships and 30,000 seafarers, Nortraship was at the time one of the world's largest shipping companies and would play a key role in ensuring delivery of supplies to the Allied forces in Europe. Norwegian tanker ships alone transported a fifth of all the fuel delivered to the Allied forces, and after the war, UK prime minister Winston Churchill claimed that the Norwegian war sailors were worth more than a million soldiers.[59]

But the sailors paid a high personal price. Seafarers were the first Norwegians to be killed in the Second World War, and already before German troops attacked Norway on 9. April 1940, 58 ships and nearly 400 crew and passengers had lost their life in what has later been referred to as the Forgotten War.[60] Before the end of the war, half the Norwegian fleet had been torpedoed and sunk, and 3,700 Norwegians and almost a thousand sailors of foreign nationalities were killed in service for the Norwegian merchant fleet.[61] Of those who did return, many were haunted for the rest of their lives by horrific memories and the trauma of years spent lying awake at night in nerve-wracking anticipation of a torpedo strike.

The official homage paid to the merchant fleet's contribution stood in glaring contrast to the reception the sailors received when they started to return to Norway after the end of the war. They did not return as a group, since many were obliged to stay on board for months and years after the war. While the military forces that had fought the Germans at home were welcomed with parades, flags and fanfare, thousands of exhausted and traumatised war sailors returned to a society that turned its back on them. When peacetime came, the Norwegian government confiscated the Nortraship fund, where the sailors' war risk allowance had been deposited during the war. Adding insult to injury, it was the sailors' 'own' Labour Party government, supported by the Norwegian Confederation of Trade Unions, of which the Norwegian Seafarers' Union was at this time the largest member, who was responsible for this gross injustice. The war sailors were left to manage on their own, and for many that meant unemployment, divorce, alcohol abuse and chronic anxiety and depression.

Norwegian society and successive Norwegian governments shamefully betrayed those who paid the highest price for ensuring that we live in a free and democratic country today.

Brimming with Confidence

When the USA was drawn into the First and Second World Wars, in both cases reluctantly, and late in the game, it was a direct consequence of attacks from the sea. The USA came to understand that it could no longer remain safely ensconced behind its vast bordering ocean regions, that it no longer had a choice regarding whether to become involved or distance itself from the conflicts and problems of the rest of the world. Not even the rich country on the abundant North American continent, surrounded by vast ocean expanses, could isolate itself from a tumultuous world. 'The stopping power of water', as it turned out, had its limitations.

This led to a pivotal recognition with far-reaching consequences for both the US defence strategy and the international landscape of the post-war period: it was not enough to secure the protection of own coastal waters and seaborne trade. The key to the USA's security and geopolitical influence lay in its ability to exercise military power on and across the ocean.

As the strategic and geopolitical consequences became evident, the country did an about-face in its approach to the world. The US doctrine of a

4. URBANISATION, COASTALISATION AND GLOBALISATION

forward-deployed defence was established, a strategy that continues to determine the country's most important military priorities. The front line of military conflict was to be in the enemy's own coastal waters. With a strategy of forward-deployment, the USA could project power all over the world and engage in combat with the fleets of other nations long before the latter could approach the coast of the USA.

The strategy required political influence and close diplomatic ties in other parts of the world. Because of the great distances across the ocean, the US fleet would need access to ports for contingency, rest, maintenance, refuelling and replenishments. The USA therefore cultivated its relations with friendly countries, and today the US Navy has bases in every part of the world. But through such agreements, the host countries also declared their loyalty to the USA. This could make them vulnerable to economic and political pressure from other countries – and from the USA. In 1947, the Central Intelligence Agency (CIA) was established to ensure that the USA had foreign intelligence and counterintelligence capabilities that could match the needs and requirements of the country's expanded international ambitions.

The USA's heightened awareness of the ocean's strategic importance is evidenced by the strange fact that, although before the Second World War no US president had served in the Navy, after the war this became essentially a requirement for anyone with ambitions of occupying the Oval Office: all US presidents of the 1960s, 70s and 80s – Kennedy, Johnson, Nixon, Ford, Carter and Bush senior had served in the US Navy.[62]

While the economies and infrastructures of all other major powers were largely destroyed during the Second World War, the Americans did not suffer attacks on their own mainland. The USA emerged from the war with its entire physical infrastructure intact and with unrivalled technology and military force. Because of the Allied countries' gratitude for the critical role the USA played in winning the war, the latter enjoyed unique international status and influence.

Brimming with self-confidence, even before the end of the war, the USA took its next step on the geopolitical stage, championing the establishment of a new 'rules-based world order' administrated and led by a new world organisation, the United Nations. 'Universal values', joint principles and mutually binding regulations would prevent war, promote international collaboration and stimulate global economic growth and prosperity. The entire system would be underpinned by new international financial institutions and massive economic aid packages to assist with reconstruction and development.

The USA wanted to secure peace – and to shape the world in its own ideological and political image.

*'The shipping container made the world
smaller and the world economy bigger.'*
Marc Levinson, American economist

5.
WAVES OF GROWTH AND PROSPERITY

About the establishment of the United Nations and the legal framework governing the ocean, and how superior naval power has since the Second World War been key to the USA's dominant geopolitical position and its role as a global police and enforcer of the rules-based, UN-led world order. About the Cold War and the final phase of the colonial era, and how maritime trade drove economic growth leading to the fall of the Soviet Union, China's emergence as an economic powerhouse and major shifts in Western policies.

Imagine

I always register a special feeling of respect when I walk through the entrance to the United Nations' headquarters in New York. For those of us who profess allegiance to the global organisation's objectives and values, the building's towering, mirror-like facades appear to be an architectonic manifestation of one of the most important advances in the modern development of civilisation. The sea-blue UN logo, featuring the globe encircled by olive branches, is a universal emblem of peace, accountability and humanism. The text of the UN Charter can be read as a ratified protocol version of John Lennon's beautiful song 'Imagine'.

But not everyone is seduced by such poetic descriptions. In many parts of the world, both the conditions and consequences of the United Nations' activities are considered far more controversial. For some, the UN represents yet another arena for cynical power politics, a system designed to consolidate the global position and influence of the West. Although the core of the United Nations' basic philosophy is to 'replace might with right', there are many who ask, 'Which might and with what right?'

5. WAVES OF GROWTH AND PROSPERITY

When people in the West speak of 'the rules-based world order', there are many outside the West who hear 'we make the rules – you obey orders!'

A Few United Nations

The new global organisation was formally founded on 24 October 1945. Originally, forty six nations, that had all declared war on Germany and Japan and had signed an agreement that formalised the Allies of the Second World War, were invited to the table.[1] According to the two introductory paragraphs of the organisation's Charter, the objective of the United Nations is 'to save succeeding generations from the scourge of war, which twice in our lifetime has brought untold sorrow to mankind, and to reaffirm faith in fundamental human rights, in the dignity and worth of the human person, in the equal rights of men and women and of nations large and small'.[2] Among the founding principles of the organisation are key provisions of what is usually referred to as the Westphalian model, after the two peace treaties that ended the Thirty Year War in Europe in 1648: that each nation state has sovereignty over its territory and domestic affairs, that there shall be no interference in another country's domestic affairs, and that each state is equal in international law.[3]

Its admirable objectives notwithstanding, most of world's countries did not take part when the principles were defined and the rules set out for the new organisation. As five more countries were invited, there were initially only fifty-one nation states united under the United Nations. Almost the entire African continent remained under European colonial rule, and the four African countries represented – Egypt, Ethiopia, Liberia and South Africa – had effectively no voice whatsoever. Most of the countries of Latin America were autonomous, but none of them had sufficient political relevance or economic heft to merit inclusion in the negotiations. This was also the case for most of Asia. At this time, Pakistan was a part of India, and both were a part of the British Empire. Indonesia was a Dutch colony and the Philippines were under US control. Vietnam, Laos and Cambodia were all French colonies and referred to by the umbrella term 'Indochina'. Japan had been vanquished and humiliated, and in two of its largest cities the rubble from history's first, and so far only, nuclear attacks was still smouldering.

China was in the grips of a gruesome civil war, although Mao's communist army had taken control of most of the mainland after Chiang Kai-Shek's Western-backed national forces fled head over heels to Taiwan. To restrict communist influence internationally, the USA and Western Allies decided, despite vehement protests on the part of the Soviet Union and others, that it was the regime on the island of Taiwan that would be the UN representative for the Chinese people.

In Europe, Germany lay in ruins, subjugated and divided, saddled with heavy financial and moral debts owed to the rest of the world. The Italians were paying the political price for their alliance with the Nazis and the East European countries were all under the rule of Stalin's iron fist.

None of the states of the Middle East were in a position to exercise any influence when the new world organisation was founded. Their marginal role and impotence became even more evident when the United Nations voted to support establishment of the Jewish state of Israel in 1948. In the USA and Europe, guilt prevailed over having failed to prevent the Holocaust, but this did not translate into popular support for an offer of resettlement for the large number of surviving Jews who had been driven out of their homes in Europe. The European 'Jew problem' was therefore sought resolved by offering them their own country. Several regions were considered, including in Uganda, Australia, Surinam and Madagascar.[4]

The major powers, however, quickly agreed to grant the Zionists' demand for the establishment of a Jewish state in the region termed Mandatory Palestine by the League of Nations. The British had at this time virtually abandoned this region and surrendered its fate into the hands of the United Nations. The latter decided to divide Mandatory Palestine into one Jewish and one Palestinian state. Those residing in the region were, however, not consulted. The 'two-state solution', which still constitutes the basis of the UN's official policy, led to the brutal displacement of several hundred thousand residents, who were thrown out of the new Jewish state of Israel. Thousands of civilians were massacred by Israeli soldiers and Jewish militias.

The partitioning of Mandatory Palestine provoked violent reactions throughout the Arab world. Almost one million Jews who were residents of Arab nations such as Iraq, Tunisia, Algeria and Morocco sought refuge in Israel when they, in turn, were driven from their homes.[5] The same year, Egypt, Syria, Iraq and several other Arab countries launched attacks on Israel. The outcome of these acts of war was that the land the United Nations had allocated to the Palestinians was almost immediately occupied by Israel and Arab neighbouring states.

The new Palestinian state did not survive its own birth.

A Stubborn Norwegian

In practical terms, it was the major victors of the Second World War – the USA, Great Britain, France and the Soviet Union – who defined all the important conditions for the United Nation' objectives, design, governance and priorities. They also secured future control by assigning themselves permanent seats and exclusive veto rights in the UN's most powerful body, the Security Council. A few decades later, the People's Republic of China would be incorporated into this important club of veto-wielding nations, referred to today as 'the Permanent Five'.

The world organisation's first secretary-general was Norwegian. Trygve Lie represented a small country without any noteworthy political influence. He came from a post as minister of foreign affairs in a government still ideologically divided between the capitalist West and socialist East. Lie was of humble origins, from a working-class district on the east side of Oslo, not far from where I grew up. Critical voices have subsequently questioned his loyalty and character, and his legacy is rather controversial. But in 1945 he was a safe 'lowest common denominator' candidate around whom the major powers could unite.

However, once he had been appointed secretary-general, the officious Norwegian started to behave more like a general than a secretary in a number of areas. He became deeply involved in the construction of the organisation's headquarters in New York. To the raised eyebrows of both his own colleagues and representatives of other member countries, he saw to it that the meeting rooms for the United Nations' three councils – the Security Council, the Trusteeship Council and the Economic and Social Council – were all designed and outfitted by Scandinavian architects. The commissions were granted to Swedish Sven Markelius, Finnish Finn Juhl and his own close friend, the Norwegian Arnstein Arneberg. Trygve Lie hereby ensured that Nordic and Western culture also left its unmistakable mark on the United Nations' aesthetic expression and architectonic design.

He would simultaneously take such a proactive role in supporting the establishment of the Jewish state of Israel that he antagonised large parts of the Arab world and effectively undermined his own position as secretary-general.

A Striking Paradox

Although the USA initiated and was the prime mover behind establishment of the United Nations and the rules-based international order, the country brazenly took liberties with the rules regarding territorial rights in the ocean, humanity's largest public commons. After the death of Franklin D. Roosevelt in April 1945, one of Harry Truman's first presidential acts was to declare that the USA would unilaterally expand its exclusive economic zone from twelve to 200 nautical miles (370 kilometres) from its own coastline.

Up until this point, the rights of countries had been based on traditional use and coastal states' legitimate need to defend themselves against military attacks from the sea. From the time the Dutch Hans Grotius's principles acquired prominence in the eighteenth century, it had been universally accepted that the ocean was to be accessible to all. In keeping with John Selden's legal counter-response, the recognised practice had nonetheless been that countries should have territorial rights within three nautical miles from own coastlines. This originated with the 'cannon-shot rule', which was based on the range of coastal artillery at the time. As the cannon range subsequently increased, the border was extended to twelve nautical miles.

When the USA took the law into its own hands, in violation of international common law, other countries also claimed expanded exclusive rights within the regions off their own coastlines. The first to do so were Chile and Peru in 1947, followed by several Arab states just two years later. They all followed the USA's example and basically claimed territorial rights over the ocean regions extending to 200 nautical miles from their own coasts. It quickly became evident that a situation of chaos and conflict was in the making, unless a more mutually binding agreement was formed.

However, time went by, and negotiations for a 'Constitution of the Ocean' did not start until the ambassador of Malta, in an address to the United Nations General Assembly in 1967, called for 'an effective international regime over the seabed and the ocean floor beyond a clearly defined national jurisdiction'.[6] Fifteen years later, afters successive rounds of negotiations, the Convention on the Law of the Sea (UNCLOS) was opened for signature, and in 1994 it entered into force. Encompassing more than three hundred articles and nine annexes, this comprehensive legal framework regulates all ocean space, its uses and resources. It is the most extensive and detailed set of rules that member states have ever successfully negotiated under the aegis of the United Nations.

The USA was at all times actively involved in the negotiations but it has never signed this vitally important agreement. However, the USA has been the most important global guarantor and enforcer of the Convention's key rules, which US officials refer to as 'customary international law'.

The Magic Capitalist Formula

After the Second World War, the USA and other leading Western countries were more than ever before convinced that capitalism and free markets constituted the most important conditions for economic growth and prosperity, and that economic growth, in turn, provided the best conditions for the development of democratic rule. Since, historically speaking, democracies had never gone to war against one another, it was also assumed that expansion of this form of governance would mitigate the risk of war.

5. WAVES OF GROWTH AND PROSPERITY

The first part of this line of thought rests on the Scottish Adam Smith's formula in *The Wealth of Nations* from 1776: capitalism and free markets engender enterprise, innovation and creativity and the ongoing interaction between supply, demand and prices serves as an 'invisible hand' guiding the production of more and better goods, at lower prices and in quantities required by society. This school of thought is also based on the idea advocated by liberal thinkers such as Karl Popper and Friedrich Hayek that economic growth and the development of prosperity will produce a larger middle class, who, with better knowledge, greater self-awareness and a stronger desire for individual freedom, in the long term will not accept subjugation to authoritarian political regimes.

This is how the 'magic capitalist formula' was designed and explained, the pillar of the West's narrative of how capitalism and free markets supposedly produce free, content and peaceful human beings. This narrative also underpins the USA's idea of itself as an exceptional country, a globally inspiring 'City upon a Hill', with a unique moral responsibility to ensure lives of freedom and well-being for human beings all over the world.

By way of this approach, the USA took the lead in supporting the United Nations and the new, rules-based international order of treaties, schemes and institutions designed to promote international trade and economic cooperation. Through the Marshall Plan, the USA also contributed significant funding for the post-war reconstruction of Western Europe.

By the end of the war, the Western countries had agreed to establish the World Bank and the International Monetary Fund (IMF), often collectively referred to as the Bretton Woods system, named after the location in the USA where the first talks took place.[7] In 1947, the General Agreement on Tariffs and Trade (GATT) went into force, a measure to reduce tariffs and minimise technical barriers on the trade of industrial goods between Western and other non-communist countries.

The collaboration agreements, funding packages and reduction of trade barriers contributed to the strong economic growth and rapid upswing in international trade of the decades that followed the end of the Second World War.

A Hyper-efficient Logistics System

While world trade constituted an estimated ten per cent of global economic value creation in the nineteenth century, and twenty to twenty-five per cent in the first half of the twentieth century, it passed fifty per cent for the first time in the twenty-first century.[8] It was first and foremost the international fleet that transported this powerful surge in world trade after the Second World War. Then, as now, close to ninety per cent of the global trade in goods was transported by ship. Shipping hereby laid the foundation and conditions for the global growth which in the course of a few decades would completely alter the world's economic, political and military balance of power.

The international merchant fleet made it possible to ship large quantities of goods between different parts of the world at extremely low costs. If you purchase a pair of jeans or running shoes for US $150 in a fancy Manhattan boutique today, the overseas transport cost from China or India is typically a mere fifty cents. Different countries and regions of the world can therefore produce what they do best and trade goods and services on the international market. This enables individual countries to focus on the production of fewer products and services while the sales revenues from the latter can be funnelled towards the import of a broader selection of whatever else might be in demand. In the United Arab Emirates, for example, most of the export income stems from oil, gas and tourism, while almost everything in the way of food, clothing, cars, computers and beach parasols is imported from other countries. The opportunities for import constitute the most important driver for each individual country's prosperity development, while export's most important function is the generation of revenue to pay for the import.

This hyper-efficient system for the international trade also means that companies can achieve economies of scale by competing in markets in other countries. At the same time, companies from different parts of the world can collaborate through extended, tightly interwoven supply chains. This reduces production costs and boosts the exchange of expertise and ideas, which in turn promotes efficiency gains, quality improvement and innovation. Each element contributes to increasing global value generation.

Revolution in a Box

The strong growth in international trade led to an increase in the number of ships and the building of vessels that were larger, faster – and more specialised. While previously it was common to use the same ship for different purposes, ships were now being constructed with hulls, cranes, decks, tanks and equipment with specific designs, depending on whether they would be used to transport coal, iron ore, wheat, cars, lumber, oil, chemicals or consumer goods. A ship was no longer merely a ship.

In 1955, Malcolm Maclean sold his large American lorry company to raise the capital needed for building the world's first container ship. When *Ideal X* set out on its maiden voyage the following year, it marked the start of a

5. WAVES OF GROWTH AND PROSPERITY

revolution in the conditions of world trade and the global division of labour. Up until this time, the loading and unloading of ships was predominantly done manually by stevedores. Port workers loaded and unloaded goods to and from cargo holds, whether this meant bicycles, sacks of coffee or crates of garments. But the introduction of containers with a standard breadth, height and length meant that cranes could be used to stack them like Lego blocks on ships. With time the containers were also equipped with refrigeration and freezer systems for the transport of fruit, flowers, medicine and fresh produce, and each ship could thereby transport a range of different products on the same voyage.

The containers could also be seamlessly transferred between ships, trains and lorries while en route from one part of the world to another. This increased efficiency and led to a substantial reduction in transport costs.

Even in their infancy, container terminals could handle twenty times more cargo per worker than traditional ports.[9] This rapidly resulted in lay-offs of large numbers of stevedores and port workers, triggering wide-spread strikes and fierce protests from powerful trade unions. But there was no going back, and the containers not only reduced the number of jobs for these groups of workers. This new invention would also fundamentally change the conditions and parameters for the global distribution of work, income, wealth and prosperity. It provided, not least, the basis for China's strong economic growth a few decades later.

The powerful growth in maritime trade also generated large, new business opportunities for companies offering maritime services such as insurance, financing and classification. Maritime research and development took off, and new, innovative navigation and communication technologies were introduced.

One for All

Despite these significant technological advances, the strong growth in seaborne trade was accompanied by an increase in the number of accidents and mishaps at sea. Ships sank, seafarers died and large oil spills from grounded tankers destroyed coastlines and marine wildlife. The international community responded by developing new and more comprehensive sets of rules for the environmental, technical and operational standards for ships and port operations, and for training, certification, wage agreements and working conditions for seafarers.

The first international technical standards for ships had been introduced already in 1912, following the tragic wreck of the *Titanic*, but it was not until the establishment of the United Nations that global conventions and effective regulations were developed. In 1958 the International Maritime Organization (IMO) was formed, the UN agency responsible for the international regulations on maritime competence, safety and the environment. The mandate of the International Labour Organization (ILO), established by the League of Nations in 1919, was simultaneously broadened to include seafarers' working conditions, salaries, paid annual leave and other terms of employment.

Through the United Nations conventions, shipping also became, despite what most people seem to believe, one of the most pervasively regulated international industries. There are few if any other industries today subjected to such a broad set of global rules and regulations relating to technical, operational and environmental standards, safety, security, competence, training, certification, working conditions, salaries and labour rights.

In many fields the regulatory standards for shipping can be low and inadequate, and in some fields exceedingly so, but they have universal application. There is one set of rules for all.

Rootless, Disloyal and Invisible

While the shipping industry's hyper-efficient transport work formed the basis for the modern world's division and organisation of economic activity, shipping companies themselves have always been at the forefront of globalisation with respect to its logic and consequences. A ship can be designed, financed, built, equipped, owned, flagged, classified, insured, manned and operated from anywhere in the world, the transport crosses national borders and an intense international competition is a key characteristic of all commercial activity.

This means that the profit margins in every part of the shipping supply chain are under constant pressure. But it also implies that shipping companies are free to choose from the entire world's abundant menu of profitable opportunities, smart adaptations and attractive locations.

In all of modern history, it is the expected return on investments, market opportunities and commercial requirements that have guided shipping companies' decisions.

Shipping is fundamentally a rootless industry, which invites national disloyalty.

As shipping's significance for global growth and prosperity rose to unprecedented levels, the industry was also moved out of sight and out of mind for the public at large. After the Second World War, many ports in central urban districts were relocated, as the warehouses, hinterland, roads and

5. WAVES OF GROWTH AND PROSPERITY

railways required for handling the flow of goods were squeezed out by an influx of housing units, shopping malls and office buildings. As a result, the ships, seafarers and port workers were no longer a part of the urban landscape of daily life.

In response to increasingly tough international competition, shipping companies also began hiring seafarers and commissioning new builds from low-cost countries. Ships were registered in offshore flag states offering the most favourable terms and formal ownership relocated exotic tax havens. The international fleet was manned by more seafarers from countries of the Global South, such as the Philippines, India and Bangladesh, and fewer from Western high-cost countries such as Norway, France, Spain, Denmark and Great Britain. The number of ships built, equipped, flagged, manned and taxed in Western industrialised nations declined, although key activities were still controlled and administered from these countries. As a result, the political and social ties between the shipping companies and their 'actual' national affiliation were eroded.

In large Western shipping nations, the industry became more invisible, alien and controversial. Shipping companies' national affiliation and social utility were obscured, and, when the industry did receive public or political attention, the focus was more often on 'social dumping' of seafarers and the tax evasion tactics of wealthy shipowners. On the other hand, the international standing and prominence of the shipping industry rose in many countries in Asia and the Global South, due to the industry's vital role in creating new jobs, revenues and developmental opportunities. Today, for example, China, South Korea and Japan are the world's largest shipbuilding countries, while the Philippines, India and Indonesia are main suppliers of seafarers to the international fleet.

Also, in the 1950s and 1960s, airline companies took over virtually all intercontinental passenger traffic and the traditional ocean liners that had up to this point carried travellers back and forth across the Atlantic and the Pacific soon went out of business. From this point on, it was Air France, British Airways and PanAm, rather than the *Queen Elizabeth*, SS *France* or SS *Vaterland*, that people associated with transoceanic journeys.

In this climate of economic growth and expanding material prosperity, a number of the passenger ship companies spotted a new and attractive business niche: the concept of an onboard holiday. The cruise industry was hereby born. For the first time in history, passengers boarded ships in pursuit of enjoyment and entertainment, rather than out of necessity.

MAD, NUTS and NATO

The UN Charter's lofty ambitions of peaceful coexistence did little to alleviate tensions between the world organisation's two most important founders. The relationship between the USA and the Soviet Union quickly became so icily fraught that the author George Orwell, as early as 1945, referred to it as the Cold War. This period of 'peace that is no peace' would have an impact on all international relations until the dissolution of the Soviet Union in 1991, and the after-effects continue to reverberate even up to the present day. The bipolar logic of the post-war period determined the two superpowers' respective spheres of power and influence, much as the papal decree had done for Spain and Portugal more than 400 years previously. Several countries attempted to remain neutral, but in practice very few were able to avoid choosing sides.

Both the USA and the Soviet Union built military, political and economic alliances to consolidate and expand their spheres of power. In 1949, the North Atlantic Treaty Organization (NATO) was formed, a military collaboration between the USA, Canada and a handful of West European countries. The core principle behind the alliance was that an attack on one member country would be considered an attack on all, and that all member states were bound to defend one another. But, while NATO is often referred to today as a purely military alliance, the organisation was initially founded on the basis of three objectives: to curb Soviet expansion, prevent military nationalism from threatening peace on the European continent, and promote political cooperation and integration in Western Europe. At this time, many of the West European nations in particular were concerned that the USA would revert to an isolationist existence behind the vast expanses of the Atlantic Ocean, which might tempt both Russia and Germany to stage new military escapades on the European continent. The organisation's first secretary general, the free-spoken Lord Hastings Ismay, said undiplomatically that NATO was created to 'keep the Soviet Union out, the Americans in and the Germans down'.[10]

In 1955, the Soviet Union countered by forming the Warsaw Pact, an alliance between the Soviet Union and seven East European countries. But, while NATO membership is voluntary and achieved by application from democratic states, those in power in the Kremlin imposed membership in the Warsaw Pact on the East European countries.

The rivalry between the USA and the Soviet Union manifested in the form of a fierce military arms race, and by the 1950s both of the countries had sufficient nuclear capacity to eradicate the other many times over. On one

5. WAVES OF GROWTH AND PROSPERITY

occasion, Albert Einstein commented, 'I don't know with what weapons World War III will be fought, but World War IV will be fought with sticks and stones.'[11]

The destructive force of nuclear weapons is so extreme that for the first time in the history of humankind, the defence strategies of the two most powerful countries were based on a hypothesis of total and mutual annihilation: a nuclear attack by one party would trigger immediate response on the part of the other, and both would go up in smoke. This strategy of mutual deterrence was tagged with the label MAD, Mutually Assured Destruction, but few found humour in the acronym's biting irony. MAD was for all practical purposes a suicide pact. Eventually, the USA and the Soviet Union developed nuclear missiles that were so precise and long-range that in theory each could knock out the other's arsenal in a surprise first strike and thereby prevent mutual annihilation. This was the origin of the term NUTS, Nuclear Utilization Target Selection, MAD's darkly humoristic counterpart. Both nations quickly understood that although the strategy might work well on paper, it would never succeed in practice.

Even in the event of a surprise first strike, wiping out an adversary's entire fleet of submarines carrying intercontinental nuclear missiles, operating covertly in the dark depths of the ocean all over the world, was inconceivable.

The threat of mutual annihilation was therefore at all times credible enough to ensure that the Cold War remained cold: the relations between the USA and the Soviet Union were rife with unabating hostility and intransigence, but neither ever risked a first strike. Neither of the two nuclear powers harboured any doubts about the destructive threat of the other, and neither wanted to risk heating up the cold war. Yet this did not prevent either of them from supporting uprisings and conflicts around the world or from becoming involved in wars by proxy.

Doomsday Machines in the Depths

The arms race between the two superpowers extended into the atmosphere and further into space, but both also built huge fleets of seagoing warships. Power in the world requires power at sea – and power at sea brings power on land.

USA made particularly heavy investments in building up a fleet of aircraft carriers. Because of the country's isolated geographic location behind the Atlantic and Pacific Oceans, the USA required long-range military fleets to assert its power and counter the communist threat in other parts of the world. At first the aircraft carriers were solely intended for use as bases for fighter planes to protect seagoing combat groups from hostile fleets or attacks launched from land. With time, however, it became evident that the carriers could be even more useful in offensive operations. During the Vietnam War, the aircraft carriers demonstrated their strategic advantage as forward-deployment bases for airborne attacks on land.

Since then, it is therefore the accompanying fleet that protects the aircraft carrier – and not the other way around.

During the Cold War, the superpowers' strategic game moved into the ocean depths. Both the USA and the Soviet Union built large nuclear-powered submarines armed with long-range, ballistic nuclear missiles (a ballistic missile is launched in a high arc and rocket-propelled as it ascends toward the highest point in its trajectory, where the engine stops firing and the forces of gravity make the missile free-fall towards the target at high speed). In 1960, the American USS *Triton* became the first submarine to complete a submerged circumnavigation of the earth.[12] While up until this time submarines had solely constituted a threat to seagoing vessels, nuclear-powered submarines were now capable of attacking cities, infrastructure, industrial areas and other land-based targets all over the world. The black, cigar-shaped hulls of today measure up to 180 metres and have crew capacity of more than 150. Yet they are almost completely silent, can remain submerged for months at a time and can hide beneath the ice on the North Pole. Hidden in the murky depths beneath the Atlantic Ocean's foaming waves and the Pacific Ocean's long swells, these submarines can navigate imperceptibly and remain completely undetected.

The battle *on* and *over* the ocean has become a battle *in* and *from* the ocean.

One of history's most nerve-wracking and potentially fateful confrontations took place beneath the surface of the ocean. In 1962, US surveillance photos revealed that the Soviet Union was in the process of establishing a military base in Cuba, just 150 kilometres off the coast of Florida. The US simultaneously discovered that Soviet ships were en route, carrying missiles to be deployed on the communist island state. The USA perceived this as such a grave strategic threat that President Kennedy issued an ultimatum, which effectively implied the start of a nuclear war if the missiles were not removed. This was the Monroe Doctrine with humanity on the line. Due to the extreme tension of the situation, people all over the world began preparing for the worst. I was only six years old at this time, but I vividly remember my mother showing me our family's emergency rations of flour, canned goods and water in the cramped storage space in the cellar of our modest block of flats.

The ships carrying the missiles to Cuba were but minutes away from the Kennedy ultimatum's 'red line', when the Soviet Union's President Khrushchev gave the order to retreat. An entire terrified world population was standing by, watching the countdown to nuclear war on a live television

5. WAVES OF GROWTH AND PROSPERITY

broadcast, when the ships abruptly turned round, the waves churning and foaming in their wake as they reset their course for home. Total annihilation was forestalled and the Cold War continued.

But what viewers could not see on television – and did not become public knowledge until several decades later – was that another drama was simultaneously unfolding underwater. Because the Soviet Union had not only dispatched surface vessels; they had also sent submarines into the waters around Cuba. The US military knew that the submarines were there, but were unaware that they were armed with nuclear warheads. On 27 October 1962, one of the Soviet submarines was detected and the US Navy attempted to force it to the surface. Armed and prepared for combat in its submerged state, the submarine was surrounded by eleven US Navy vessels. Unable to communicate with the outside world and with depth charges detonating on all sides, the desperate and terrified Soviet crew on board the submarine *B-59* were convinced that they were at war.

The exact details of what happened next are somewhat fuzzy. There does, however, appear to be little doubt about the fact that the submarine captain, under extreme duress, issued the order to fire a nuclear warhead torpedo with an explosive force equivalent to the bomb dropped on Hiroshima. Before the order could be executed, however, the two other high-ranking officers on board were required to give their authorisation and unlock their respective launch codes. One of the officers approved the order, while the other refused. Thirty-four-year-old Vasili Arkhipov had noticed that the depth charges were being detonated on the starboard and port sides of the submarine in alternation. He understood quite correctly that this meant that the US warships wanted to force them to the surface, rather than blow up the submarine. After a heated argument between the captain and Arkhipov, *B-59* finally rose to the surface and made itself known to the US forces. Without further ado, the Soviet submarine was then escorted out of the waters around Cuba.[13]

The young Arkhipov's cool, analytical thinking essentially prevented the start of World War III. One man's personal integrity, courage and quick-wittedness saved the planet. Not in a Hollywood movie, but in the everyday reality of the Cold War.

This would be one of several known incidents in the modern era during which a global apocalypse was averted by random events. For the first time in history, nuclear weapons, command systems and rapid response times implied that misunderstandings, human error and technical failures held the potential to wipe out large numbers of the world population.

Colonial Collapse

In the decades following the Second World War, waves of decolonisation swept across Africa and Southeast Asia. The war had drained the colonial powers' economic, military and administrative resources, and the support for colonisation was also dwindling in many European countries. In many of the colonies, a greater national self-awareness had also emerged , and they were often themselves the main drivers of liberation.

Kenan Malik writes:

> Today, decades after decolonization, it can be difficult to grasp the degree to which a brutally naked form of imperialism dominated the globe. It can be difficult also to conceive the depth of often violent opposition to imperial rule, from mass strikes to national liberation movements. 'Decolonization' is often imagined as a process by which the colonial powers 'granted' independence to their colonies as a benevolent gesture. The decades of fierce struggle, and the ferocious suppression of such struggles, have been largely erased from public consciousness and left to moulder in archives and historians' texts.[14]

On the basis of widely disparate ideological arguments, both the USA and the Soviet Union employed rhetoric condemning coercion, tyranny and oppression. The USA made decolonisation a condition for economic post-war recovery aid to Europe under the Marshall Plan.[15] Both the USA and the Soviet Union also mobilised substantial economic, military and diplomatic resources as a means of gaining the allegiance of the newly autonomous nations. The locations of many of these countries were of strategic importance, and many also had vital natural resources. This was the case of Africa in particular, with its large reserves of gold, diamonds, copper, cobalt, nickel, uranium and oil.

The USA wanted to set an example by taking the lead, and in 1946 recognised the Philippines' independence after almost fifty years under US rule. The same year, the inhabitants of Indochina went to war to oust the oppressive French colonial rulers. The communist guerrilla leader Ho Chi Minh led the charge of the revolt. The war of independence continued for eight years, until France and the other major powers agreed on the division of Indochina into four parts: Cambodia and Laos were recognised as independent nations. The remaining territory was divided into a communist North Vietnam and a South Vietnam supported by the West, pending reunification after a democratic election. This election, however, never took place.

5. WAVES OF GROWTH AND PROSPERITY

The USA and the Soviet Union were both vying for political power and ideological influence in every part of the world, and the US strategic thinking was based on the 'domino theory': if one country falls to the communists, the neighbouring nations will follow.[16] In 1950, the US invested considerable military resources in an attempt to throw Kim Il Sung's communist-backed occupying forces out of South Korea. When France threw in the towel in Southeast Asia in 1954, the USA entered the fray once more to prevent the communists from strengthening their position in the region. The Vietnam War would continue until 1975, during which more than one million people had been killed and wounded. When American soldiers were forced to flee Saigon (today Ho Chi Minh City), it marked the USA's first humiliating defeat as a military superpower.

Although France withdrew from Indochina, it was not willing to relinquish control over Africa's largest country, Algeria, and its substantial oil and gas resources. Algeria's fight for independence lasted for seven years and was one of the most gruesome on the African continent. By the time the French were thrown out of the country in 1962, an estimated one million had been killed and more than two million had lost their homes.

In Indonesia, the Netherlands never succeeded in regaining control after the surrender of the Japanese occupying forces towards the end of the Second World War. The country, with its more than 17,000 islands and population of seventy million, achieved autonomy in 1949.

Released from the straitjacket of colonial rule, in many places internal conflicts and civil wars erupted. In the period prior to India's independence in 1947, the antagonism between Hindu and Muslim populations became so violent that the British colonial authorities decided to divide India in two. In the course of just a few weeks, more than two million Muslims were forcibly displaced to areas in the east and the west, where they were given their own country, Pakistan, 'the land of the spiritually pure'. This incited violent protests and riots, and more than 100,000 people were killed. It quickly became obvious that a single Pakistan divided in two by a larger, Hindu-dominated India, was a hopeless proposition. After a war with India in 1971, East Pakistan was made a separate country, named Bangladesh, while West Pakistan became what we today recognise as Pakistan. The relation between India and Pakistan remains strained to this day, while India is fraught with large and growing antagonisms between the country's Hindu majority and Muslim minority.

Following its liberation from Belgium in 1960, long-term tensions between many of Congo's two hundred ethnic groups continued to smoulder until they burst into flames in a series of violent civil wars in the late 1990s, the impact of which continues to haunt the country. The figures are uncertain, but most estimates indicate that, since that time, between four and five million people have been killed, the greatest number of fatalities in any armed conflict since the Second World War.[17]

Just a few years before civil war broke out in Congo, the Hutus attacked the Tutsis in Rwanda, one of the smallest and most densely populated countries in Africa. The country was consolidated first as a German, and later Belgian colony, until it gained independence in 1962. The Hutu people made up the majority in Rwanda, but, even during the early colonial period, the Germans fostered the notion that the Tutsis were racially superior to the Hutus. For generations this engendered economic disparities and social antagonisms that laid the foundation for the genocide of 1994, during which one million Tutsis and moderate Hutus were massacred in the course of just one hundred days.

Nigeria ceased to be a British colony in 1960, along with several other African countries. Today there are more than two hundred ethnic groups, the majority with their own language, living within Nigerian borders defined by European colonial powers. Nigeria is Africa's most populous country and has the continent's largest economy, but the majority of the more than 200 million inhabitants live in abject poverty. The large natural resources, including oil and gas, could have formed a basis for well-being and prosperity, but the country has been consistently beleaguered by political mismanagement, endemic corruption, terrorist attacks, criminality and rivalries between tribes, clans and ethnic groups.

When Angola gained its independence from Portugal in 1975, a gruesome and devastating war broke out between communist and nationalist guerrilla groups. Like several other armed conflicts in Africa and Southeast Asia, the war in Angola was as much a proxy war and part of the Cold War as it was a civil war between the Angolan people. The Soviet Union and the USA – the latter together with the apartheid government of South Africa – contributed to prolonging the war and the suffering of the Angolan people by supplying each side with large amounts of weapon and equipment. More than a half million people lost their lives, and to this day people are killed or maimed by the landmines that were strewn in fields throughout the country during the almost thirty-year-long war. On the streets of the capital, Luanda, it is impossible not to notice the many residents who are missing an arm, leg or an eye.

From the end of the Second World War and up to the new millennium, more than one hundred countries in Africa and Southeast Asia were decolonised. When Great Britain handed Hong Kong over to China in 1997 and Portugal relinquished Macao in 1999, this also marked the return of the final

5. WAVES OF GROWTH AND PROSPERITY

parts of the Chinese Empire that had been occupied during the Century of Humiliation.

The colonial era left deep, permanent scars. The majority of the current ethnic, sectarian, religious and political tensions and conflicts in Southeast Asia, the Middle East and Africa stem from the burdensome legacy of Western colonialism.

Immigration and Assimilation

For many of the former colonies, the path to independence and autonomy was convoluted and thorny. After centuries under the rule of foreign powers, the newly independent states were suddenly required to stand on their own two feet. They endeavoured to fast-track political, economic and social development processes, which in the West had been cultivated and refined over the course of many generations. The results were as diverse as the countries' cultural and historical starting points. Some took rapid strides towards freedom and progress, while others collapsed into chaos and civil unrest. In many countries liberal democracy gained a foothold, while in others kleptocracy, the dictatorship of thieves, took root.

Many European nations were initially open to welcoming immigrants from the former colonies. Until the early 1960s, people from all of the more than fifty countries in the British Commonwealth, as the former empire had now been renamed, were relatively free to settle and work in Great Britain. Following the partitioning of India, a large number of people from India and Pakistan migrated to the British Isles. The Netherlands opened its borders for immigration from Indonesia and Suriname, and France for immigrants from Algeria.

This produced new and unfamiliar issues in European countries, because racial prejudice was alive and well and the integration policy immature and flawed. The prevailing attitude was that the new citizens were to be *assimilated*, not *integrated*. Although this may sound like sociological hair-splitting, the difference is much like being asked by a partner to become more 'like him' as opposed to being 'with him'.

Many of the immigrants ended up on the bottom rungs of the social ladder in their new homelands. Their difficult situation was exacerbated with the abatement of economic growth and the rise in unemployment in the 1970s and 1980s, which added further fuel to tensions and hostilities between immigrants and host populations.

In present-day France, some of the greatest social problems are found in immigrant communities in, for example, the satellite towns in the north-east of Paris and some of the *arrondissements* in the northern districts of Marseille. The unemployment, violence and crime statistics here top national averages, while the communities land in a solid last place on tables depicting living conditions, life satisfaction and hopes for the future. As demonstrators set fire to cars, threw stones and fought with the police in French suburbs in 2005, they shouted in unison 'We are here because you were there!', a slogan coined by the Sri Lankan author Ambalavaner Sivanandan.[18]

A Poverty Trap

Throughout the last half of the twentieth century, uprisings, regional feuds and civil wars continued to prevail in large parts of the resource-rich African continent. Widespread corruption hindered or delayed economic growth and the development of democracy in many African countries. Authoritarian and quasi-democratic regimes were kept in place by ample and often covert funding from foreign, international corporations that helped themselves to the continent's abundant natural resources. Most production took place under the direction of the companies themselves, while executive or skilled positions were predominantly filled by individuals recruited from headquarters mostly located in Western countries. The dangerous, exhausting and poorly paid jobs were given to the locals. Hundreds of thousands of often under-age children were sent down into cramped, dark and dust-filled mines equipped with little more than picks, crowbars, spades and headlamps.

For the greater majority of Africa's population, the abundance of natural resources became more of a curse than a blessing. Funds that could have been invested in social development, infrastructure, health and education became instead the accumulated wealth of dictators, directors and investors.

This glaring disparity in the distribution of income and wealth, and the one-sided harvest of natural resources, also undermined the prosperity and economic development of many African countries. Economic growth is not created merely by doing more of the same thing in the same way. It requires doing more of the same thing in more efficient ways, or doing something else that generate more value added for the man hours, money and resources spent. Economies must constantly innovate, reinvent and reposition themselves in order to grow; they must work smarter, quicker, better – and with other things. It is the ability to change and adapt that creates torque in the economic growth engine, and the greater the adaptability, the more quickly progress is achieved.

But innovation, adaptation and repositioning require knowledge, skill, ideas, infrastructure and investments, all of which were systematically neglected and quashed in many of the newly independent African nation states. The lack of investment in education, health care, welfare schemes and infrastructure also meant that only a small fraction of the population took part in the organised economic value creation. It was thus not long before the populations' trust in the governing authorities deteriorated. This made it difficult for many of the low-income African countries to pull themselves out of the poverty trap.

Consequently, large parts of the African continent have remained poor economic performers. Today approximately two thirds of the world's extreme poor are living in sub-Saharan Africa.[19] Also, two thirds of the fifty-four states on the African continent remain in the category of 'the world's least developed countries' (LDCs).[20] By 2024 only three of these thirty-six African countries – Botswana, Cape Verde and Equatorial Guinea – had managed to graduate from this category since the United Nations established it in 1971 as a measure used to design development aid policy.[21] In several of the remaining thirty-three countries, political mismanagement, corruption, violence and war continue to perpetuate poverty and suffering for large portions of the population.

Then, the African continent has also witnessed remarkable improvements in some important areas, such as education, health and child mortality. While still higher than on other continents, and varying significantly between the African countries, the overall child mortality rate has steadily declined over the past seven decades due to improved health-care services, improved nutrition and better access to clean drinking water.[22] There are also great differences between the African states in their social, economic and political developments, and several have made major strides towards improved economic growth, such as Kenya and Ghana (which were never LDCs), and Uganda, Tanzania and Rwanda, which in 2024 were all candidates for graduating from the LDC list.[23]

The Curse of the Landlocked

But even with more targeted investments in people and infrastructure, many of the impoverished African countries would be fettered by the lingering chains of colonialism.

Large parts of the basic infrastructure which is needed for economic and social development, had been developed to serve the interests of the European colonial powers, not the African continent itself, writes Walter Rodney:

> [Roads and railways in Africa] had a clear geographical distribution according to the extent to which particular regions needed to be opened

up to import-export activities. Where exports were not available, roads and railways had no place. The only slight exception is that certain roads and railways were built to move troops and make conquest and oppression easier... All roads and railways led down to the sea. They were built to extract gold or manganese or coffee or cotton. They were built to make business possible for the timber companies, trading companies, and agricultural concession firms, and for white settlers. Any catering to African interests was purely coincidental.[24]

Also, due to the random delineation of national borders by the Berlin Conference of the 1880s, almost one third of the African states do not currently have access to the ocean.[25] Forty per cent of the population of sub-Saharan Africa is landlocked, and this alone creates unique challenges.

Comprehensive empirical studies describe how landlocked countries have consistently less trade, lower economic growth, more poverty and overall, a less auspicious social development than states with a coastline.[26] The annual value creation per inhabitant of landlocked African countries is half that of other countries on the continent.[27] Of the sixteen states in Africa that do not have a coastline, thirteen are on the UN list of the world's poorest countries and, on average, African landlocked countries trade thirty per cent less than the coastal states on the same continent.[28] One study of Africa states that 'there is a temptation to say that the more a nations' borders are constituted by water, the more successful that nation becomes in all respects'.[29]

Landlocked countries are often less able to profit from shipping's cost-effective routes to the rest of the world, and this constitutes a key disadvantage. For instance, the cost of sending a container from a landlocked country on the African continent is typically twice that of countries with a coastline.[30] Landlocked countries are thereby often unable to leverage the full extent of their competitiveness in export markets, and imported goods also become more expensive.

The UN Convention on the Law of the Sea (UNCLOS) does, in fact, give landlocked countries the right to uninhibited and customs-free transhipment of goods through neighbouring countries to and from the coast. However, in practice, it is the political stability, infrastructure and good will of these countries that determines how this system functions.

Tiger Economies

While many African countries remain suppliers of raw materials, several of the countries in the eastern and southern regions of Asia were better positioned, historically, culturally and politically, to reap the benefits of the global trade economy which flourished in the decades following the Second World War.

Under British colonial rule, Hong Kong had already become an Asian financial hub. Japan was enjoying the fruits of the industrial foundation it had built after the Meiji Restoration. After the Second World War, both the international community and Japan itself imposed restrictions on its military activities, as a result of which most of its resources could be poured into economic growth and development. From the mid-1960s, Singapore developed a thriving economy under Lee Kuan Yew's autocratic but capable regime. At the same time, President Park Chung-hee implemented a large-scale modernisation programme in South Korea with the help of economic aid from the USA and Japan, both of whom had an interest in countervailing the communist influence of the Soviet Union and China. In Taiwan, Chiang Kai-shek launched a comprehensive economic growth and industrial development programme through use of China's gold reserve, which the nationalists had taken with them when they fled the mainland.[31] The Taiwanese economy continued to flourish under the protective hand of the USA and with the help of generous US aid packages that provided funding, investment, technology and military support.

These Asian countries therefore had an advantageous position as global economic growth continued to accelerate throughout the 1960s, 1970s and 1980s. The volume of trade expanded, the ports became more efficient, and, not least because of containerisation, the costs of intercontinental freight shipping were further reduced. This led to powerful economic growth in Japan, South Korea, Taiwan, Hong Kong and Singapore. The four latter countries, often referred to as 'the tiger economies', would also inspire extensive economic reforms in countries such as India, Indonesia, Thailand, Malaysia – and China.

Cadillacs and Coca-Cola

A new twist in the Cold War occurred when the Berlin Wall was erected in 1961, and the Soviet Union in the subsequent years installed an 'Iron Curtain' stretching from the north to the south of Europe. At the same time, China wove a 'Bamboo Curtain' around its own borders. While the communist countries walled themselves in with their oppressive regimes and state five-year plans, 'the free world' expanded on the back of the post-war recovery and a virtually explosive upsurge in international trade.

The post-war decades were also the golden era of the Western middle class. Millions of new jobs were created, income levels rose and prosperity grew.

In the USA, the 'American dream' became a reality for increasingly more of its citizens. The 1950s and 1960s were the era of Elvis, James Dean, Marilyn Monroe and John Kennedy, the era of Cadillacs, Marlboro Man, Coca-Cola and the American moon landing. The USA was at the peak of its cultural and ideological influence, for which the political scientist Joseph Nye would later coin the term 'soft power'. The American lifestyle, the country and the culture came to constitute an attractive global brand.

Sixty years later, the nostalgia and memories of this period would find expression in the slogan 'Make America Great Again!' By mining the dream of what once had been, Donald Trump would generate political clout for his politically and socially divisive democracy demolition project. *Again!* – not *Great* – is the dominant theme of his bombastic message.

The powerful economic growth of Western industrialised nations led to higher expectations on the part of the population for better schools, roads and health care. There was also growing public demand for welfare schemes and social security nets that could better provide for the sick, unemployed and elderly. More workers joined trade unions, which in turn grew in size, power and influence.

Extensive reform programmes were introduced in the USA in the 1960s, including assistance for poor and less privileged citizens. Also, in 1965, African Americans where granted full voting rights. It was, however, the European countries who in the decades after the war built what we today associate with the modern welfare state. Schemes were introduced for old age pension, disability, sickness benefits, childcare subsidies, social welfare and unemployment benefits. Public schemes and services, such as free schooling and health care, were broadened, and laws passed ensuring wage guarantee schemes, minimum wage, holiday leave, maternity and paternity leave and working environment regulations.

As welfare schemes expanded in scope and prevalence, the need for state revenues grew apace. In 1954, France was the first country to introduce a general value-added tax (VAT) on goods and services, and all Western countries raised or introduced new taxes and fees. In the 1960s, for example, the marginal tax on high-income brackets was fixed at seventy to ninety per cent in countries such as the USA, Great Britain, and France, while corporate profits were taxed at rates ranging from thirty to fifty per cent.[32][33][34] In the mid-1970s, Western industrialised countries channelled on average the equivalent of one third of their gross domestic product via public budgets.[35]

The welfare schemes and redistribution of income and wealth created better lives for more citizens, but often entailed significant hikes in corporate tax rates and salary costs.

5. WAVES OF GROWTH AND PROSPERITY

When Asian countries began making their presence felt in the modern-day international economy, this would backlash in a way few had anticipated – and even fewer had provided for – when the schemes were first introduced.

The Iron Fist of the Oil Club

In the aftermath of decolonisation, some of the largest oil- and gas-producing countries – Iran, Iraq, Kuwait, Saudi Arabia and Venezuela – joined forces in 1960 to found the Organization of the Petroleum Exporting Countries (OPEC). Other countries such as Qatar, Libya, Angola, Nigeria and the United Arab Emirates later became members, and today this club controls around eighty per cent of the world's known crude oil reserves (although the members stand for a smaller percentage of the production).[36] OPEC is effectively a cartel that restricts competition and promotes cooperation between oil-producing countries, geared to extract even larger revenues from the world market.

The effect of the OPEC cooperation was a rise in oil prices that was far greater than otherwise would have been the case during the 1960s, but it was not until 1973 that the organisation really demonstrated its power. It did so in the wake of Egypt's and Syria's attack on Israel during the Jewish Yom Kippur holiday. Israel responded in kind, and the resulting war lasted for only six days, but the Arab member countries of OPEC stopped oil exports to the USA, Canada, Japan and the Netherlands, all of whom sided with Israel in the conflict. The oil boycott remained in force until the following year, when the USA convinced Israel to withdraw from the Egyptian territories it had occupied during the Six Day War. In the interim, the oil prices had risen sharply, and the increasing revenues were funnelled into the pockets of wealthy oil sheiks and authoritarian regimes.[37]

The 1973 oil crisis introduced a new chapter in the history of oil, power and strategic interests. A few years later, the Iran-Iraq War inflicted another oil price shock on the global economy. From 1978 to 1980 prices continued to rise, increasing by almost forty per cent in the period leading up to 1990.[38] From this point on, it became evident that 'the black gold' gave the oil-producing states income, power and global influence wholly disproportionate to their population size and previous historical achievements.

Our Son of a Bitch

While popular uprisings in the 1970s led to the overthrow of dictatorships and the installation of democratic institutions in Portugal, Spain and Greece, the USA's strategic interests hindered and delayed the development of democracy in Central and South America. For the USA it was critical to ensure that the Soviet Union did not establish a foothold in Cuba or acquire other allies in Latin America. The USA therefore supported democracies where it could and dictatorships where it 'was obliged'.

Freedom, democracy and human rights were tabled in the struggle to keep communism out of the US backyard in the south. The Monroe Doctrine continued to dominate strategic thought, and under both Republican and Democratic administrations the USA lent economic, political and military support to brutally oppressive, right-wing military dictatorships in countries including Chile, Argentina, Brazil, Panama, Venezuela, Nicaragua, Uruguay, Paraguay, Mexico and Colombia. Of one of these dictators, Nicaragua's Anastasio Somoza, President Franklin D. Roosevelt allegedly said, 'He may be a son of a bitch, but he's our son of a bitch.'

Across Latin America, popular resistance nonetheless gradually gained momentum throughout the 1960s and 1970s, and in some countries guerrilla forces and rebel groups took up arms against the military juntas. The juntas were also challenged by the economic problems that followed in the aftermath of the oil price shocks of the 1970s. While Latin American countries struggled under the burden of large debts, the price increases generated huge incomes for the oil-producing countries of the Middle East.

To support the faltering military juntas, the USA helped channel funds from the Middle East into loans to its Latin American allies. As a result, the already considerable foreign debt of several Latin American countries increased more than ten times over in the 1970s. The military regimes were therefore able to continue business as usual until rising inflation in the USA and Western Europe induced major interest rate hikes in international markets. Buckling under the untenable strain of the rise in interest rates, sixteen Latin American countries went bankrupt and were obliged to seek aid from the USA and the IMF to manage their debt.[39]

The social unrest and popular discontent with the military regimes escalated in step with the rising interest rates and growing inflation throughout what is often referred to as Latin America's 'lost decade'. It was a period fraught with bloody and dramatic revolutions and counter-revolutions, but by the mid-1990s, most of the military juntas had been ousted and replaced by democratically elected governments.

Rust Belts

With greater knowledge, more experience, improved skills and a large injection of Western investment capital, the companies of East and South

5. WAVES OF GROWTH AND PROSPERITY

Asia quickly became reliable, cheap and efficient suppliers of the global production chains. Neither was it long before their expertise enabled them to compete with Western companies on price, quality and creativity. In East Asia, owners and investors were spared having to contend with high taxes, costly regulations and powerful unions. The workers had lower wages, longer working days and fewer rights, but in return this provided opportunities for millions of people to secure a better life and a brighter future for themselves and their families.

At this time the Western countries still had the largest economies and controlled the most important markets. But since maritime freight shipping was so efficient and inexpensive, much of the labour-intensive industrial production in the West was moved to East Asia. Meanwhile, East Asian investors took advantage of the emerging market opportunities. Cars from Japan, electronics from South Korea and clothing from Taiwan quickly acquired large international market shares, squeezing out American and European competitors who had not yet cut their labour costs sufficiently. Some of the latter companies responded by offshoring their production to East Asia, others by extensively rationalizing and automating their domestic production processes.

For the first time, Western industrial workers began to experience the ramifications of competition with millions of disciplined and ever-more proficient East and South Asians who were now a part of the global economy. The most labour-intensive segments of industrial production were the first and hardest hit by the rising competition. In the West, shipyards, steelworks, mining companies, car factories and clothing manufacturers went bankrupt. Others downsized, automated or relocated operations to the other side of the globe. Several Western economies failed to adapt quickly enough to replace the jobs that were lost. When companies were relocated or shut down, the foundation for economic value creation, tax revenues and prosperity was also reduced. Millions of industrial workers, the core of the Western middle class, suddenly found themselves without income, work or hope for the future. Unemployment rose, costly and ambitious welfare systems were under pressure, and economic policy was put to the test.

In the USA, the formerly prosperous industrial regions of the Midwest and north-east deteriorated into 'rust belts' of low economic activity and proliferating social problems. The economic growth in Europe also slowed in the course of the 1970s, and the flow of tax revenues was reduced. Many governments experienced difficulties servicing the large state loans used to finance ever more costly and ambitious welfare schemes.

The British economy was with time in such a sorry state that the United Kingdom – which only fifty years before had been the ruling global superpower – was obliged to seek help from the global community to stay afloat. In 1976, the country became the first country in the West to swallow its pride and apply for a loan from the IMF.

The Actor and the Iron Lady

It was at this point that Ronald Reagan, the B-list actor from Hollywood who gained A-list status as a US president, and the British prime minister, the 'Iron Lady' Margaret Thatcher, entered the arena as the most prominent Western politicians of the day. Reagan's election-winning slogan was 'Let's Make America Great Again!' The two leaders had vastly different personalities and backgrounds, but they found common ground through the struggle to bolster market forces.

According to their narrative, the West could only restore economic growth by removing trade restrictions and reducing company costs tied to taxes, wages and public regulations. The West had no choice but to become more like the Rest if it wanted to recoup international market shares. Both of them also wanted to limit the power of government, Reagan famously stating that he had always felt that 'the nine most terrifying words in the English language are "I'm from the government and I'm here to help"'.[40]

In both the USA and Great Britain, taxes were therefore cut for companies, investors and high wage earners. Banks and financial institutions were granted greater latitude. Regulations restricting trade, investment and the movement of capital were relaxed, and both president Reagan and prime minister Thatcher attacked the powerful labour unions.

The dramatic shift in economic policy is clearly evident in the official statistics. Wage growth for ordinary labourers stagnated, while business executives and investors helped themselves to increasingly larger pieces of increasingly larger pies.

American historian Timothy Snyder writes:

> Between 1940 and 1980, the bottom 90 % of American earners gained more wealth than the top 1 % did. This condition of growing equality was what Americans remember with warmth as the time of American greatness… Since 1980, 90 % of the American population has gained essentially nothing, either in wealth or income. All gains have gone to the top 10 % – and within the top 10 %, most to the top 1 %; and within the top 1 %, most to the top 0,1 %, and within the top 0,1 % most to the top 0,01 %.[41]

5. WAVES OF GROWTH AND PROSPERITY

When companies were permitted to trade in large-scale international markets, produce at lower costs and in practice pay as little in taxes as they saw fit, it was the owners, investors and directors who walked away with the spoils.

The economic growth with time recovered in the USA and Great Britain, but in both countries the economic and social disparities continued to grow while income and wealth was concentrated in fewer hands. So when governments boasted of economic growth, some people whose lives did not improve ironically commented that 'it's not my GDP'.

Eurosclerosis

On the European continent, economic growth was now so debilitated that it was referred to as 'eurosclerosis'. The effect of Reagan's and Thatcher's reforms was to increase the competitiveness of US and British companies over European competitors, and this came on top of the growing challenges from the 'tiger economies' and other countries in East and South Asia.

The countermove of the European Economic Community, the precursor to the European Union, was to prescribe a remedy called 'the Four Freedoms'. In 1987, the member states formed an agreement guaranteeing the free movement of goods, services, capital and labour within the European community. In this way, the member states attempted to pull themselves up by their bootstraps through creation of a European economic growth zone, an 'internal market', and enhance their competitive edge within an increasingly tough world market.

The member states recognised that, when they integrated their markets, they must at the same time introduce supranational rules and institutions that could secure level competitive playing fields, good working conditions and social protections. If not, companies and investors would simply move back and forth between those member states offering the most favourable terms and conditions. This would foster new problems that individual countries could not resolve on their own. The Four Freedoms and the internal market were therefore in 1992 consolidated by the Maastricht Treaty, which established the tighter and more wide-reaching economic and political cooperation embodied by the European Union.

Since then, further and more comprehensive transfers of power and authority from the member states' popularly elected assemblies to the EU's central bodies have been implemented. The internal market is based on a logic which implies that the European Union inevitably seeks to expand and elaborate its decision-making authority on issues ranging from approval schemes, industry standards, government subsidies and competition law to research, education and trade. The EU's common system of laws and regulations, the so-called *acquis communautaires*, now contains more than 100,000 ordinances, directives and other legal acts that apply to all member states. In addition to this, the latter must also adapt domestic legislation to assure compliance with the EU regulatory system.[42]

In a number of the EU member states, from the start there has been considerable, and in many places growing resistance to this transfer of power. For many, the term 'the Bureaucrats in Brussels' acquired with time a derogatory connotation, becoming a symbol of how market forces and 'the global elites' undermine 'national self-determination' and democratic processes. This resentment was brought to bear in 2016 when the British voted to leave the European Union on the force of vague notions about 'taking back control' – although nobody was fully able to explain what that exactly meant or understand how it was to be achieved.

A Westerly Wind

Throughout the 1970s and 1980s, the Japanese and West German economies continued to grow, bolstered by US support and increasingly unrestricted access to international markets. At the same time, both countries, on the losing side of the Second World War, were obliged to limit their military capacity. They were thereby spared having to shoulder the economic burdens of large military budgets and a costly arms race.

The military costs in particular were among the most important factors that ultimately led to the fall of the Berlin Wall in November 1989 and the collapse of the Soviet Union two years later. It was an inferior economic system and sluggish economic growth that eventually forced the Soviet Union to its knees. The inefficient Soviet planned economy was unable to absorb the rapidly rising costs of the arms race with the USA and it could not withstand the pressure of economic globalisation and the free world's superior growth momentum. Neither could the communist system prevail against the growing pressure for reform from its own citizens. The majority in countries under communist rule had long understood that the Wall had been built to lock the population in – not to keep the capitalist world out.

In the West, the fall of the Berlin Wall and dissolution of the Soviet Union were celebrated as the ultimate triumph of democratic ideals and the principles of the market economy. A Westerly wind swept across the world, and there were few political or ideological forces that could slow down the continued liberalisation of world trade. The idea took hold that if the principles of a market economy were good for countries and regions, they

5. WAVES OF GROWTH AND PROSPERITY

must be even better on a global scale: larger markets would provide more opportunity and increased economic growth. Two hundred years after he wrote his manifesto on market liberalism, the reach of Adam Smith's ideas extended into both former socialist states and the far-left factions of democratic countries. Even the Chinese Communist Party knew how to read the signs of the times and introduced increasingly comprehensive liberalisation measures to transition to a market economy.

Gradually the restrictions on international trade, investment and financial transactions were eliminated, and in 1995 the World Trade Organization (WTO) was born. The WTO was entrusted with the task of administrating the existing General Agreement on Tariffs and Trade (GATT) and a new system of agreements on intellectual property rights and trade of services. Today the WTO regulations cover ninety-eight per cent of all international trade in goods and services, and only a dozen or so small states are not members of the organisation.[43]

Unlike the EU's internal market, this new global organisation did not, however, produce a set of supranational regulations and establish strong institutions to ensure level competitive playing fields, minimum environmental standards and good working conditions for employees. In practice, the WTO was based on a consensus regarding the reduction and elimination of a number of trade restrictions, yet without effective mechanisms for enforcing or ensuring compliance with the regulations on the part of individual countries. Neither were mechanisms put into place to secure a more reasonable and fair distribution of income, wealth and economic growth.

Ping-Pong Diplomacy

The Cultural Revolution of the 1960s drained China of human talent, economic resources and political allies. The world's most populous country more or less self-isolated behind the Bamboo Curtain, and the Chinese population was forced to live an ideologically dogmatic and, materially speaking, bleak existence. Mao's dream of a powerful China that would assume leadership in the communist world gradually gave way to the fear that the Soviet Union would launch an attack on his impoverished country. In March 1969, skirmishes broke out between Soviet and Chinese troops in the border regions near Manchuria. The fighting continued for more than six months before the conflict froze into yet another cold war within the Cold War.

After this, Mao was more concerned about the military threat from the regime's ideological cousins in the Kremlin than the destructive influence of the capitalist adversary on the other side of the Pacific. And of course, it wasn't long before the US government and intelligence community understood this.

What could be seen from Washington was not only an opportunity to drive a wedge between the two communist giants. The annual value of trade between the USA and China was a mere US $100 million, and the USA was confident that with increased trade and economic development a Chinese middle class would emerge, which with time would give rise to reforms and democracy, opening up the ideologically insular country.[44] In the eyes of the USA, the road to a democratic China ran through economic growth and it would only be a question of time before the Middle Empire became a part of the free world.

But Mao, the son of a farmer from a poor village, was ideologically steadfast and had no interest in capitalism or democracy. A softer relation with the USA was for Mao exclusively a means of consolidating his domestic political power, expanding his communist project and strengthening China's economic and military prowess.

All the same, a diplomatic process had been set into motion, and when in 1971 a team of US athletes was invited to play ping-pong in Beijing (at the time Romanized as Peking), it was the first sign of a thaw in the relations between the two countries. This was the first official US visit since the People's Republic of China was founded and it was immediately followed by a series of secret talks and the massaging of unofficial contacts, which would come to be known as 'ping-pong diplomacy'.

To further grease the wheels of the process, in the autumn of that year the USA agreed to allow the People's Republic of China a seat in the United Nations. Since Taiwan and the People's Republic of China were entangled in the 'One China' narrative's fundamental premise that there can only be one legitimate representative for the Chinese people, Taiwan was thrown out of the United Nations. The USA nonetheless refused to formally recognise the People's Republic of China and continued to uphold diplomatic ties with the government in Taiwan. A series of hectic developments culminated in President Nixon's famous handshake with Chairman Mao in Beijing in February 1972. Their meeting was brief; the almost eighty-year-old Mao had been struggling with his health for a long time, but it represented a historic turning point. The two heads of state agreed to initiate a political dialogue to take the first steps towards an economic collaboration and full normalisation of relations between the two countries. They also agreed to disagree about Taiwan.

The USA hereby found its way out of the diplomatic tug of war between Beijing and Taipei, which to this day remains a painful source of foreign policy strain for the USA.

5. WAVES OF GROWTH AND PROSPERITY

Little Boss with a Capital B

Although Chairman Mao looms large and alone on the summit of China's modern history, it was Deng Xiaoping who modernised China. While for the remainder of Mao's lifetime little of significance transpired in the relation to the USA, Deng knew how to avail himself of the benefits of the USA's extended hand and ongoing offer of economic cooperation. The chain-smoking little man, measuring around 150 cm in his stocking feet, became China's powerful reformer and perhaps the world's most important individual in the modern era. Under his leadership, China was ushered into a period of growth and rising prosperity that is unprecedented in the history of the world.

Deng was a hardcore communist, but also an inveterate pragmatist. One of his most famous statements was: 'It doesn't matter whether a cat is black or white, as long as it catches mice.'[45] For him, what mattered were the results. Deng also had a belief in the importance of private initiative and personal rewards in generating vigorous growth. In a deviation from Mao's ideology of equality, he shocked his party comrades by claiming that one should 'let some people get rich first'. Even communist China needed someone to take the lead and pave the way for growth and prosperity.[46]

The rigid economic system was revitalised and decentralised, so more decisions could be taken at lower levels. Although the state retained ownership of the large and strategically important companies, private individuals were now allowed to start and own enterprises. The market economy hereby made its first, faltering entrance into modern China. Laws were adapted, the legal system modernised and 'economic zones' set up with conditions designed to stimulate export and foreign investment. Deng and his political leadership were unbothered by the historical irony lying in the fact that several of the economic zones were established in the very coastal cities that were forcibly opened for trade by Western countries during the Century of Humiliation.

On the contrary: Deng immediately understood the large economic opportunities that would be created should China succeed in exploiting its trade connections with other countries.

Eggs in Two Baskets

With Deng in the driver's seat, the USA saw an opportunity to strengthen bilateral relations and rewarded Beijing with long-coveted diplomatic recognition. On New Year's Day in 1979, the two countries signed an agreement on the establishment of formal diplomatic ties. The 'One China' narrative's underlying premise that had earlier pushed Taiwan out of the United Nations also implied removal of the country from the US diplomatic registry. The USA moreover phased out its military bases in Taiwan. The following year, China achieved status as 'Most Favored Nation', on the force of which Chinese export to the USA would no longer be burdened with higher customs duties or less expedient conditions than those of any other country that receives preferential treatment.

But the USA did not want to put all of its eggs into the Chinese mainland basket. The US Congress therefore approved the 'Taiwan Relations Act', which authorised the sales of arms to the government in Taipei and stated that the United States should 'maintain the capacity... to resist any resort to force or other forms of coercion that would jeopardize the security, or the social or economic system, of the people on Taiwan'.[47]

China's collaboration with the USA added to the strain on the relationship between the Kremlin and Beijing, and towards the end of the 1970s, a quarter of the Soviet Union's ground troops and a third of its air force were stationed along the Chinese border. The Soviet Union was an ongoing source of concern and anxiety for Beijing, because at this time China's military forces predominantly consisted of a large, outdated and poorly equipped army of farmer recruits. When Soviet forces invaded Afghanistan in 1979, this did little to mitigate China's concerns. China responded by cooperating with the USA in the deployment of more than thirty listening stations on Chinese territory to monitor the movements of Soviet troops. China also invested heavily in the development of advanced weapons technology and ramping up its professional military forces on land, sea and in the air.

The USA, which still had its main strategic focus on the Soviet Union, contributed funding, technology, and technical expertise to the build-up of China's military power.

A Growth Miracle

Deng's reforms and the opening of China for trade with other countries worked like a fuel injection into the country's growth engine. From the late 1980s, the country's economic development surpassed anything ever before seen in the history of the world.

Initially, it was the access to Western markets, first and foremost, that of the USA, which provided traction for China's economic growth and subsequent emergence as a global superpower. The world's most populous country

5. WAVES OF GROWTH AND PROSPERITY

received economic support, trade benefits, military aid and advice from the world's richest and most powerful country, and before long international investment capital from a number of Western countries was flowing into China. From the late 1980s until the early 2010s, China's exports to the USA increased hundredfold.[48] In the same period, the reduction in the number of Chinese citizens living below the poverty line was twice the size of the US population.[49] When the Cold War came to an end in 1989, the American economy was 1,500 per cent larger than the Chinese. Today that number has dropped below forty per cent.[50] Measured in purchasing power, which takes into account that the prices and costs in the two nations differ, in the mid-2010s, China surpassed the USA as the world's largest economy.[51]

The unprecedented growth in the export of goods, transported by ship, constituted the first phase of China's impressive industrial development and, in a short period of time, the country became the world's largest exporter of finished goods. The revenues were invested in the construction of domestic infrastructure, housing and factories at a scale and pace never before seen in any other country, let alone in China. The Chinese economy grew at a record rate, and this in turn triggered a demand for the import of large quantities of raw materials, semi-fabricated goods, coal, oil, iron ore, food and consumer products.

This also meant that China's growth and prosperity became increasingly more dependent on other countries – and on the global web of maritime trade routes.

> *'For the first time since the 17th century, the majority of the world's most affluent population segment live outside of Europe and North America.'*
> Professor Mike Martin, London School of Economics

6.
A NEW GLOBAL TRIANGLE

How maritime trade has facilitated economic growth and the division of labour between different continents, left parts of Africa trapped by poverty and natural resource extraction, lifted hundreds of millions of people out of poverty in East Asia and landed tens of millions in unemployment and despair in the West. How the sea lanes and merchant ships constitute the backbone of the modern, global economy and the worldwide web of cross-boundary supply chains. How this has fundamentally altered the conditions for national economic policy and expanded the divide between the Rich and the Rest. How the risks and vulnerability of this system became evident during the 2008 financial crisis and the Covid-19 pandemic, and how the world's economic and geopolitical centres of gravity are shifting.

How to Square a Circle

When as a young economist graduate, I started working at the Norwegian Ministry of Finance in the mid-1980s, rapid changes in the conditions for Norway's economic policy were underway. It was becoming increasingly difficult for a small, wealthy country to shield itself from the international market forces. The foreign exchange rate on the Norwegian kroner was locked against other important currencies, and cross-border capital transactions were subject to strict regulations. The banks were obliged to comply with strict rules on loans, which included a cap on the interest rate they were permitted to charge. The latter was fixed by the Ministry of Finance in the form of so-called 'interest rate declarations'. Many of the banks invented quite creative solutions for circumventing this cap, such as introducing more frequent compounding periods and charging a range of additional administrative fees.

6. A NEW GLOBAL TRIANGLE

The Ministry therefore decided that the interest rate declarations would be based on the 'effective interest rate', which meant that interest paid, administrative fees and payment dates would be taken into account. The task of creating the formula for the new scheme was delegated downward through the bureaucratic ranks, until it landed on my desk as the youngest consultant. Advanced mathematics was not required, but the formula was nonetheless a bit complex. When the completed formula was finally sent to the banks, it was not long before it appeared in the humorous column in one of the country's largest newspapers, headlined 'How to Square a Circle'.

There was nothing wrong with the maths. But the complex formula served as a telling example of how difficult it had become to cling to an old-fashioned monetary policy in a world where there were increasingly fewer restrictions on the cross-border movement of capital, goods and services.

Chinese Containers

At this time, the Chinese growth engine had sputtered to life and this had a direct impact on seaborne trade. From this point on, the maritime routes between China and the rest of the world served not solely as gigantic conveyor belts, but also as chain drives in the global economy. Throughout the 1990s and until the financial crisis of 2008, trade in goods and services grew from 39 to 61 per cent of global gross domestic product,[1] and global freight shipping grew almost twice as rapidly as the world economy.[2]

The majority of China's exported goods were packaged in standardised containers, and the Chinese containers became the foremost symbol of the tremendous activity in the world economy. The number of vessels in the international merchant fleet increased at a record pace, as did the size and cargo capacity of the ships. In 1979, the year the USA officially recognised and established diplomatic ties with China, the largest tanker ship of all time was launched, TT *Seawise Giant*. The 500-metre-long vessel had cargo capacity for 700 million litres of oil, equivalent to the capacity of more than 20,000 large fuel tank lorries. In the late 1980s, the largest container ship of the time, *President Truman*, had capacity for 4,500 standard containers on each trip (a 'standard container' is a twenty-foot equivalent unit (TEU), commonly seen on semi-trailers and lorries). When the *Algeciras*-class of container ships was launched thirty years later, the load-carrying capacity had increased to 24,000 containers. If these containers were to be put after each other like a chain, they would extend almost 150 kilometres.

Simultaneously, the ships' hulls and propellers were modified to minimise drag and water resistance, the engines were made more efficient and the navigation systems more advanced. The international and intercontinental transport of goods by ship thereby became even cheaper, faster and more energy-efficient.

A Global Triangle

Given the rapid growth of China and other East and South Asian economies towards the end of the twentieth century, the world economy gradually began to resemble a global triangle: Africa was the world's supplier of raw materials, Asia was the world's factory of finished goods and the West was the world's shopping mall.

At this time, Africa continued to derive most of its income from the export of natural resources to the rest of the world. Even today, unprocessed raw materials, such as crude oil and minerals, constitute more than two thirds of the continent's export income. The trade between African countries is modest in scale. Less than twenty per cent of African export goes to other African countries, while more than sixty per cent of the Asian and European export goes to other countries on their own continents.[3] The African continent has not yet come close to realising its large economic potential to the benefit of its population.

China and several other countries in East and South Asia underwent a demographic shift similar to that which took place in Europe and the USA in the nineteenth century, but within a much shorter timeframe and on a much larger scale: people migrated from agriculture to industry and from rural areas to the cities. There was a decrease in the number of farmers and an increase in the number of factory workers. Since almost all transport in the international trade of goods was by ship, the factories and affiliated activities were located near the coast. Because of this, millions of people relocated from inland to the coastline, and the large coastal cities, most of which had historically grown out of international ports, became even larger. The countries transitioned from agrarian to industrial societies in much the same way that the Western part of the world had done almost two centuries earlier, but on a much bigger scale and in a much shorter period of time.

Industrialisation also engendered a burgeoning need for financial, technical and legal expertise, and knowledge about production processes, accounting, sales, marketing and business management. Several hundred million people were sent to school to receive education, training and further upskilling. This paved the way for China's and other South and East Asian countries' advancement to the first division of the global economy. They went from

6. A NEW GLOBAL TRIANGLE

being mere suppliers to eventually outperforming Western companies in several of the most specialised and cutting-edge niches of the world's supply chains.

At the same time, the countries' own economies and markets underwent expansion. China and other South and East Asian countries generated their own, domestic demand. Gradually, this also made them more economically dependent on one another, and less dependent on the West.

Confirmation Bias

In spite of the strong growth in China and other parts of Asia, up until the financial crisis of 2008, Western countries nonetheless had the largest economies, the majority of the investment capital, the best industrial expertise and the leading universities and research communities. Some of the largest and best-known companies in the world today were established at this time. Companies like Apple, Amazon, Microsoft and Google (Alphabet) were all listed on the stock exchange in the 1980s and 1990s.

In the West, China's rapid industrial development and the booming growth in the global economy were interpreted as yet another confirmation of the superiority of the market economy and liberal societies. Not even the financial crisis that swept through a number of Asian countries in the late 1990s served to rattle this conviction. The global growth was also reinforced by an abundance of technological inventions and innovations. The internet, mobile phones and SMS messages were born, along with the MP3 player, PlayStation and Nintendo 64. The robust creativity leveraging the opportunities of the internet also fuelled the expansive growth of the dotcom wave in the late 1990s.

The technological advances produced new and dramatic shifts in the global distribution of labour and economic growth. Once China and other low-cost nations in South and East Asia began outperforming millions of Western industrial workplaces, they wasted no time in moving into the service sectors, bolstered by favourable conditions under the WTO agreement. Companies could now employ low-salary, skilled workers in different time zones all over the world to offer round-the-clock service internationally. If you were in Madrid and wanted to book a flight from New York to Shanghai, it could very well be an employee in India who answered the phone when you rang Qatar Airways. If you were a Nigerian oil worker employed by a French company operating on the Brazilian continental shelf, a human resources department in Lithuania could be responsible for handling the administration of your paycheque.

But the period leading up to and following the turn of the millennium was not solely characterised by powerful growth, an increase in trade, groundbreaking investments and technological revolutions. With fewer regulations and greater leeway, the banks' ingenuity and creativity in inventing new financial instruments to increase lending and revenues also grew.

Chain Reactions

For many people, financial 'derivatives' are as difficult to understand as they are tempting to use, and, especially in the USA, large financial bubbles began to accumulate. Ordinary families could qualify for oversized short-term consumer loans against anticipated future income and collateral in housing values that were expected to rise. Local politicians could finance costly campaign promises by taking out large loans rather than raising taxes. The new financial instruments were, however, so complex that even the banks' own customer service advisors did not always fully understand them. But the advisers did understand that their own salaries and compensation packages were tied to the commissions on new loans. This produced a rapid and destructive merry-go-round based on a large increase in lending, skyrocketing housing prices and unmanageable municipal deficits.

The financial bubble ballooned and finally burst when Lehman Brothers, one of the largest banks in the USA, collapsed on 15 September 2008. The bankruptcy sent shockwaves around the world. It was only at this time that it became evident how many of the new financial instruments were in reality elements in a huge pyramid scheme, the underlying values of which were in no way proportionate to the size of the loans, and where existing loans had often been used as collateral for new ones. The stock exchanges of the world plummeted in a free fall, hundreds of thousands of companies went bankrupt, millions of people lost their homes and jobs, and the world's tightly interwoven banking system teetered on the brink of collapse.

The crash of the stock exchange in 2008 triggered the worst economic crisis the world had seen since the 1930s. International trade and maritime transport collapsed, serving as seismographs for the economic earthquake that shook the world. Global growth and trade screeched to a halt. In the course of a few short autumn months in 2008, the shipment of cars from Japan to Europe was halved. The day rates for large vessels dropped by more than ninety per cent: ships that sailed for US $200,000 per day in August were offered US $4,000 per day in October.

The financial crisis proliferated in an extended chain reaction throughout the global economy. International capital flows dried up immediately.

6. A NEW GLOBAL TRIANGLE

Equivalent to around fifteen per cent of the world economy before the crisis, the international flow of capital dropped to just two per cent in the subsequent years.[4]

But the consequences would extend far beyond financial and commercial realms. The crisis also sent shockwaves through social and political structures, international relations and major power politics.

The tide had turned – and the world would never be the same again.

Bin Bags and Cardboard Boxes

The Western nations were, in relative terms, hardest hit by the financial crisis. It was the West that had the wealthiest and most advanced economies, the most close-knit financial systems and the highest borrowings on the part of public authorities, companies and private individuals. The West also had the largest share of its population in paid labour, the most expensive welfare systems – and the most inflated financial bubbles. When the crisis hit, the ripple effects were therefore more powerful and pervasive in Western countries.

Tax revenues diminished abruptly when companies went bankrupt and people lost their jobs. At the same time, government spending rose dramatically when the authorities had to contribute massive sums to unemployment benefits and measures to prevent the bank systems from collapsing altogether. Countries could not borrow their way out of this crisis, because in the financial markets both the capital and trust were gone. Many governments were therefore obliged to resort to heavy-handed austerity measures, slashing spending on welfare schemes, construction projects and public procurement. This produced a vicious cycle which dragged the economic activity further downward, while unemployment spiked.

In the USA the banks put entire housing districts up for sale in an attempt to cover losses on defaulted loans. Thousands of middle-class families were evicted and found themselves suddenly on the streets carrying everything they owned in suitcases, bin bags and cardboard boxes. In Spain, bank representatives went door to door delivering legal notices of eviction to people who were unable to pay their bills. In Greece the state itself went bankrupt, and shortly thereafter one out of every four adults was unemployed, while half the country's twenty-year-olds had neither studies nor a job to go to. In British port cities, Portuguese suburbs and French *banlieues*, secure lives and permanent employment were replaced by a daily existence of uncertainty, fear and frustration.

Millions of people lost not only their jobs, homes and incomes – they also lost their identities, self-esteem and pride. They lost the contents, sense of direction and fixed points of reference for their lives. They lost hope and belief in the future and their social status deteriorated. They fought to swallow their humiliation when faced with the welfare systems' rigid rules and meagre benefits.

Inward and Backward

The first and foremost priority of the Western governments was to rescue the banks, secure the financial systems and keep interest rates down, in hopes of stimulating investments and consumer spending. Massive amounts of public funds were allocated to saving the banks, while brutal cutbacks were made in public services for people in crisis. The large banks' losses were made a public issue, while the losses and problems of ordinary people were privatised.

The economic setback rumbled through the world's supply chains, spreading to every corner of the globe. The crisis unfolded with such force that ten years would pass before the economic activity of the countries in Europe fully recovered. The USA rebounded more quickly, but throughout the entire Western world, the crisis produced higher unemployment rates, greater social dissatisfaction and deeper political divides, challenging citizens' trust in their governments and each other.

The crisis also contributed to the reinforcement of underlying nationalist and protectionist trends in Western countries. In the part of the world that had been the driving force of globalisation, more people turned against it. The part of the world that had laid the foundation for the capitalist market economy and international trade system increasingly turned inward. Strong political forces in the countries that had exploited this system to become the richest and most powerful in history wanted to turn their backs on the rest of the world.

The Western countries were no longer able to shoulder the consequences of the international order they had designed and fostered. The West's uncontested position as leader of the global economy was about to founder.

Role Reversal

China was far better equipped than most countries to handle the worldwide crisis. The Chinese economy remained more loosely enmeshed with the international financial system than Western economies, and it was still less advanced and less automated. The powerful leadership of the Communist Party could also mobilise the entirety of the country's large population

6. A NEW GLOBAL TRIANGLE

and vast resources to keep the wheels turning. Because of its authoritarian system, the regime was spared having to squander time and energy on political opposition and critical media. The government could focus all its attention and resources on pulling the country out of the economic crisis.

While since the early 1990s China's double-digit economic growth rate had been fuelled by export to the West, it was now domestic demand which powered the huge economic engine. When the world markets shrank, the Chinese government responded by pouring even more money into large-scale construction projects for roads, railways, power plants, airports, harbours, suburbs, schools and universities. After a brief setback, growth continued, and in the years following the financial crisis and until the beginning of the 2020s, China alone stood for more than a third of the world's economic growth.[5] In three short years, from 2011 to 2013, the country consumed as much concrete as the USA had done throughout the entirety of the twentieth century.[6]

But China simultaneously showed the world that it was capable of much more than casting concrete for bridges, buildings and roads. In international rankings of the ten top universities in the world, around half of the institutions are now located in China.[7] China has also acquired leading international positions in a number of the most advanced fields of research and industrial development, such as biomedicine, artificial intelligence, robotics, superconductors, quantum computing and advances in space. Chinese astronauts were for example the first to land successfully on the far side of the moon. Having essentially started from scratch at the beginning of the new millennium, China now produces more high-impact scientific papers than the USA and the EU countries combined.[8] China produces electric cars, aeroplanes, spacecraft, 5G systems and smartphones that are wholly on a par with, and in many cases superior, to Western products.

The massive economic activity in China following the financial crisis stood in striking contrast to the West's anaemic growth. This led to a seismic shift in trade relations, economic power and geopolitical influence. At the turn of the millennium, eight out of ten countries traded more with the USA than with China. Just twenty years later, seven out of ten countries are trading more with China. Today China is the most important trade partner for more than 120 countries, and, for the first time in history, the USA and China are global superpowers – at the same time.[9]

For more than thirty years, the Chinese growth engine has continued to grind indefatigably onward at a growth rate between two and three times greater than that of Western countries. In China, India and Southeast Asia, hundreds of millions have been able to work their way out of poverty, and in the same period the Asian middle-class population has increased by more than one billion. Today, China's middle class is already larger than that of the USA.

Dramatic changes have also occurred in the global distribution of the one tenth of the world population with the highest income. In the early 1990s, seventy per cent of this group came from Europe or the USA, but today the majority are from Asia. Now, 'for the first time since the 17th century, the majority of the world's affluent population lives outside Europe and North America', British sociologist Mike Martin writes, adding that '[t]his squeeze has largely been at European expense: the US and Canadian share of the global top 10 per cent has declined much less.'[10]

The Rich and the Rest

Since the early 1990s, globalisation has generated a shift in the world's economic centre of gravity from the north-west towards the south-east. Also, until the Covid-19 pandemic virtually paralysed the world in 2020, the economic disparities between rich and poor nations were on the wane but saw a resurgence after the pandemic because the wealthy nations recovered more quickly.[11]

Over the past decades, however, the economic disparities within most countries have increased, not least in the Western world. This is due to the fact that a small number of individuals have become exceedingly wealthy while many more people have not experienced any significant increase in income, assets or prosperity. In the USA, for example, the number of 25- to 34-year-olds living with their parents or grandparents tripled from 1970 to 2020.[12] Average life expectancy in the USA has also declined significantly over the past few years, undoing over two decades of progress.[13]

The contrasts become glaring when we consider globalisation's super-winners, the large investors and capital owners. They can reap dividends from expanding markets, and at any time funnel their investments to those parts of the world where they expect the highest returns, the lowest tax burdens and the cheapest labour costs. Globalisation has also fostered a situation of competition between countries endeavouring to retain their own companies and attract those of other countries, such as through offers of favourable tax schemes. In the period from 1995 to 2021, the average corporate tax burden for companies included in one of the most widely quoted global stock market indexes (MSCI ACWI IMI) was more than halved. In the same period, the average return on these companies' equity capital increased by more than fifty per cent.[14] Today, a billionaire is running or the principal shareholder of seven out of ten of the world's biggest corporations.[15]

6. A NEW GLOBAL TRIANGLE

In the Western countries we must look one hundred years back in time to find differences in income and wealth corresponding with that today, and the disparities are on the rise in almost every country around the world.[16] The ten richest people on earth (all of whom are men) now own almost as much as the poorest half of the world population combined.[17] The number of super-rich individuals is growing the most rapidly in Asia, and even in communist China the disparity gaps are large and widening. One report shows that the richest tenth of the Chinese population owns more than half of the country's accumulated assets.[18] There are more dollar billionaires in Beijing today than in any other city in the world.[19]

In the West it is the USA that has the greatest income and wealth disparity, and the inequalities continue to grow. As we have seen in Chapter 5, over the past forty years, almost the entire increase in the country's value creation has gone to those already at the top of the American social pyramid. Professor Daniel Markovits at Yale Law School describes how this leads to self-perpetuating and widening economic disparities and declining social mobility:

> The rich and the rest now work, live, marry and reproduce, and shop, eat, play, and pray differently and in largely separate social worlds … Today, the rich and the rest each lead lives that the other could hardly recognize and cannot understand.[20]

In this way, the widening disparities do not merely produce greater social divides and political polarisation within countries; they also create a new dynamic across national borders. When it comes to world view, ideology, attitudes, lifestyle and preferences, the rich and privileged in New York, Paris, Moscow, Shanghai, Dubai, Cape Town and Rio de Janeiro often have more in common with one another than with their fellow countrymen.

Previously the great divide was between the West and the Rest. Now it is equally pervasive between the Rich and the Rest.

Shock and Disbelief

In the years following the financial crisis, radical shifts occurred in the Western political landscape. Weak economic growth, tight budgets, high unemployment rates and large income disparities fostered polarisation and diminished trust across political and social divides. In almost all Western democracies, the support for centre-leaning and compromise-oriented parties dwindled, and in many countries, the traditional governing parties were crippled. Political extremists from both ends of the political spectrum joined forces in a rejection of globalisation, fear of immigration and contempt for the socio-economic and political elites. This rejuvenated the 'horseshoe theory' in the discussions of political scientists. At the same time, reactionaries, radicals and populists eagerly exploited this window of opportunity by fanning the flames of people's fears, anger and frustration.

The force of these movements was far greater than most people and political pundits were able to comprehend. In the summer of 2016, Europe reacted with shock when the British voted to leave the European Union. Just a few months later, the world gaped in stunned disbelief as Americans opened the doors of the White House to a spectacularly unsuited and hugely incompetent man with a dubious past, who has since also become a convicted felon. Through the election of Donald Trump as the president of the USA, 'democracy has lost its beacon, Europe has lost its guarantor, and the world has lost its referee', commented French political scientist Dominique Moïsi.[21]

Quaranta Giorni

The world community was again put to the test when the inhabitants of Wuhan, the most populous city in central China, began falling ill and dying from an unknown virus in December 2019. Soon thousands were infected, the hospitals in Wuhan filled beyond capacity, and the dead and dying found lying on the streets. The Covid-19 virus spread from China to one country after the next at a rate that led the WHO to declare the illness a pandemic: a global epidemic. Countries all over the world closed their borders and, only a few weeks later, one third of the world population was in total or partial lockdown. The world stock markets immediately plummeted, companies went bankrupt and the unemployment rate skyrocketed. Millions of people died, health systems were overloaded and economies ground to a halt. Governments and central banks responded by doling out funds in an attempt to maintain economic activity, financial stability and social calm. The situation was so extraordinary that the terrorist organisation the Islamic State (IS) advised its members to refrain from carrying out attacks in Europe – out of fear of being infected by the virus.

Aircrafts were grounded, lorries piled up at border crossings and the vulnerabilities of the seaborne transport system and the close-knit global web of supply chains became glaringly evident. When the borders were closed, the world fleet of merchant vessels could no longer carry out its key role in the international trade of goods. Ships were denied docking, the ports of Los Angeles and Shanghai refused to accommodate cruise ships carrying infected passengers, shipyards could no longer provide maintenance and repairs, and bunkering and refuelling became difficult. Flight cancellations, travel restrictions and strict quarantine regulations rendered international crew changes virtually impossible. This immediately caused a snarl of difficulties for the 350,000 port calls and 200,000 crew changes that normally took place every month around the world.

By the late summer of 2020, almost half a million seafarers were trapped on board ships. These maritime transport workers, whom we seldom see, but upon whom we all depend in our daily lives, were not permitted to go ashore, although eventually many of them had been on board for a year or more. A number of seafarers developed serious mental health problems due to isolation, uncertainty and because they missed their loved ones. For some the burden was more than they could bear and they committed suicide by jumping into the ocean. During this period, I co-chaired, together with Stephen Cotton, head of the International Transport Workers' Federation (ITF), and Guy Platten, head of the International Chamber of Shipping (ICS), a task force made up of relevant United Nations agencies, the EU Commission and other important stakeholders, which provided recommendations on how the international community was to handle the critical situation faced by seafarers.[22] As a result, in the autumn of 2020, the UN secretary-general advised member states to relax the restrictions on shipping. This was first and foremost out of consideration for the health and well-being of seafarers, but also because of the critical role they play in keeping the world economy up and running.

When seafarers were finally allowed to disembark, they were sent directly into quarantine, where yet again they spent weeks in isolation, a containment measure dating back to the Black Death of the mid-fourteenth century. The word 'quarantine' stems from the Italian term *quaranta giorni*, signifying the forty days ships from infected regions were required to remain in isolation before they could call at the Venetian port of Ragusa (today Dubrovnik in Croatia).

Almost 700 years later, Italy would be one of the countries hardest hit by the pandemic – and the first country to place its entire population in lockdown.

Panic Hoarding

The Covid-19 pandemic spotlighted the systemic risk inherent to the dense, finely meshed global weave of cross-border cooperation and interdependencies. The international supply chains had driven economic growth upward, costs downward and development forward. But the system was unable to withstand closed borders and shuttered ports.

A short time after the Covid-19 virus broke out, the sea-based supply lines were on the verge of being throttled. A backlog of empty containers accumulated in the Western world because ships weren't permitted to transport them back to China. The production of cars from Volkswagen, Ford and Fiat ceased because the supply chains broke down. Apple halted production of smartphones and laptops on all continents.

The risk of delays and breakdowns in the delivery of critical staples such as food, fuel, medicines and medical equipment nourished the fear of goods shortages in the general population in many countries, and triggered outbreaks of panic hoarding. In Australian supermarkets people fought over pasta, toilet paper and packages of nappies, and in the USA the queues of people trying to stockpile canned goods, weapons and ammunitions grew. It was suddenly easy to see how short the distance from law and order to anarchy and chaos can be, even in well-organised, affluent societies.

The trust between fellow citizens were put to the test – as was international solidarity.

Vaccine Battle

International relations were challenged as countries began competing over access to Covid-19 vaccines and medical equipment. When vaccines came onto the market on the heels of a record-breaking production period, they were not distributed on the basis of epidemiological, humanitarian or mutually agreed criteria. The argument that 'nobody is safe until everyone is safe' fell upon deaf ears.

The principle of 'survival of the fittest' prevailed, and the richest countries elbowed their way to the front of the queue. For the poor countries of the Global South, the perception of the West's lack of solidarity was confirmed and reinforced. Now it was the USA, Great Britain, Germany, France, Norway, Sweden and Canada 'First!' All of these countries gave their own populations the first dose of the vaccine and several subsequent rounds of booster doses, before they permitted dispatch of supplies to the developing nations in the Global South. In the summer of 2022, almost eighty per cent of the population of the USA and the European Union had been fully or partially vaccinated, while in Africa the number was just above twenty per cent.[23]

The fight over vaccines was not just a health issue. It also became a symptom of more fundamental weaknesses in international systems and institutions. In the global crisis, the United Nations failed to mobilise definitive authority and effective coordination. The UN secretary-general had little to offer beyond compassion and heartfelt appeals. The European Union suspended several of

6. A NEW GLOBAL TRIANGLE

the regulations which made the Union a community, and for the first time, borders between member states were closed. It took a full six months for the EU to agree on measures to assist the hardest-hit members of the Union.

The impotence of international institutions provided additional fuel to strong nationalist and protectionist trends that were already pervasive in many Western countries. It was the nation states, not the international and supranational bodies, that confirmed their relevance, strength and legitimacy. It was the nation states alone that could close their borders and protect their citizens. They alone were able to effectively mobilise resources and adequate measures to combat the pandemic.

Covid-19 also opened the door to further strategic manoeuvres and positioning on the part of the major powers. Before the European Union was able to rally, China and Russia had already sent specialists and medical equipment to Italy in response to its desperate pleas for help. The Trump administration's inept and egregious mismanagement of the pandemic provided the Chinese regime with ample opportunities to gain the upper hand on the world stage.

While in the USA the number of fatalities rose and the illness spread, President Xi declared, prematurely and erroneously, that China had succeeded in eradicating the virus. By March 2020, the world media was releasing photos of a smiling Chinese president on a well-staged tour of triumph through the streets of Wuhan, surrounded by applauding health care workers and visibly grateful supporters. The US administration had failed its citizens completely; China's leadership, on the other hand, was perceived as unifying, competent and capable.

With time, as the vaccine programmes were rolled out, life in large parts of the Western Hemisphere slowly began to normalise. In the summer of 2022, Europe and the USA had emerged from the crisis, although another year would pass before the WHO formally declared that the Covid-19 pandemic had come to an end also in other parts of the world.

A Warped Narrative

In February 2022, just as the pandemic had begun to release its grip in Europe, the first large-scale invasion on this continent since the Second World War took place. On the pretext of unsubstantiated claims that NATO countries were threatening Russia's security and spouting a warped narrative about the urgent need to 'denazify' their brother land, Putin's regime sent military forces over the Ukrainian border in an attempted full-scale invasion. Just weeks later, thousands of people had been killed or wounded, cities, factories, roads and homes destroyed, and more than ten million Ukrainians were internally displaced or had sought refuge in other countries. But the Russian attack forces quickly hit the wall. The invasion was poorly planned and badly executed, and the Ukrainian resistance proved far stronger and more resilient than the Russian regime had anticipated.

Russia's invasion of Ukraine immediately throttled global supply lines due to acts of war and international sanctions. The regions around the Black Sea are some of the world's largest breadbaskets, and Russia and Ukraine are also two of the world's major exporters of oil, gas, minerals and fertilisers. Since the invasion, mines and attacks at sea have transformed the transport of goods across the Black Sea into a dangerous undertaking.

Raw materials scarcity and rising prices on food, electricity and fuel – while the pandemic continued to rage in large parts of the world – set the international stage for famine, economic setbacks, social unrest, increased international tensions and geopolitical rivalries. The uncertainty induced stock market declines and rising inflation all over the world. Meanwhile, many state treasuries had already been critically depleted by the financial crisis and the pandemic. In the Western world, government debt increased to constitute a larger share relative to gross domestic product than at any time since the Napoleonic Wars.[24]

A Broader Palette

In most countries, there was no doubt about who was the victim and who was the assailant in the senseless, brutal war between Russia and Ukraine. The United Nations was nonetheless paralysed, because Russia vetoed all UN Security Council decisions condemning the attack and calling for the withdrawal of Russian military forces.

Russia was, however, unable to prevent an Extraordinary General Assembly of the United Nations from passing a resolution condemning the invasion. The resolution received the support of a large majority of the member states, but the landscape behind the vote count was more complex. The Western countries gave the resolution their unanimous support, while Russia received the votes of a small number of autocratic states.

However, almost every other member state in the world abstained, and the countries in support of condemning Russia only represented one seventh of the world population.[25] The motives of those who did not choose sides were as divergent as the countries themselves. China attempted to assume the role of a global broker, putting itself above the conflict. For economic and political reasons, a number of countries felt their interests were best served by following China's lead. Also, many Asian and African countries have close economic and military ties with Russia, many are large-scale importers of Russian grain and raw materials, and at the time, the majority were still wobbling from the aftershocks of the pandemic. Most of India's oil imports come from Russia, and in Brazil the agricultural sector is heavily dependent on artificial fertilisers produced in Russia.

Moreover, in several parts of the world, the invasion of Ukraine is considered an issue Europeans must resolve on their own. There may be sympathy for the Ukrainian population, but many of these countries are all too familiar with war, death, suffering and atrocities on their own soil or in their own neighbourhoods. Neither did it go unnoticed that Europe welcomed the Ukrainian refugees with open arms, while families fleeing war, violence and hostilities in Afghanistan, Syria, Yemen or Eritrea encountered barbed wire and armed border patrols – or drowned in the attempt to cross the Mediterranean or the English Channel.

Importantly, there also remains an acute awareness of the legacy of colonialism and its injustices in the Global South. Further bitterness was added to these historical grievances when the USA and countries in Europe stockpiled Covid-19 vaccines for their own populations during the pandemic, leaving the rest of the world to its own devices. Neither does it restore good will that by the mid 2020s, more than ten years have passed since a US president went to the trouble of visiting any African country south of the Sahara.[26]

For many of the countries in the Global South, the UN resolution on the invasion of Ukraine was therefore not solely a matter of principle, humanitarian issues or international law. It was a part of a broader palette of political trade-offs and embedded motives of self-interest and historical grievances, and an allergic sensitivity to what is perceived as Western double standards, moralism and racism.

Yet again it is the West against the Rest.

A Perfect Storm

Simultaneously, even darker and more menacing clouds were accumulating all over our planet. The world was not only in the midst of a pandemic and a geopolitical crisis; the consequences of the global warming and environmental degradation were becoming blatantly visible. The extreme weather events, floods, droughts, heatwaves and forest fires during the first half of the 2020s were the worst ever recorded.[27] Nature is striking back.

And once again, as so often before, the ocean constitutes a critical part of the problem – and of the solution.

> *'The huge unending ocean. Indifferent to everything and able to obliterate everything. The ocean, in its indifference, forgives everything. Ancient, irresponsible, inhuman.'*
> Pär Lagerkvist, *Pilgrim på Havet (1962)*

7.
THE OCEAN STRIKES BACK

About the impact of global warming, pollution and litter on marine life and ecosystems, and about nature's 'tipping points'. How the ocean-climate nexus, the mutually reinforcing interaction between a warmer atmosphere and a warmer ocean, leads to more intense drought, flooding, heatwaves, wildfires and extreme weather events, and how a rising global sea level is threatening lives, livelihoods, food production, infrastructure and cities.

Boeing 737

A split second of hesitation and he was forced to retreat. The strapping American paratrooper did not succeed in exiting the aircraft until his second attempt. When he was finally airborne, his goggles were blown off. The second paratrooper was thrown hurtling across the aisle, where he crashed into the toilet door on the opposite wall. As the third man up, I witnessed how the first two had fared. Yet I was still unprepared for the powerful force of the wind when I exited the Boeing 737, at an altitude of 12,000 feet. It was the early 1980s and we were a small group of Norwegian and American paratroopers doing parachute training from a jet passenger plane. Although the speed of the aircraft had been slowed to under 300 km/h, the wind pressure was so intense that it took strength and determination just to get out the door.

These memories sometimes come back to me when I watch TV reports of devastating hurricanes around the world, such as when Dorian swept ruthlessly across the Bahamas one day in the early autumn of 2019. It was the most powerful hurricane ever recorded in this region. For almost two days, the massive low-pressure system hovered over the islands, stirring up persistent winds at speeds higher than those we had experienced from the door of the jet plane.

Alan's tears: the Earth seen from the Moon
(*Photo: NASA*)

The West and the Rest: the East India company destroying
Chinese war junks in 1841, during the First Opium War
(*Painting: Edward Duncan*)

Shipping, the backbone of the global economy
(*Photo: Shutterstock*)

More people live in cities, and most cities are by the coast
(*Photo: Mark Lehmkuhler*)

Three tons in three hours: with WWF on clean-up campaign in the Oslo Fjord
(*Photo: Fredrik Myhre, WWF*)

We are consuming the eco-system supply of almost two planets
(*Photo: Shutterstock*)

The ocean-climate nexus is the most important driver of global warming
(*Photo: UNHCR / R. Rocamora*)

The global fishing fleet is more than double the size needed to catch the amount that the ocean can sustainably support
(*Photo: Wikimedia*)

Most of the coral reefs will be gone by the middle of the century if we continue like we do today (*Photo: Shutterstock*)

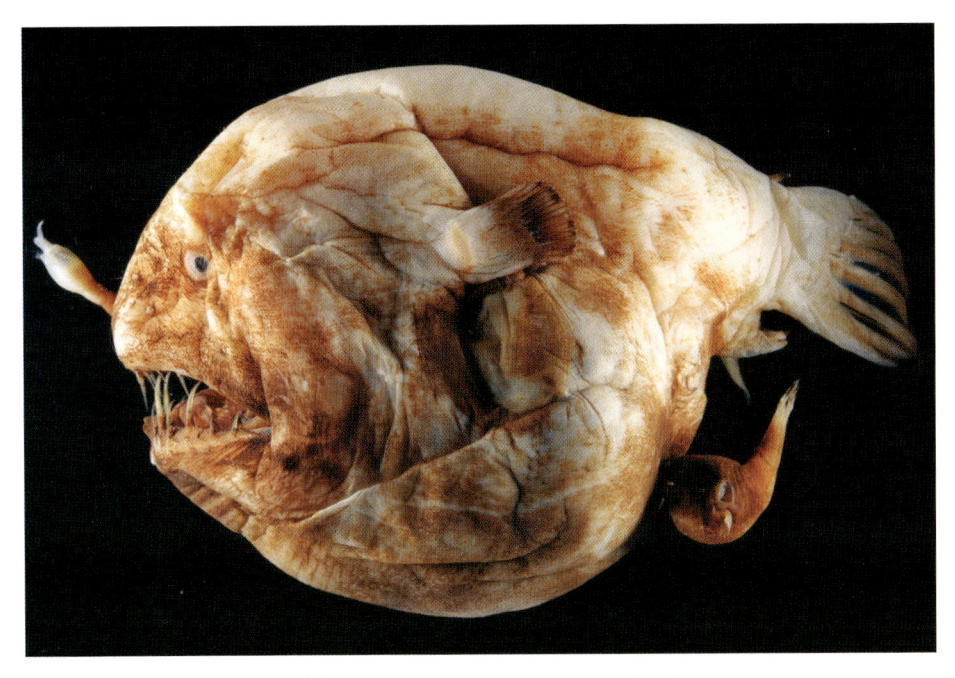

World's worst sex? Deep sea angler fish
(*Photo: Shutterstock*)

Worldwide, the volume of farmed fish has surpassed landings of harvested wild fish (*Photo: Jo Halvard Halleraker*)

Extensive marine bioprospecting is going on to produce food, pharmaceuticals and cosmetics from the lower trophical levels (*Photo: Shutterstock*)

Watts from winds off shore
(*Photo: David Dixon*)

Yara Birkeland: The world's first fully electric and autonomous container ship (*Photo: Ørnulf Rødseth*)

Ten thousand passengers and crew coming to a local community near you
(*Photo: Shutterstock*)

Deep sea mining for minerals? No tiptoeing around on the ocean floor
(*Photo: Adobe Stock*)

Attacking the arteries of world trade: burning oil tanker struck by Houthi missiles in the Red Sea in 2024 (*Photo: EPA-EFE*)

Geopolitics heating up in the cold North: a US nuclear submarine in the Arctic (*Photo: Defense Visual Information Distribution Service*)

Keynote opening speech at the World Ocean Week 2023, Xiamen, China
(*Photo: Erik Giercksky*)

With colleague Erik Giercksky, left
(*Photo: Jian Yang*)

With UN Secretary-General António Guterres (*Photo: UNGC*)

Ocean Advisor meets Aquaman: with Jason Momoa on beach in Cascais, Portugal (*Photo: UNGC*)

Too complicated for a prime minister… (*Photo: Author*)

Certifying Norwegian Prime Minister Jens Stoltenberg, right, and son Aksel for scuba-diving (*Photo: Lill Haugen*)

Safely back on shore after being almost trapped under the ice (*Photo: Author*)

Weightless in the blue space outside the Phi Phi Islands, Thailand (*Photo: Kon-Tiki, Ao Nang*)

My crowd: the generation that must take responsibility where my own has failed (*Photo: Kristin Kristoffersen*)

7. THE OCEAN STRIKES BACK

Wetter and Wilder

Violent gusts of wind tore roofs off buildings, overturned cars and snapped tree trunks in two. Hurricane Dorian moved slowly, barely at walking pace, which gave it ample time to absorb moisture from the ocean and dump cascades of torrential rains onto land. In the course of just a few hours, the gigantic pump system of natural forces released almost a ton of water onto every single square metre of these islands. At the same time, towering waves washed over the flat landscape. The waves rode on top of a storm surge that caused the local sea level to rise by almost seven metres. The storm surge is the hurricane's faithful companion, as the winds push the ocean towards land.

Just a few weeks later, it was Japan's turn when Hagibis, the most powerful typhoon the country had experienced in fifty years, carried the waves of the Pacific inland on its destructive journey. Several hundred people lost their lives, thousands lost their homes, and hundreds of thousands were evacuated. Fields and pastures were flooded, and the entire centre of Tokyo, the world's most populous city, was shut down.

Dorian and Hagibis are part of a clear and consistent trend of powerful wind systems that are becoming larger, wilder and wetter, and they have since been followed by even more devastating events. Due to global warming, these massive winds are likely to move more slowly and wreak far greater damage. They are part of a pattern of stronger and more destructive torrential rains, snowstorms, flooding, mudslides, forest fires, heatwaves and droughts. The number of climate related disasters has tripled over the past thirty years.[1]

What until now have been referred to as 'once in a century events' are, by the end of this century, expected to occur with such frequency that they will be redubbed 'once in a decade'.[2]

It is a deadly trend that will become even more dangerous and dramatic with time, as the global temperature increases. The consequences of extreme weather events are also compounded because more people are relocating to areas closer to the ocean. Every week, one and a half million people throughout the world move from rural areas to cities and most large cities are located by the coast.[3]

The highly populated parts of the low-lying delta regions of Southeast Asia are particularly vulnerable when the ocean strikes back, 'indifferent to everything and able to obliterate everything'.

THE OCEAN

A Costly Contribution

The warming of the atmosphere is now happening ten times faster than the average rate of warming after the eight cycles of ice ages that has occurred over the past 800,000 years.[4] This is also the first time the global temperature has risen as a direct consequence of human activity. While previously it was climate change caused by natural phenomena that affected the quantity of greenhouse gases in the atmosphere, it is now human-made greenhouse gas emissions that are driving climate change. Our modern society emits far greater amounts of these gases than the earth's natural processes are able to absorb. Roughly speaking, only half of human emissions are absorbed by the ecosystems on land and in the ocean. The rest remain in the atmosphere, hugging the earth like a thick blanket.[5]

In Chapter 2 we looked at the unique characteristics of water, including its capacity to hold heat. It is because of this heat absorption capacity that the earth remains habitable, even though global warming is in overdrive. As we discussed, the top three metres of the ocean alone hold as much thermal energy as the entire atmosphere, and since the Industrial Revolution the ocean has absorbed ninety per cent of all human-made surplus heat.

The downside of the ocean's critical contribution to curbing atmospheric warming, however, is that the ocean itself is growing warmer. Never before have ocean temperatures been as high as they are today, neither on the surface nor in the deeper ocean zones. The average surface temperature is now almost 1.5°C higher than one hundred years ago, and in some areas and time periods, the increase is far greater.[6] The warming of the upper parts of the ocean is expected to continue to an extent that may be impossible to reverse for hundreds of years into the future.[7]

The already high and rapidly rising temperature contributes to a vicious cycle with devastating consequences.

Little Boy and Girl

The warmer the ocean, the more moisture is released into the atmosphere. Simultaneously, the warmer the atmosphere becomes, the greater its capacity to retain moisture. For each degree Celsius of air temperature increase, the atmosphere can hold seven per cent more water vapour.[8] The water vapour in the atmosphere, like other greenhouse gases, also contributes to global warming because it forms an insulating blanket around the earth. Water vapour is in fact the earth's most abundant greenhouse gas, and even more potent and consequential than carbon dioxide.[9] While 'climate change

7. THE OCEAN STRIKES BACK

deniers' and sceptics make frequent reference to this, they unfortunately overlook the simple fact that increased amounts of water vapour in the atmosphere are a result – not a root cause – of global warming.[10]

The water vapour has an effect on the wind systems that is comparable to that of high-octane fuel in an engine. Large amounts of energy are released when the vapour in the atmosphere is condensed into drops of water. This additional energy causes the air to rotate more quickly and the wind systems to expand. Hurricanes, typhoons and cyclones (the terms are based on the world region in which the storms occur) today contain forty per cent more water vapour than they did in the middle of the last century. This is one of the reasons why the American National Aeronautics and Space Administration (NASA) predicts that the strongest North Atlantic hurricanes, those with wind speeds of 200–300 km/h or more, will arise twice as often in this century as in the last. The coastal regions in tropical zones will be hit especially hard and with the greatest frequency, precisely because it is the combination of heat and humidity that provides the energy source for these destructive weather systems.

As the wind systems become stronger and larger, they also move more slowly. This is because global warming impacts the interaction between the high-flying jet streams and the lower layers of the atmosphere (which is also why there is now a greater risk of serious turbulence on intercontinental flights, having already cost the lives of several passengers).[11] The wind systems must move through air masses with more 'hilly terrain' than previously, which slows them down. Both the wind and the rain – or the snow if it is cold – can therefore cause even greater damage. Since the middle of the last century, the forward speed of hurricanes, typhoons and cyclones has decreased by ten per cent, and in the Northern Hemisphere the decrease is even twice that.[12]

The destructive interaction between rising ocean temperatures and an increase in atmospheric water vapour also compounds the forces of the weather phenomenon known as El Niño, which is defined by periods of abnormally high surface-water temperatures in the eastern Pacific Ocean regions near the equator. El Niño often causes more extreme precipitation in South America, and more drought and heatwaves in Southeast Asia and Australia.[13] Since El Niño often starts at Christmas time, people who live on the coast of South America have for centuries called the phenomenon El Niño, meaning 'the little boy', in reference to baby Jesus. Its counterpart is La Niña, or 'the little girl', and refers to periods of lower-than-average surface water temperatures. El Niño and La Niña events occur at four- to seven-year intervals, and each event can last for up to two years at a time.

Although these phenomena occur in the Pacific Ocean, both El Niño and La Niña are so powerful that they can affect the weather all over the planet. Now global warming is providing both with added force.

The rising ocean temperature caused by El Niño can also impact heavily on fish stocks in the South Pacific. In Chapter 2 we saw how the Humboldt Current along the coast of Chile and Peru is carrying nutritious water feeding some of the world's most productive marine ecosystems and fish stocks, such as the anchoveta. But El Niño often leads to dramatic reductions in the anchoveta populations because the strong winds are weakening the massive upwelling of nutrient water in this region. This causes the anchoveta to aggregate in smaller areas, where they are easily overfished, creating challenges for the sustainable management of these fish stocks. Since wild-capture fishing of anchoveta is a significant contributor to food supply, livelihoods and feed production for aquaculture, this can also create serious regional ripple effects.

Atlantis 2.0

Sea-level rise is one of the most reliable and robust indicators that global warming is taking place. The gradual rise of the global sea-level basically 'cancels out' all random and short-term variations in atmospheric temperatures.

Since the beginning of the twentieth century, the average global sea level has risen by around twenty centimetres.[14] In recent years, approximately one third of this stems from the expansion of the seawater due to heat, called thermic expansion.[15] The rest is the result of meltwater from glaciers on land, first and foremost the melting of the large glaciers in the Himalayas, Greenland and Antarctica.

The global sea level is rising now almost three times as quickly as it was at the start of the twentieth century, and it is no longer inconceivable that the global sea level will have risen to one metre, or even more, by the end of this century.[16][17] It is worth remembering that, if we go some 120,000 years back in time, when temperatures globally were around 2 to 3°C higher than today, and recalling that 2°C represents the upper goal of the Paris Agreement, the global sea level was between four and ten metres higher than it is today.[18]

Even a far more modest increase will have huge consequences for the world population and society. One report estimates that a gradual sea-level rise of 'only' sixty centimetres can reduce the global economy by one fifth.[19]

The consequences of a given rise in the global sea level will vary in different parts of the world. For the inhabitants near the fjords of Norway and

7. THE OCEAN STRIKES BACK

in New Zealand, where the mountainsides often plunge from great heights directly into the ocean, an increase of ten centimetres will have little practical significance. But the consequences are much more dire for the several hundred million people living in the low-lying deltas of Southeast Asia, where such an increase will make large regions uninhabitable. In countries such as Bangladesh and Myanmar, the salty ocean water will flood villages, fields, pastures, factories, infrastructure and groundwater reservoirs. In Thailand more than eighty per cent of the population reside in vulnerable coastal cities. Many of these regions lack the options and resources required to build their way out of the problems.

In other parts of the world, such as Singapore, Tokyo, Venice, Rotterdam and New Orleans, longer, higher, thicker and more sophisticated walls, sluices and drainage systems are being constructed as protection against rising sea levels. In the Netherlands, almost two thirds of the population already live one metre below sea level, so here it is only the dykes that can prevent inundation and obliteration of large portions of the country.

But it is probably the state of Kiribati in the Pacific Ocean that will suffer one of the first and most dramatic consequences of sea level rise. This tiny island state has neither the money nor the physical parameters to construct a form of protection from the ocean. The government has therefore purchased land in the neighbouring state of Fiji, over 2,000 kilometres away, to ensure 'migration with dignity' for its population.[20] Kiribati can be the first nation state obliged to displace its entire population because the country is drowning in the ocean. The country is suffering the fate of the Atlantis in slow motion.

Not because the island is sinking – but because the ocean is rising.

The Doomsday Glacier

As if these scenarios were not dismal enough, there is also a risk that the global sea-level rise may evolve far more quickly and be much greater than scientists formerly anticipated. This is first and foremost due to uncertainty about the melting of glaciers and ice shelves. Ice-covered areas function as reflectors and refrigerators, and their atmospheric cooling effect is reduced through the accelerated melting of ice. The world's glaciers are now losing almost one third more snow and ice annually than was the case just fifteen years ago.[21]

When the glaciers in the Himalayas, Antarctica and Greenland melt, the water will also run off land and into the ocean. Consequently 'more water is poured into the bucket', which will cause the global sea level to rise. In the summer of 2019, the Greenland ice sheet alone lost more than one billion tons of ice – every single day![22] The Greenland ice sheet is so large that, if it were to melt entirely, this would cause a global sea-level rise of seven metres. In such a case, the centres of the majority of the largest cities in the world would be under water.

The melting of the Greenland ice sheet will also affect the major ocean current systems of the North Atlantic. Fresh water from the melted ice will dilute salt content and weaken the mixing of warm surface water and cold water deeper down in the northern part of the Atlantic. This will diminish the force of the northern link of the Atlantic Meridional Overturning Circulation (AMOC), which, as we have seen in Chapter 2, contributes to the temperate climate enjoyed by the populations of Scandinavia and Northern Europe. Recent research indicates that this process has evolved further and more rapidly than previous projections have suggested. AMOC also appears to be more capricious and 'binary' than previously believed; the changes may not necessarily be gradual and linear. Scientists therefore no longer rule out the possibility that the weakening of the AMOC in the course of this century will be on a scale that can dramatically alter the climate of Scandinavia and other parts of Northern Europe.

On the opposite side of the planet, the land area of ice-covered Antarctica equals that of the USA and Mexico combined. The ice sheet here is on average more than two kilometres thick, and this barren continent is the home of more than ninety per cent of all the ice on our planet. If this ice sheet were to melt in its entirety, the global sea level would rise by sixty to seventy metres.[23] In that case, the entire world map would have to be redrawn.

A melting of the ice sheets of the Himalayas, Greenland and Antarctica would generate a sea-level rise that would vary in different parts of the world. Because both of the latter two ice sheets are so massive, they also create gravitational forces that impact nearby ocean regions. Paradoxically, this can lead to sea levels rising less in the areas where the melting of ice occurs, because the gravitational force will diminish as the ice disappears. As a result, the ocean will rise even more in other parts of the world.

While the gradual melting of the glaciers on land has already caused an annual sea-level rise that can be measured in millimetres, a sudden breakage of ice off the Antarctic ice sheet can raise sea levels by several dozen centimetres within months. The impact can be understood as similar to the way the water in a bathtub will rise as you lower your body into it. There is a growing concern among scientists that, as warmer ocean water undermines the Antarctic ice sheet, large chunks of ice can suddenly break off and slide into the ocean. Even though subject to uncertainty and discussions among scientists, the scenario of 'marine ice-cliff instability' has been a focus of increasing attention and substantial research in recent years, not least with an eye to predicting the potential impact of such sudden break-offs.[24]

The Thwaites Glacier in particular has come under the watchful eyes of scientists. The glacier is the size of Great Britain and more than one kilometre thick. Should it break off and slide out into the ocean, the global sea level would rise immediately by more than sixty centimetres.[25] The Thwaites Glacier also appears to function as a brake pad for the masses of ice behind it. Were this section to break away, it could cause large portions of the Antarctic ice sheet to slide out into the ocean.

It is not without reason that the Thwaites Glacier has been given the nickname 'The Doomsday Glacier'.

Archimedes' Principle

While Greenland and Antarctica are large areas of land covered by ice, a large percentage of the ice-covered Arctic regions are found in the ocean. Here the melting will not have a direct impact on the sea level. Why this is the case is perhaps not intuitively evident, but we find the explanation in Archimedes' principle, which states that the upward buoyant force of a body immersed in fluid is equal to the weight of the fluid the body displaces. When the ice melts, it becomes seawater, with the same volume as the water it formerly displaced as ice. The sea level therefore remains the same.

The melting of ice in the Arctic nonetheless has a large impact on sea-level rise, ocean currents, climate change and global weather systems. When the ice in these regions melts, this decreases the area of white surfaces on land and at sea made up of snow and ice, which reflect sunrays back into the atmosphere. At the same time, melting ice produces an expansion of the heat-absorbing areas of dark water, mountain and marsh surfaces. The latter is what scientists call the 'albedo feedback', a self-reinforcing feedback loop which accelerates warming of both the air and the ocean. When the Arctic ice melts, this frozen lid covering large deposits of methane gas on both land and sea disappears, releasing this hyper-potent gas into the atmosphere. Methane has a much shorter atmospheric lifetime than carbon dioxide, but its impact on global warming is more than thirty times greater.[26]

Because of the albedo feedback loop, the temperature of the ocean and the air is rising four times more rapidly in the Arctic than in other parts of the world.[27] The air above the Arctic is now more than four degrees warmer than it was a century ago, and in less than two decades the mean temperature in the summer has gone from sub-zero to above zero.

The rapid changes in the Arctic regions can be illustrated by a timeline from my own family. Since the birth of my two eldest daughters, Christine and Cecilie, less than forty years ago, half the summer ice in the Arctic Ocean has disappeared. When our family's four teenagers, Silja, Mathilde, Theodor and Ferdinand, turn thirty, all the summer ice may very well be gone. In the course of this brief period of my own family's life, the Arctic region will be reset to more than 700,000 years back in time, which was the last time this ocean region was free of ice.

Ocean Rainforests

While a warmer ocean leads to more extreme weather, melting ice and rising sea levels, all of which pose a threat to human life and society on land, rising ocean temperatures also harm marine life. Since it is the upper layers of water that absorb the most heat, this is where the temperature increases are the fastest and largest. It is also here that we find the majority of marine life, and many of these life forms are extremely vulnerable to the rapid temperature increase taking place.

Some of the most serious and dramatic consequences affect the coral reefs, which live in both cold and tropical waters. As far as we know, the world's largest cold-water coral reef is 'Røst-revet', found off the coast of Lofoten in northern Norway. But the great mother of all coral reefs and the largest and most well known of them all is, of course, the Great Barrier Reef, on the east coast of Australia. This is also the world's largest living organism, apparently so large that it can be seen by the naked eye from the moon. The Great Barrier Reef is a gigantic, oval cascade of colour, light, shadows, life and movement.

But the coral reefs are not only spectacular to observe; they also play a wholly unique role in the ocean's ecosystems. They are of such critical importance for sustaining marine life, photosynthesis in the ocean and carbon capture from the atmosphere that they are often called 'the ocean's rainforests'.[28] Although the coral reefs cover less than one per cent of the ocean floor, a quarter of all the fish in the ocean spend all or parts of their lives by these reefs.[29]

The coral reefs also make important and direct contributions to modern society. The fish inhabiting the areas near and around the reefs provide food and employment for several hundred million people. In many coastal regions the coral reefs function as effective breakwaters. Tourists wishing to experience these colourful and spectacular colonies of coral animals also represent a thriving industry. It is estimated that the accumulated value of the coral reefs' ecosystem services amounts to almost US $10 trillion annually.[30]

7. THE OCEAN STRIKES BACK

But now coral reefs are being destroyed all over the planet by thermal stress, pollution, acidification and litter.[31] Throughout the 2010s and the first half of the 2020s, one seventh of the world's coral reefs were eradicated. The demise of coral reefs is so grave in particular because most of them will not be able to survive temperatures above 30°C. If we don't succeed in curbing global warming in line with the goals of the Paris Agreement, it is estimated that between seventy and ninety per cent of the coral reefs could be gone by the middle of this century.[32]

Gender Discrimination

Global warming creates more intense and long-lasting heatwaves both on land and in the ocean. 'Marine heatwaves' can in the course of a few days or weeks kill coral reefs and other marine life forms. In recent years, ocean surface temperatures have been increasingly recorded at 38°C or higher, which is the upper limit of what most marine life can tolerate.[33] In the summer of 2024, the extreme high temperatures produced what some scientists have likened to 'wildfires underwater'.[34]

It is nonetheless the gradual, enduring increase in temperature that has the greatest consequences. We find an example of this exactly where land meets the sea, on beaches where sea turtles lay their eggs in protective nests of sand. The sex of these small foetuses is determined by the sand's temperature, and if the temperature around the egg rises above 31°C, it is almost guaranteed that the baby turtle will be female. Due to global warming, in recent years ninety-nine per cent of newborn turtles in Australia have been female.[35] If this development continues, upholding a natural, viable balance in the turtle population will prove impossible.

As the temperature of the ocean rises, many types of fish will gravitate towards cooler waters. Scientists can already document the occurrence of a parallel displacement of fish stocks to habitats in the north and south. Fish are leaving tropical waters to seek out more temperate zones, also closer to the cool, nutritious and (still) richly oxygenated ocean regions around the Arctic and Antarctica. In the North Sea, herring and cod are migrating north, while anchovies and sardines are moving in from the south. At the same time, mackerel are leaving the North Sea, moving into the Norwegian Sea and all the way up to Svalbard.

In the Arctic Ocean, as the Polar ice cap melts, the edge of the ice moves north. The nutritious zones surrounding the ice are the habitats of an abundant and diverse marine life. Because of this, other wildlife, such as polar bears and seals follow the ice edge as it moves.

A Carbonate Competition

In addition to absorbing heat, the ocean helps curb global warming by absorbing a quarter of the carbon dioxide emitted into the atmosphere. This takes place through a chemical interaction between the carbon dioxide in the atmosphere and salt ions in seawater. A portion of the carbon dioxide is used in photosynthesis, whereby plankton converts carbon dioxide into oxygen and energy. Another portion of the carbon dioxide is propelled downward by ocean currents such as the AMOC, into the ocean depths, where it can be stored for hundreds of thousands of years. It is because of this storage mechanism that the ocean is able to hold almost sixty times more carbon dioxide than the atmosphere.[36] However, since carbon dioxide dissolved in water produces a weak acid, increased absorption of carbon dioxide leads to increased ocean acidity and a decrease in pH values, which indicate that the water is becoming more alkaline. This is compounded by the release of methane from the ocean floor, caused by not least by melting of ice in the polar regions, where the gas interacts with water to produce carbon dioxide. Over the past few decades the ocean has become more than twenty-five per cent more acidic, and scientists have now recorded the lowest level and the fastest decline in the ocean's pH values in almost 20,000 years.[37]

The greater the amount of salt that is depleted through this process today, the less there will be in the future to absorb carbon dioxide.[38] This is because, as discussed previously, the salt in the ocean comes from the gradual erosion of volcanic stone on the ocean floor, or the run-off from weathered stone on land. These erosion processes take place over the course of millions of years, which means that the amount of available salt in the ocean in practice is fixed and limited.

Much of marine life relies on calcium and carbonate from the salt in the ocean to build skeletons, shells and coral. These animals have little chance of holding their own in the increasingly fierce competition with greenhouse gases, in particular because their habitats are also being altered and destroyed by rising ocean temperatures, deoxygenation, pollution and increased acidification.

We can gain an understanding of the consequences by comparing today's species with samples collected during the HMS *Challenger*'s research expedition in the 1870s. The thickness of many of the shells of today is only one quarter of those found back then.[39] Marine shellfish have thinner shells and they have become more fragile and more vulnerable, and this has an impact on the entire aquatic food web.

7. THE OCEAN STRIKES BACK

Modern society's use of fossil energy sources does not 'only' lead to the emission of vast amounts of carbon dioxide, methane gas and other harmful greenhouse gases. The emissions also contain large quantities of sulphur, nitrogen and soot particles. These are gases and substances that are not fully absorbed by the atmosphere. They can be carried by the weather across great distances, but for the most part are deposited in the areas close to the emission source.

The ocean reacts to many of these substances in the same way it does to carbon dioxide, and this affects photosynthesis, the ecosystems and marine species. Since almost half of the earth's population and even more of the economic activity are located near the coast, most of this 'acid rain' lands in shallow water in coastal regions. This is also where we find most of the marine life.

It is also in the coastal regions that the impact of litter and harmful run-offs from land on ecosystems and marine life is the greatest.

Respiratory Difficulties

Chemical processes dictate that the warmer the water temperature, the more the water's capacity to absorb and hold gases is debilitated. The rising ocean temperatures therefore also lead to a reduction in the content of life-sustaining oxygen in the ocean. All marine life is dependent upon oxygen and since the middle of the last century, the oxygen content has been reduced by more than one per cent.[40]

In many coastal regions, the oxygen content of the ocean has been reduced even more due to alien nutrients, sewage and hazardous chemical runoff into the ocean from agriculture, industry, households and human activity on land. Along the coast of the USA, the marine life in two thirds of the oceanic regions has been moderately or critically damaged by run-off from land.[41] Especially nitrogen and phosphates foster what is called 'eutrophication', a human-made fertilisation of the ocean that contributes to the destruction of finely calibrated aquatic food webs. Eutrophication, in combination with warmer water, often leads to extensive bloom of algae and plankton, which consumes oxygen and depletes other marine life. Bloom sometimes manifests as large, colourful areas on the water surface at the mouths of wide rivers, or close to agricultural land. As the layer of algae on the surface expands, it forms a lid that prevents sunshine from reaching the marine flora and fauna below. As the sunlight is diminished, photosynthesis occurs more slowly and marine plants die. Eutrophication can also kill different types of bacteria on the ocean floor that produce vital vitamins for fish, shellfish and corals.[42] In several locations, the production of oxygen in the water can be so depleted that entire regions become uninhabitable 'dead zones'.

There are hundreds of such dead zones in the ocean, and the number is growing. Today, we find the largest dead zones off the coast of Oman and in the Gulf of Mexico.

Fortunately, it has proven possible to rescue and resuscitate such zones if we respond quickly enough.

Three Tons in Three Hours

The ocean is not only the human race's most important public commons – it also, unfortunately, serves as our largest rubbish bin. For generations we have thought of the ocean as being so vast that it can contain, hide and withstand just about anything. As recently as in the 1970s, every winter a number of Norwegian municipalities would place adverts in local newspapers informing readers that they could dispose of scrapped buses, lorries, cars, tractors or other heavy machinery by placing these at designated sites on the fjord ice, where it would all sink into the depths when the ice was melted by the spring sunshine. If you wanted to discard an old refrigerator, a bicycle, sewing machine or a stove, dumping it in the fjord was fine. Out of sight, out of mind.

Even today, the idea lingers that pollution is not a problem as long as it is distributed throughout the ocean: 'a solution to pollution is dilution'. The international regulations under the UN International Maritime Organization (IMO) have still not banned the dumping of sewage from ships in the ocean – as long as it is done 'at a sufficient speed' and 'far enough' from land.

Luckily, more people are beginning to understand that we can't use the ocean as a public rubbish bin. Plastic waste in the ocean in particular has attracted a lot of attention. When a dying Cuvier's beaked whale washed ashore on the west coast of Norway in 2017 and its stomach proved to be full of plastic bags, it attracted international attention. The plastic had paralysed the whale's breathing, digestion and blood circulation, and the creature was put down to end its suffering. But, despite its tragic fate, it nonetheless had a more meaningful demise than many other members of its species. Photos of the dying whale became a powerful symbol of our collective negligence. A short time later, local campaigns were organised to clear away plastic and other litter on beaches all over Norway. Together with a group of scuba divers, I took part in one such nationally televised clean-up initiative at the invitation of the World Wildlife Fund. In less than three hours, we gathered more than three tons of rubbish from the seabed

7. THE OCEAN STRIKES BACK

of the fjord just outside of Oslo, Norway's capital. Norwegians live in one of the most affluent and well-regulated countries in the world, and we like to pride ourselves that there are few nations with stricter environmental regulations or a more conscious and active relationship to nature.

All the same, three tons in three hours.

Plastic waste is a massive and growing problem. The annual production of plastic equals the weight of the entire world population. Nearly one fifth of the world's oil production goes to the manufacturing of plastic products for use in everything from automobiles, computers and home appliances to clothing, packaging materials, insulation and fishing nets.[43]

It is estimated that less than ten per cent of the plastic produced gets recycled, and that more than a fifth of plastic waste worldwide is either not collected, improperly disposed of, or ends up as litter in refuse dumps or in nature.[44] A great deal of this waste eventually finds its way into the ocean by way of rivers or runoffs from land, but large quantities of plastic waste are also dumped from ships, oil rigs and fishing vessels.

Until recently, scientists estimated that every year nine to thirteen million tons of plastic were deposited in the ocean. It now turns out that the amount may be twice that and continues to grow.[45] In the time it takes you to read this page, another fifty to a hundred tons of plastic waste will have ended up in the ocean.

Once plastic has reached the ocean, it begins a virtually endless journey. Only a fraction – an estimated five per cent – is washed back onto land.[46] That means that, for every plastic bag you find on the shoreline, there are another nineteen bags in the ocean that you are unable to see.

Of the plastic waste that remains in the ocean, only around one per cent is floating on the water surface. Large quantities of this waste accumulate into patches in the gyres of the major ocean currents. The largest of them all is the Great Pacific Garbage Patch, located in the northern Pacific Ocean. While difficult to measure precisely, it probably already covers an area three times larger than France, and there are indications that it is still expanding. Every year, hundreds of thousands of seabirds and marine mammals perish because they are trapped in such floating garbage patches around the world.[47] Many get stuck and drown, while others are strangled or die from intestinal blockage because they mistake the plastic for food.

I once met a yachtsman who told me that, on a crossing of the Indian Ocean, he had seen palm trees swaying in the waves. At first, he thought it was a mirage, but as he drew closer, it turned out to be real. The palms tree had put down roots and grown on a huge patch of plastic waste, a floating artificial island. The trees drew sustenance from the organic materials that had attached to the plastic.

The Smaller, the Worse

The biggest problem, however, is the plastic we cannot see. It is estimated that more than ninety per cent of the plastic waste in the ocean sinks to the bottom, where it will remain virtually for ever.[48] With time, this plastic is broken down into smaller and smaller pieces. The decomposition time varies, depending on whether the waste is plastic bags, cigarette butts, contact lenses, bottles, gloves, fishing nets or laptop covers, and generally ranges from a few decades to several centuries. The fragments eventually become microplastic (smaller than five millimetres) and nano particles (smaller than 0.001 millimetres). These plastic fragments are so small that they can be absorbed into the marine food chain. Today the quantity of micro- and nanoparticles of plastic in the ocean is so great that scientists are often more surprised when they don't find plastic in fish and other marine animals than when they do. Even in the Arctic Ocean, far away from the largest cities and the most important sources of pollution, scientists often find plastic in the bellies of cod, king crabs and seabirds.

When plastic is absorbed into the food chain, it also finds its way to our own plates. One study estimates that every week on average the amount of plastic we ingest corresponds with the amount in a credit card.[49] An Italian study showed that three out of every four healthy, breast-feeding mothers included in the study had microparticles of plastic in their breast milk.[50]

Scientists remain uncertain about how much damage this can inflict on the human body, but there is considerable evidence suggesting that the smaller the particles, the greater the damage they can do. 'Micron-sized plastic particles can enter the bloodstream and the liver. When the particles are nanometric, they can apparently pass into the body tissues,' claims Tanja Kögel from the Institute of Marine Research in Bergen.[51] A compounded risk of life-threatening blood clots or serious tissue and genetic material damage is then close at hand. The tiny particles of plastic also bind to environmental toxins such as PCB and other hazardous chemicals, which in turn can have serious ramifications if absorbed by the human body.

It is basically impossible to remove plastic and other waste once it has landed in the ocean. There is only one solution to this problem: prevent it from ending up there in the first place.

The problem of plastic in the ocean must be solved on land.

7. THE OCEAN STRIKES BACK

A Painful Experience

The noise beneath the surface of the ocean has increased in step with world trade ever since steam ships replaced sailing vessels in the late nineteenth century. Larger ships, more powerful engines and heavier traffic currently produce human-made background noise at far higher volume than ever before.[52] Some readings show that the noise level has increased more than thirty times over in the past fifty years.[53]

Large areas of the marine soundscape are affected by the noise of engine vibrations, rotating propellers and the wave slap caused by the large hulls of moving vessels. Especially intense is the sound from the air bubbles created by the tips of propeller blades, the technical term for which is 'cavitation'. The noise is similar to the sound of whiplashes. But the most dominant and disturbing encroachment is the low-frequency sounds that travel great distances and can be heard many kilometres away from the shipping lanes.

I was made painfully aware of this one day when I was diving by an island in the middle of one of the narrow straits of the Oslo Fjord. Cruise ships, cargo vessels and ferries pass through the area on their way to the port in Oslo. For people on land, it might appear as if these ships glide in and out of the fjord virtually without a sound, but looks can be deceiving, because below the surface they actually produce a deafening commotion. On this occasion, we were diving at around thirty-metres' depth when a low-frequency pounding noise began increasing in volume. Eventually the noise was so powerful that we could feel the vibration of the sound waves in our bodies. It did not continue for long, maybe a couple of minutes, but it was an extremely unpleasant experience. When we later surfaced, we were told that the noise had come from the ferry to Kiel as it sailed by. Although it had passed at a safe distance, sound waves travel more quickly and further in water, so it felt like the ferry had been right on top of us.

It doesn't take much imagination to understand how invasive, frightening and disruptive this deafening noise must be for underwater wildlife. Many marine animals have extremely sensitive receptors and communicate with their offspring and other members of their species using sophisticated forms of acoustic signalling. There is solid documentation demonstrating that the impact of noise pollution on marine wildlife is stress, illness and increased mortality. The noise inhibits the animals' communication, weakens their immune system and negatively impacts reproduction. It has been proven that chronic noise exposure can reduce growth in shrimp, and lead to cellular mutations in lobsters.[54] When a noise becomes sufficiently invasive and persistent, marine animals will often seek out other habitats, either because the noise disturbs them or because the species below them in the food chain disappear.

Just like human beings, underwater animals can be startled and frightened by sudden or loud noises. Military sonars used to detect submarines can affect the hearing of marine animals within a radius of several thousand kilometres, and following large-scale naval drills an increase in stranded whales has been observed.[55] It is also well known that seismic trials in conjunction with oil explorations, during which air is fired down into the water column, can frighten fish, seals and other marine mammals. This is particularly disruptive and harmful during spawning and mating seasons. The explosions can also cause frightened fish and other marine creatures to rise to the surface too quickly. Like scuba divers, they can then suffer cerebral haemorrhages or decompression sickness, otherwise known as the bends that we discussed in a Chapter 2.

Wave Movement

Wind and weather permitting, I like to go paddling in my sea kayak on the Oslo Fjord. Sometimes I will paddle across the fjord to the idyllic islands on the other side. Then I must cut through the sea lanes of the large ferries and cruise ships, which requires a bit of additional caution. But it is not only the risk of being run over that can be a bit annoying; the waves and whirlpools that form in the ships' wakes can also be problematic. The tall waves are easy to spot, but the flat whirlpools created by the huge propellors can be surprisingly powerful. The first time I experienced this, it was undeniably a bit stressful, because I was paddling alone and almost capsized in the busy sea lane near the port of Oslo.

Waves and whirlpools caused by shipping traffic create challenges, not only for recreational paddlers but, more importantly, for marine life forms worldwide. When these powerful movements in the water travel towards shallow shorelines, they can destroy marine habitats and local ecosystems. It is difficult for fish and crustaceans to thrive, hatch eggs and raise offspring in areas where at any time they can be washed ashore by waves from passing ships. The waves also carry sand, sludge and other bottom sediments that foster harsh growth conditions for coral reefs, mangroves and marine plants.

Many scientists are also concerned about how the waves and whirlpools from increased shipping traffic can accelerate the melting of ice in the Arctic Ocean. When the Arctic ice melts, it does not solely mean that the ice pack shrinks. The ice cover also becomes thinner: indeed, in the course of a little more than one century, the thickness of the summer ice has been halved.

7. THE OCEAN STRIKES BACK

Because of the thinner summer ice, ships can navigate through larger ice-covered areas without need for assistance from icebreakers. When a ship passes through the ice, the hull breaks open broad, kilometre-long gashes in the ice cover, while waves and whirlpools also shatter ice floes, opening up areas in the water that are far wider than required for passage. This will in turn produce larger dark surfaces that absorb heat. The increased shipping traffic in the north will thereby also compound the already destructive albedo feedback.

Invasive Species

The tens of thousands of merchant vessels that keep the wheels of the global economy in motion do not solely transport cargo along the coast and between continents. They also carry ballast water that is pumped in and out of tanks in the hull to adapt the weight of the ship to the amount of cargo, ensuring that the vessel remains balanced and stable in the water at all times. While previously iron or stones were used for this purpose, each of the ballast tanks on large, modern ships contain several million litres of water. The water is pumped into the tanks and transported across large distances before being discharged in another area, perhaps in a wholly different part of the world. Often ballast water contains micro-organisms, algae, fish and crustaceans, all of which can be alien species in the habitats where they are emitted. Also, various aquatic organisms often accumulate on ships' hulls, referred to by experts as 'biofouling'. This can have serious consequences for the local marine environment.

A comprehensive system of international regulations under the IMO has therefore been developed to prevent harmful and unintended spreading of these 'invasive species'. It has nonetheless proven difficult to avoid the mixing of marine animals and organisms between coastal regions and continents. According to statistics from the Norwegian Maritime Authority, every year an estimated ten billion tons of ballast water is transported between ports, and that 'at any given time there are approximately 7,000 species in a ship's ballast tank water'.[56]

International trade and globalisation have led to an ecological reunion of the continents – also under water.

Coastal Cultivation

All over the world people move to coastal regions to live, work and spend their holidays. Today almost half of the world population lives less than 100 kilometres from the coast, and both the number and percentage continue to grow. This implies, for better or worse, that increasingly more people live where the ocean meets civilisation, in the important and fragile interface between land and sea. More people can therefore take pleasure in everything the ocean has to offer, but at the same time more people are exposed to the ocean's power and whims – and vice versa.

Not only are more people moving closer to the coast – they are also settling in concentrations of cities, towns and holiday complexes along the coastline. It is estimated that, by the middle of this century, eighty per cent of the world population will be living in the 700 largest cities, and that these will generate ninety per cent of the global economic activity.[57] In principle, this can be good for both sustainability and economic growth. In cities, the companies are close to the consumers and people live in proximity to their places of work. Electricity, public transport and urban spaces can therefore be utilised more efficiently. Urban environments can be smart, green and productive – if they are designed accordingly.

However, more people and larger cities will increase the already sizeable pressure on the marine ecosystems in coastal zones, where most of both the large cities and marine life are found. Large-scale construction of cities and infrastructure almost always leads to degradation of natural areas, reduction of biodiversity and destruction of ecosystems. Every year, for example, several billion tons of sand, gravel and sediment are taken from the sea to build houses, roads and other infrastructure.[58] This dredging and excavation activity releases greenhouse gases, degrades water quality, disturbs marine habitats and damages biological diversity.

Unfortunately, care and consideration for mangrove forests, wetlands, seaweed, coral reefs and marine life rarely carry sufficient weight in competition with the construction of new housing, hotels, office buildings, quay facilities, motorways, railways, airports, factories, power plants, beachfront promenades, beaches or small-craft harbours. Nature has no voice in this context, no voting rights or financial leverage with which to push back against people's expectations, politicians' ambitions and investors' calculations.

The UN Sustainable Development Goal 14, 'Life Below Water', is exclusively dedicated to measures for protecting, preserving and restoring the marine ecosystems and life under water. Here we find goals such as conserving coastal and other areas to ensure that marine animals and plants have peaceful havens where they can live and reproduce without suffering the adverse effects of human activity. Such protected areas, often referred to as Marine Protected Areas (MPA), include both marine sanctuaries and areas with active, sustainable management of marine life. Today, these protected zones constitute less than ten per cent of the world's oceanic regions, but

7. THE OCEAN STRIKES BACK

the target agreed in the Kunming-Montreal Global Biodiversity Framework adopted in 2022 is that they should cover thirty per cent by 2030, often referred to by insiders as the '30 by 30' target.[59][60]

Despite this, blasting, drilling, digging and construction continue along the coast in many parts of the world. Such large areas of the earth have been affected, changed and destroyed by human activity that scientists are now describing it with a specific geological term: the Anthropocene Epoch, or 'the human impact era'. This is also the epoch during which humankind's reckless overconsumption and abuse of the ocean has been effectively wiping out marine life forms and destroying crucial underwater ecosystems. At the same time, global warming is causing ice melt, sea-level rise and the movement of water from the poles to the equator, which is even slowing down the earth's rotation on its axis. The change is imperceptible in our daily lives, but the ultraprecise atomic clocks that control the world's navigation systems and financial markets must from time to time be adjusted by one second to provide for this.[61]

Yet it is precisely the ocean that offers some of the largest and best opportunities for combating global warming, preventing the destruction of nature and biodiversity, and securing a sustainable future for a growing world population.

> 'Then Jesus directed them to have all the people sit down in groups on the green grass. So they sat down in groups of hundreds and fifties. Taking the five loaves and the two fish and looking up to heaven, he gave thanks and broke the loaves.'
>
> Mark 6:39–41

8.
BLUE GROWTH FOR A GREEN FUTURE

About the vitally important role of the ocean in the green transition, and how nature-based marine solutions and maritime industries alone can contribute a full third of the emission-cuts needed to reach the goal of the Paris Agreement. How the ocean can ensure a more prosperous and secure future for a growing world population, and why the blue economy is expected to grow more rapidly than the global mainland economy in the coming decades.

Colourful Contrast

We were attending the climate change conference COP26 in Glasgow in the autumn of 2021, and my good colleague Erik and I were taking a quick lunch break. Dressed in dark suits and white shirts and with accreditation cards on a lanyard hanging around our necks, we were indistinguishable from the thousands of other delegates slowly streaming between plenary discussions, bilateral meetings and random pull-aside chats. The contrast to the young woman being interviewed at the next table could hardly have been more pronounced. She was wearing, as always on such occasions, her characteristic headwear and traditional colourful dress.

Time magazine has named Hindou Oumarou Ibrahim one of the most prominent women in the fight against climate change.[1] She is a geographer, leads a number of community-based organisations and is a highly sought-after public speaker. She comes from a poor goat-farming family and grew up by Lake Chad, the main water source for more than forty million people.

8. BLUE GROWTH FOR A GREEN FUTURE

Because of drought and release of water vapour caused by climate change, this lake – previously one of the largest in the world – has shrunk to one tenth of the size it was in the 1960s.[2]

Since so much of the water is gone, many people who live in the areas surrounding Lake Chad can no longer make a living from the sale of meat, mangos, sesame seeds or cashew nuts. They must keep most of the food they produce to survive, and the men travel to the cities in search of employment. In the absence of the men, it is the women who assume responsibility for the home and family. They must procure food and water and take care of both their children and ageing parents. Many of the men rarely or never come back, because they can't afford to, because they can't find work or because they become involved with other women in the cities.

Hindou attends international conferences all over the world to represent the voices of these people. She speaks about how climate change and freshwater scarcity is not only a threat to crops, people's livelihoods and life itself, but also destroys familial relations and social structures. While young people demonstrating against climate change in Western countries take to the streets to save their future, Hindou's sisters who live around Lake Chad are fighting for their daily lives.

Day Zero

Freshwater scarcity is one of the most imminent and severe consequences of global warming. A report by the UN Food and Agriculture Organization (FAO) of 2020 states that 'the annual amount of available freshwater resources per person has declined by more than 20 per cent in the past two decades'. The report elaborates:

> [T]his is a particularly serious issue in Northern Africa and Western Asia, where per capita freshwater has declined by more than 30 percent and where the average annual volume of water per person barely reaches 1,000 m^3, which is conventionally considered the threshold for severe water scarcity.[3]

More than two billion people still lack access to safe drinking water and one quarter of the world population lives in regions that are already experiencing 'extremely high water stress'.[4] [5] In China, most of the provinces in the north, home to forty per cent of the population, is at or beyond the threshold defined by the United Nations as 'water scarcity'.[6] Megalopolises such as Sao Paolo, Jakarta, Bangalore and Chennai are already teetering on the brink of 'Day Zero', the moment when freshwater reservoirs are so depleted that they do not cover the population's basic needs. In 2018, it started to rain just a few days before Cape Town was about to run out of freshwater.[7] The Global Commission on the Economics of Water has stated that, due to decades of faulty and deficient management, 'we now face the prospect of a 40 per cent shortfall in freshwater supply by 2030, with severe shortages in water-constrained regions'.[8]

Water scarcity may seem like a paradox when we consider images of our shimmering, blue planet. Yet only three per cent of all the water on earth is freshwater, and half of this is bound up in snow and ice, the majority in the Arctic, the Antarctic, Greenland and the Himalayas. Global warming melts ice and releases freshwater from glaciers, but the better part of this meltwater flows directly into the ocean. Only 0.5 percent of all the water found on earth is therefore available as fresh water for human use or consumption.[9] One third of this is found in reservoirs, so-called aquifers, which are located under the ground.

Global economic and population growth increases the total need for freshwater, while the consistent trend of urbanisation means that more people are dependent upon the same freshwater sources. While in 1960, New York City and Tokyo were the only two cities in the world with more than ten million residents, today there are more than thirty such cities.[10] This is causing the rapid depletion of one-half of the world's groundwater reserves, on a scale that renders natural replenishment impossible. Intensive groundwater pumping, combined with the increased weight of more infrastructure and taller buildings, is also causing cities to sink. Mexico City is sinking at a rate of almost a half-metre per year and in China a quarter of a billion people are living in cities that are sinking several millimetres annually.[11] This entails potentially huge costs related to the consequent destruction of buildings and infrastructure, especially in coastal cities that are also exposed to the impacts of sea-level rise. This is the case in Indonesia, for example, where the process of relocating the capital from Jakarta to the new city of Nusantara, located on higher ground in Borneo, is already underway. As the city of Jakarta sinks, this produces cracks in bridges and buildings, while sea-level rise causes saltwater intrusion of underground drinking water. While the formal transition of the capital was celebrated with music and parades on the Independence Day, 17 August 2024, no one could avoid noticing the serious backdrop of the event.[12]

Global economic growth and the rise in prosperity have introduced changes in dietary habits all over the world, which in turn increases the highly water-intensive production of meat and dairy products. Also, a number of countries advocate the use of biofuels in planes and cars as a measure to combat climate change, even though the production of such fuel consumes seventy to 400 times more water than the petrol and diesel it replaces.[13]

Simultaneously, industrial development, agriculture, litter and pollution all contribute to the destruction of valuable drinking water sources. On top of this, in many parts of the world the available freshwater resources are poorly managed due to ineffective irrigation systems, faulty water pipes and obsolete water management schemes.

At the same time, global warming is driving an increase in the release of water vapour from lakes, more frequent desiccation of riverbeds, an acceleration in the hydrological cycle and radical changes in the precipitation patterns on earth. In many places there is more rainfall in sparsely populated areas, and less rainfall in densely populated areas. Tropical and temperate regions are experiencing more extreme precipitation, while central parts of Africa, India, Southeast Asia and the USA are experiencing longer and more intensive periods of drought, resulting in freshwater scarcity and failing crops. Global warming is simultaneously causing a sea-level rise, which causes salt water to penetrate groundwater reserves in densely populated areas, especially in Southeast Asia.

The interaction of these factors makes clean and accessible freshwater a scarce commodity that is increasingly in demand, costly and strategically important.[14]

In the years ahead, freshwater scarcity may prove to be one of the most explosive topics in international relations. A United Nations report elaborates that, 'while approximately 40 per cent of the world's population lives in transboundary river and lake basins, only a fifth of countries have cross border agreements to jointly manage these shared resources equitably'.[15] The report cites how this can compound the already substantial tensions between many countries in the Middle East and states further that 'Africa remains especially vulnerable to interstate tensions relating to water: 19 out of 22 states studied suffer from water scarcity, and two-thirds of the continent's freshwater resources are transboundary'. Another report by the United Nations maintains that 'water stress has important implications for social stability, and water deficits can be linked to 10 per cent of the increase in migrations worldwide'.[16]

Along similar lines, the Leadership Group on Water Security in Asia warns that reduced access to safe and stable freshwater sources 'will have a profound impact on security throughout the region'.[17] The group outlines the domino effects of increased scarcity and unstable access to freshwater, which include destroyed crops, unemployment, reduced earnings, large-scale migration and heightened political and military tensions.

Adding to these potential conflicts is the prospect of countries resorting to more frequent and extensive 'cloud seeding', a geo-engineering technique manipulating clouds to make it rain. Already, this has caused international tensions, such as in 2022 when Iran accused Israel and the United Arab Emirates of 'stealing their rain'.[18] As global warming is producing more intensive heatwaves and prolonged periods of draught, and since there are no international agreements in force preventing countries from manipulating clouds in their own air space, several states have stepped up their research efforts in this field. According to the Geopolitical Monitor, there are today about fifty states which manipulate clouds to ensure 'ordered' rain.[19]

These practices may produce new international tensions, and also raise more fundamental questions about 'who owns the rain'. Moreover, like several other fields of geo-engineering, cloud seeding is also considered highly controversial between experts because of its potential harmful effects on the environment and the hydrological cycle.

Water from Water

While sea-level rise threatens large freshwater reserves, the extraction of freshwater from the ocean can provide an important contribution to increasing water supplies, distributing risk and mitigating international tensions. Well-developed technologies that produce freshwater by removing salt from seawater and brackish water already exist, and today they account for approximately one per cent of global fresh water supplies.[20] These desalination processes are, however, still quite expensive and energy-intensive, and most of the electricity they consume is sourced from fossil fuel-powered plants. It is therefore not surprising that today we find a large portion of the world's total capacity for desalination in the oil-rich and desert-dry countries of the Middle East and in small island states without alternative water sources, such as Malta and Cyprus.[21] Desalination plants are nonetheless in operation in more than one hundred countries worldwide. A number of coastal states, such as South Africa, are also in the process of expanding the capacity of their plants, not least because renewable energy production makes it both cheaper and more environmentally friendly. A report from 2020 shows that use of energy from solar parks and wind power, combined with modern battery technology, can cut costs by more than one half.[22] For desalination to be sustainable, it is however necessary to address the large discharges of toxic brine and other waste from these processes. Today, most of these end up back in the ocean, where they often cause great damage to local, marine ecosystems.

There is also significant activity taking place in the mapping of geological formations of freshwater found under the ocean floor several places in the world. Like aquifers on land, these reserves were produced by the large amounts of ice that were trapped beneath thick layers of sediments after the last ice age around 10,000 years ago. When the ice pack receded, these pockets of fresh water remained under the ocean floor. Researchers have known about the existence of such reserves for a long time, but it has only been in recent years that we have come to understand their actual scope. In 2019, US scientists discovered one 200 metres beneath the ocean floor in an area extending from New Jersey to Massachusetts on the US east coast. It is estimated that this pocket alone contains 2,800 cubic kilometres of freshwater, which is enough to fill more than a billion Olympic-size swimming pools.[23] With time, as more such reservoirs are discovered around the world, this can provide new solutions for water-stressed societies – and introduce new areas of collaboration and conflict between states.

All the same, the potential of desalination and aquifers will not be the ocean's most important contribution to reducing the world's critical and increasingly pressing fresh water shortage. A far more important contribution lies in replacing food production on land that is both water-intensive and harmful to climate and the environment, with more sustainable production of food from the ocean.

Land-Based Waste

The harvest of ocean resources is providing food for hundreds of millions of people. Nonetheless, food from the ocean constitutes only one to two per cent of our total calorie intake, and less than one sixth of the protein consumed by the world population.[24]

Today, more than two billion people still suffer from famine or malnutrition, while the world population is expected to increase by more than three billion in this century. To eradicate famine and meet the needs of population and prosperity growth in the period leading up to the next century, total food production must increase by almost fifty per cent.

But we cannot continue to produce food as we do now. On a global basis, agriculture is responsible for seventy per cent of the freshwater consumed by humans.[25] Since many developing nations export their agricultural products, this also gives rise to a large 'virtual export' of the costly liquid from poor to rich countries.[26] We can get a sense of the dimensions this implies by the fact that ten to twenty tons of freshwater are needed to produce one kilo of red meat, three to five tons for one kilo of rice, and around one ton for every kilo of avocados, soy and wheat. The production of the one egg you eat for breakfast on the weekends alone requires close to 200 litres of fresh water.[27]

Land-based food production is one of the largest causes of global warming and the loss of nature and biodiversity.[28] Worldwide, more than a fifth of greenhouse gas emissions is caused by agriculture, forestry and other land use.[29] Moreover, agriculture accounts for eighty per cent of global deforestation, and the keeping of livestock alone occupies land equalling twice the total land area of India and China combined.[30][31] Meat and dairy products are also inefficient nutritional sources. They represent less than one fifth of human calorie intake but occupy more than four fifths of all cultivated land.

Moreover, most countries are experiencing large and increasing challenges related to procuring a healthier diet for their populations. In both rich and poor countries, weight gain and obesity have reached virtually epidemic proportions. The percentage of obesity in adults has doubled since 1990, while it has quadrupled in children and adolescents.[32] There are now more people in the world suffering from obesity than from starvation and malnutrition.[33] This has led to an upsurge in obesity-related illnesses, such as cancer, diabetes and cardiovascular diseases. Illnesses of this nature reduce quality of life and life expectancy for those affected, and they are extremely costly for society at large.

Blue Fields

The first and most obvious measure for meeting the world's food supply challenges must be to secure a more effective and just distribution of the food that is already available. There are already enough calories being produced to feed the entire world population. Second, we must reduce the large amounts of waste that occurs in the parts of the world that have too much food. It is estimated that as much as one third of global food production is discarded, spoiled or rots.[34] Also, we must produce healthier food, and last but not least, develop new strategies to ensure that food production becomes more sustainable and resilient in the face of climate change.

Many of these strategies lead to the ocean, which must become a significantly more important source of food in the decades ahead, as extensively documented by the FAO in its biennial flagship reports on the State of World Fisheries and Aquaculture:

> Today, aquatic systems are increasingly recognized as vital for food and nutrition security. But more can be done to feed a growing and more urbanized population. Because of their great diversity and capacity to supply ecosystem services and sustain healthy diets, aquatic food systems represent

8. BLUE GROWTH FOR A GREEN FUTURE

a viable and effective solution that offers greater opportunities to improve global food security and nutrition today and for generations to come.[35]

Moreover, sustainable aquatic food production holds the potential to generate less emissions to air, have lower environmental impact and consume less energy and freshwater. It can also contribute to reducing the need for cultivating land and alleviating further encroachment of wildlife habitat. Although climate risks are also affecting aquatic food production, producing more food in the ocean can help mitigate the risk to overall food supply caused by production downtime and crop failure due to extreme weather events, drought, flooding and rising sea levels. Since most of the world's states have coastlines, they can harvest locally from their own 'blue fields'. This will also lead to improving their preparedness and security of food supply in periods of international tensions or trade disruptions caused, for instance, by wars or a new pandemic.

Moreover, fish and other seafood contain fatty acids, proteins and other health-promoting nutrients that can reduce malnutrition and obesity. Public health authorities often advise eating fish at least twice a week, but in most parts of the world, the actual consumption of fish is far less.

Wild-caught fish have traditionally been the most important food source from the ocean, but fish stocks are facing growing pressure and degradation due to overfishing, poor management and global warming. Today it is estimated that two thirds of the world's wild fisheries are managed in a more or less sustainable manner, while the remaining third are overfished due to exigent and irresponsible exploitation.[36] According to a report by the international High Level Panel for a Sustainable Ocean Economy, illegal, unreported and unregulated (IUU) fishing accounts for twenty per cent of the world catch and more than twice that in some areas.[37] The report further states that 'the global fishing fleet is two to three times larger than needed to catch the amount of fish that the ocean can sustainably support'. The FAO considers IUU to be 'one of the greatest threats to marine ecosystems due to its potent ability to undermine national and regional efforts to manage fisheries sustainably as well as endeavours to conserve marine biodiversity', stating further that 'IUU fishing therefore threatens livelihoods, exacerbates poverty and augments food insecurity'.[38]

A Conflict Commodity

Also, rising ocean temperatures are about to push fish stocks from the increasingly warmer tropical waters towards the temperate and cooler waters nearer to the Arctic and Antarctica, threatening further access to these resources vital to the sustenance and livelihoods for hundreds of millions of people in the Global South.

Migration of wild fish stocks, caused by global warming, could be an increasingly important driver of international conflicts and already growing geopolitical tensions. Wild fish has all the essential characteristics of a conflict commodity, as it is a limited resource that for many countries is important for food security, livelihoods and economic prosperity. And, unlike oil, gas, coal, iron ore, critical minerals, forests, ground water, arable land and other natural resources, wild fish is the only strategic resource and economic commodity of major significance that moves by itself between national jurisdictions. Through history this has over and again caused international tensions and brought countries to the brink of war. Even traditionally friendly neighbours like Iceland and the United Kingdom have faced each other off militarily over these issues, and it is estimated that one quarter of all militarised conflicts between democratic countries during the Cold War were over fish stocks.[39] In recent decades, the number of such conflicts has surged, not least those involving China and Russia.

But, even with better management, and if we were able to solve the problems of IUU and find peaceful solutions to conflicts between countries, harvesting of wild fish can only provide a modest contribution to the challenges of meeting the world's future food requirements. The truly great and important potential lies in three other areas. One of these is more efficient supply chains. It is estimated that as much as a third of wild-caught fish is lost somewhere along the way from capture to consumer. The second is more sustainable industrial farming of fish, crustaceans and algae. The third is to move further down the 'trophic levels', so we do not solely harvest from the top of the marine food chain.

Caged and Seasick

Farming of fish, crustaceans and algae such as seaweed can be done with little or zero harmful emissions to air, without occupying precious land areas or excessive use of increasingly scarce freshwater resources. Unlike land-based food production, aquaculture facilities are not vulnerable to drought, flooding or torrential rainfall. Most existing aquaculture facilities are, however, exposed to extreme winds and harmful algae blooms resulting from global warming. Therefore, there is extensive research and innovation going on to develop aquaculture facilities that are even more resilient against the consequences of climate change, including pens that can be submerged if necessary to mitigate the risks of powerful storms or increasing water temperature.

Fish farming is now one of the fastest-growing sectors within the global food industry and is expected to acquire large market shares over the course of this century. While globally wild fish landing has stagnated since the early 1990s, the sale of farmed fish has doubled in the same period. Worldwide, the volume of farmed fish produced has now surpassed that of harvested wild fish.[40]

The largest fish-farming facilities are currently found in Asia. Asian manufacturers are responsible for ninety per cent of world production and China alone stands for more than half of this.[41] In this part of the world, freshwater farming of different types of carp is the most common, while saltwater farming of fish and crustaceans constitutes the largest world market share. But despite the many benefits of fish farming, large challenges remain with regard to ensuring ecologically sustainable and ethically responsible operations. Also, there are currently no overall, global regulations in existence which specifically address aquaculture or the environmental impact of this industry, even though there is guidance issued by the Organisation for Economic Co-operation and Development (OECD) and organisations such as the independent, non-profit Aquaculture Stewardship Council, established in 2010 by the World Wildlife Fund for Nature (WWF).

To address the challenges of the aquaculture industry, the construction and location of fish farms must be adapted so they do not harm and degrade the surrounding environment. In the largest fish-farming nations outside of Asia, such as Chile, New Zealand, the Faroe Islands and Norway, the facilities are typically located close to the shore or in narrow fjords. The confined waters of the fjords protect the farms from the strong waves and currents of the ocean, while proximity to land simplifies logistics, operations and maintenance. But the location close to the shore or in the fjords also poses a series of challenges, such as the emission of large quantities of excrement, infection of wild fish and other impact on local marine ecosystems, and conflicts with commercial traffic, tourism and recreational activities.

To offset some of these drawbacks, closed cages have been developed that can be located in the fjords, close to shore or further from land. Experiments have been done using for example the hulls of large vessels as pens, but here unexpected problems have arisen: the vessel's movement has basically made the fish seasick. Large cages for fish farming have been constructed that can be situated offshore, far out at sea. The idea is to avoid conflicts with other activities and reduce problems with parasites, lice and diseases by submerging the fish in deeper waters, while ocean currents will dilute the discharges of excrement and other organic waste (here, again, the idea is that 'a solution to pollution is dilution'…). But, in practice, these systems have also met with unexpected problems. When submerged in unaccustomed ocean depths, salmon for example become agitated and frighted by the darkness of deeper zones.

Other aquaculture companies are building onshore fish farms, and these are often located close to cities to shorten the distance to markets. But these onshore fish farms occupy land areas, consume large quantities of electricity to create water currents and require far more building materials than offshore alternatives. Several of the natural benefits of sustainable farming in the ocean are thereby squandered.

Moreover, if fish farming is to be ecologically sustainable and ethically responsible, the practice of crowding too many fish in cramped facilities must be phased out. Overcrowding puts stress on the fish and causes illness, suffering and high mortality rates. Frequent examples of this are found in Norway, the world's largest salmon-farming country, and where the export revenues generated are so large that they have earned salmon the nickname 'pink gold'. Although the Norwegian salmon is generally known to be of high quality, every sixth farmed salmon dies of gill disease, virus, skin lesions or lice before slaughter.[42] Since salmon lice and other parasites have become more resistant to traditional medication, the use of 'cleaner fish' – such as corkwing and goldsinny wrasse and lumpfish – has virtually exploded. For these species, the mortality rate is even higher: almost all the cleaner fish are dead by the time the salmon are slaughtered. Salmon also obviously suffer when their water is heated by lice treatment, as practised by some fish farms, and there have been instances of salmon being virtually boiled alive during such treatments.

It should be obvious to anyone that both the salmon and cleaner fish suffer under these conditions. We have previously addressed how fish, like human beings and other animals, can experience stress, fear and pain. From this perspective, a great deal of what takes place in these farming facilities is nothing short of animal abuse.

Highest Price for the Poorest

Another major challenge faced by fish farming is how to 'close the production cycle' so the fish-farming industry itself produces its own fish feed. The pellets that are now sprinkled over the cages are like small packed lunches full of nutrients, most of which come from wild fish or plant-based oils. Around one sixth of all wild fish caught in the world today are used as feed for farmed fish.[43] In an attempt to address this problem, the development of new types of feed based on omega-rich algae is being explored. Research is also being done on protein-rich feed alternatives for fish from a range of larvae and insects.

8. BLUE GROWTH FOR A GREEN FUTURE

But the aquaculture industry is not only facing challenges related to animal welfare and ecological sustainability. The current practice is also ethically and morally problematic because much of the feed comes from wild catch fisheries, some of which can be illegal, unreported or unregulated (IUU), and often in areas of the world where the human population is suffering from poverty and food scarcity. For example, sardinella, a small, herring-like type of fish that is caught off the west coast of Africa, has been for many years been used to feed farmed Norwegian salmon.[44] The demand from Norwegian salmon manufacturers has pushed up the price of sardinella, making this important and nutritious food source more expensive and therefore less accessible for the poor local population.[45] In recent years, the industry itself has launched several programmes aimed at reducing the dependence on wild catch fish and improving sustainability and transparency in the supply chain.[46]

But, much remains to be done. It is still the poor people in the Global South, not the guests of luxury restaurants in Paris, New York and Tokyo, who are paying the highest price for Norwegian farmed salmon.

Food for Thought

When we speak about seafood, many of us often mean some type of marine meat, such as fish, shrimp, crab, lobster or scallops. But one of the most important and perhaps most exciting alternatives for increased food production is that of algae, which constitutes the largest biomass on earth.[47] There is also considerable interest in how algae can contribute to the world's growing demand for feed, fertiliser, fuel, medicine, packaging, cosmetics and health products. Algae farming does not occupy valuable land areas, and neither does it consume freshwater, fertiliser or feed. In this case nature and the marine environment take care of everything.

Algae is a blanket term for tens of thousands of species, from microscopic, single-celled micro algae to macro algae measuring several metres in length, such as seaweed and kelp. Whether large or small, they are all of critical significance to the ocean's ecosystems. They produce the majority of the oxygen in the ocean, regulate acidity and serve as habitats and food sources for the entire marine food chain. Algae have an extremely rapid growth rate; micro algae can triple its biomass in one day and some kelp species grow more than a half-metre daily.[48]

Today the farming of algae, especially in the form of seaweed, is first and foremost an Asian enterprise. More than ninety per cent of the global production takes place in Asia, and the largest producing nations are China, Indonesia and the Philippines. Here we find everything from small, family-run manual-harvest farms to large, industrial production facilities. The costs of establishing such farming facilities are relatively modest, as are the risks of down time and crop failure. More than two thirds of the Asian production goes to food, but the market for this in other parts of the world remains quite small. However, the consumption of plant-based food is on the rise in many parts of the world, and especially in more affluent countries. This is due first and foremost to greater knowledge about the health benefits along with growing awareness of the harmful consequences of land-based food production for the climate and the environment.

There are also exciting opportunities to be found in macro algae farming as a measure to reduce global warming. Seaweed can absorb at least five to ten times more carbon dioxide than an equivalent area of forest land. Research is now being done to further develop this nature-based method for carbon capture and storage. An alternative explored is drying algae on land through a process that does not release carbon dioxide. Research is also being done on submerging algae to depths of more than a thousand metres, where the pressure is so great that the carbon dioxide can be stored for hundreds of years. But questions and uncertainty remain about where – and whether – the algae will remain in the depths, and if it will disturb or destroy the fragile deepwater ecosystems. Also, large-scale farming of macroalgae to absorb greenhouse gases can run the risk of becoming a kind of 'geo-engineering' which might have unanticipated and serious consequences for marine life and ecosystems.

Moreover, to increase food production from the ocean, extensive research is being done on the possibilities of moving even further down and further out in the marine food chains. The food chains are actually 'feeding chains', since the species at lower levels are eaten by the species in the levels above. These are called 'trophic levels' and today humans consume predominantly fish from the upper levels of these chains. Scientists are now exploring the possibilities for increasing total food production by taking shortcuts between the trophic levels. This can mean that in the future our daily diets will contain larger quantities of, for example, krill and microalgae that have not taken a detour through cod or mackerel.

Research on marine food chains is a part of the large and exciting field known as 'marine bioprospecting'. Here we find, *inter alia*, some of the most innovative and exciting segments of the medical research front.

8. BLUE GROWTH FOR A GREEN FUTURE

Hardy Creatures

The majority of those of us who have tried know that the pressure on our ears increases as soon as we start swimming downward in water. Once, when I was diving along a pipeline, one of my eardrums burst at only a few metres' depth because I failed to offset the pressure quickly enough.

It is the weight of the water column that creates this pressure and the increase of the latter is directly proportional to depth. Already at ten metres' depth the pressure is twice what it is on the surface and, as we have seen in Chapter 2, from around fifty metres the pressure is so great that it alters the characteristics of air. When divers descend to depths greater than this, they must use special gases because ordinary air becomes 'poisonous'. At the average depth of the ocean, the ambient pressure is 400 times as high as on the shoreline. This is comparable to releasing sixty tons, or the weight of more than ten fully grown elephants, on one of your hands.

There are few species able to survive the severe pressure or the nutrient-poor, dark and extreme environment of the abyssal ocean depths. On the other hand, those able to thrive have wholly unique and specialised characteristics. Both deep-water bacteria and marine species higher up on the food chain have developed genetic combinations and chemical processes that do not exist anywhere on land. It is precisely these traits which make the hardy marine creatures of such interest to scientists. Scientists have also discovered a number of unique DNA sequences and rare chemical compounds in species living in shallower water that can be used in the production of new types of antibiotics and medicines to treat illnesses ranging from cancer and Alzheimer's disease to asthma, diabetes and arthritis.

It is not only medical scientists who understand the exciting opportunities of these unique traits. Marine plants, microbes and animals are to a growing extent being used as ingredients in the food industry, as well as in the cosmetics industry for the production of powders, lipstick, hair care products and skin creams.

Watts from Waves…

Land-based hydropower has traditionally been the dominant source of renewable electricity, and up to the end of the 2010s produced as much electricity as all the other renewable energy sources combined.[49] The term 'hydropower' is actually misleading. In a physical sense it is more correct to speak about 'power from gravity', for that is what causes the water to descend and thereby propel the rotor blades on the turbines that produce electricity. In the ocean, on the other hand, it is accurate to speak about 'hydropower' because there we can produce electricity from water as it moves downward, upward, sideways and in circles in the form of waves, currents or tidal waters.

The theoretical potential is great, and it is estimated that, with today's technology, kinetic energy derived from the ocean's movement can cover more than twice the world's demand for electricity.[50] Synergy effects may also be derived from co-location solutions, such as electricity supply for ocean-based fish farming that is sourced from this type of energy production.

In the late nineteenth century, the tireless Thomas Alva Edison was already experimenting with using the ocean currents as a source of energy production, when he installed a propeller-like contraption on the ocean floor off the coast of Florida in an attempt to convert the forces of the Gulf Stream into electricity.[51] Since then, however, less attention has been devoted to the possibilities for harvesting the large energy potential found in ocean currents, waves and tides, than to offshore wind and solar. There are many power plants in operation around the world propelled by the water's kinetic energy, but 'ocean power' has not yet become truly fashionable.

There are also two other, slightly more exotic power sources in the ocean that can be exploited, although the technology required for industrial upscaling of these has not yet been fully developed. One of these is the production of electricity from the differences in salt content between parts of the water, such as where a river empties into the ocean. The process, which is called osmosis, leverages differences between the chemical pressure of water molecules in freshwater and saltwater. Another possibility is to exploit temperature differences in the water to produce electricity, such as between the warm surface water and the cold depths of the ocean. Since the difference in temperature must be at least 20°C, production of such thermal power is best suited for tropical regions, where surface water temperature is high.

...and Winds

In recent years, wind and solar parks on land and offshore have eclipsed hydropower as the most important sources for the world's production of renewable electricity. In 2023, wind and solar and other renewables accounted for almost a third of the world's electricity production, and the two also experienced a far more rapid growth than fossil-based electricity generation.[52] The International Energy Agency (IEA) estimates that the 1.5°C goal of the Paris Agreement will require worldwide installation of capacity for 2,000 gigawatts (GW) of offshore wind power generation. In 2023, total capacity was just slightly above 70 GW, but this nevertheless represented a tenfold increase in ten years.[53]

8. BLUE GROWTH FOR A GREEN FUTURE

On a global basis, Great Britain, Germany, the Netherlands and Denmark have led the technological development, while China is installing the greatest offshore wind power capacity. Over the past few years, China alone has installed more such capacity than all other countries combined.

Barring innovations, even with the current, off-the-shelf technology, offshore wind power alone can theoretically cover the entire world's electricity consumption. Because the technological development is advancing rapidly, the costs of offshore wind power over time are declining more quickly than for onshore facilities.[54] It is no longer unreasonable to expect that the price of electricity from offshore wind parks in several locations can match or be lower than that of electricity generated by onshore parks. This is because each individual offshore turbine can be larger, and thereby generate more electricity. The offshore winds are generally more stable, which allows for more efficient exploitation of the capacity of each turbine. Also, it is easier to find suitable sites for large wind parks offshore than onshore, and thereby better exploit the benefits of economies of scale in terms of design, installation and operation.

But the most important difference is still that offshore facilities often offer greater opportunities for reducing the impact on the environment and biodiversity and for reducing potential conflicts with other activities and stakeholder groups over the use of land. Wind parks both on land and offshore, like all other power plants, will have an impact on the surrounding environment. These environment-specific costs are seldom taken into account in a realistic manner when considering the investment proposals for such projects. For onshore wind parks, developers will often boast of zero-emission electricity, attractive employment opportunities and increased local tax revenues. But they are rarely as outspoken about the fact that onshore facilities entail extensive encroachments on nature and the occupation of large tracts of land for the construction of roads, installation zones, power lines and concrete foundations. Onshore parks often damage local ecosystems, disturb local animal life, destroy traditional ways of life and degrade beautiful natural areas. Every year, for example, tens of thousands of birds die from collisions with wind turbines. The long blades, the swept area of which is often equivalent to two football pitches, also generate invasive, low-frequency sounds that disturb both animal life and human beings, even from great distances. The extensive encroachments on nature also release large quantities of the carbon captured and stored by trees, bushes, heather and marshes. When the natural surroundings have been destroyed, they can no longer provide carbon capture or other important ecosystem-related services. Since the loss of nature and biodiversity are among the greatest challenges we are facing in the fight to arrest global warming, onshore wind parks can effectively compound aspects of the very problem they are designed to solve.

Offshore wind parks will also have a significant environmental footprint, but this can often be mitigated through use of more flexible location and design options. Moreover, if carefully planned, offshore parks can also be designed to serve as sanctuaries for fish and other marine life, providing shelter from trawlers and maritime industrial activities. Some have also suggested locating algae-farming facilities contiguous to offshore wind parks.

While wind parks located offshore offer more flexibility and options to reduce territorial conflicts and environmental impact, connection to onshore grids for distribution of the electricity generated will pose challenges in much the same way as land-based wind farms. Whether situated offshore or onshore, wind parks (and solar) will usually require building new and extensive power grids, and offshore facilities will also need landing sites for cables, land acreage for transformers and port facilities for construction and maintenance. All of these will potentially present challenges to local communities, environment and biodiversity.

But, on the whole, if all external consequences were to be included, often more would weigh in favour of installation of wind parks at sea rather than on land.

Sun, Salt and Strategic Monopolies

Floating solar farms are still more costly and less common than offshore wind parks, but here as well the technology is undergoing rapid development. Like wind turbines, offshore solar panels avoid a number of the challenges faced on land, in particular challenges related to the destruction of nature and conflicts over land use.

Solar farms, like wind power parks, generally require access to sizeable tracts of land. If for example the entire electricity supply for New York City were to be sourced from solar farms, even with today's most advanced solar panels, the land area needed for the farms would be seven times the size of Manhattan.[55] On hot days and during heatwaves, the panels' efficiency can be reduced by up to one fifth and the installations must therefore often be oversized to compensate for this.[56]

Many of these challenges can be addressed more simply by installing solar panels on lakes or on the surface of the ocean. The water surface is already flat and, on the ocean, also large. No large-scale encroachment on land is required to clear sites for the farms, and that means diminished potential for land-use conflicts. Since the water naturally cools the floating panels, they

8. BLUE GROWTH FOR A GREEN FUTURE

are also less vulnerable to production loss due to heat. Floating solar farms do require anchorage, chains and power cables that will have an impact on the underwater environment, but they also can make a positive contribution by offering a refuge for marine creatures and plants. On lakes, solar farms can help reduce surface water evaporation on hot days and provide sufficient shade to prevent toxic algae bloom in drinking water.

Moreover, the salt in the ocean could play a part in revolutionising the production of solar cell panels. Research is being done on the possibility of replacing the highly toxic and costly cadmium chloride, which is used in panels today, with magnesium chloride. The latter substance is found in seawater, it is non-hazardous, and the cost of extraction is a fraction of that of cadmium chloride. It has already proved possible to use this substance to produce solar panels offering equal energy efficiency.[57] Also, Chinese scientists have made exciting and promising progress in substituting the conventional silicon cells with perovskite thin films to enhance the performance of solar panels and electric grids.[58]

The usage areas for electricity from solar and wind power can be quite different. Solar panels are easily installed on homes, cottages and car roofs, while wind turbines are best suited for large-scale facilities. But the two are more closely related than most of us might think: wind power is an indirect form of solar energy. Winds arise because the sun creates temperature differences on earth and cold air occupies the space vacated by rising warm air. As long as the sun continues to shine, the wind will always blow.

This also illustrates the fascinating common denominator for all types of renewable electricity production, whether from the sun, water, wind, ocean currents, osmosis or differences in water temperatures: they all extract power from the energy generated through nature's endeavours to restore balance and maintain equilibrium.

I think this ought to inspire philosophical reflections that transcend the industrial talking points of electricity production.

Since most countries are coastal states, and nearly half the global population lives closer than 100 km from the sea, the ocean can offer accessible, affordable and short-travelled electricity to large cities and rural communities all over the world. Like renewable electricity produced on land, this may represents a 'democratisation' of access to clean, secure and inexpensive energy. In real terms, energy prices are as high today as they were in the 1970s. This is mainly due to two factors. One is that it is harder to reduce unit production costs in energy systems based on fossil fuel than in those based on renewables. The extraction of oil and gas, whether offshore or on land, usually requires large-scale, costly and time-consuming engineering projects with limited scope for efficiency gains. Renewable electricity, on the other hand, is based more on the logic of industrial mass production, exploiting economies of scale to drive down costs. Hence, over time, technological advances and development will lead to such a large supply capacity that the price of 'the last kilowatt hours' of electricity will be pushed down towards zero.

The other factor is related to cooperation and competition. A handful of the countries controlling most of the world's oil and gas resources, and a limited number of large international companies in charge of production and distribution, have been able to cooperate or act in a concerted manner to keep electricity prices high. The industrial mass production of renewable energy is different, in that it is based on industrial competence and commercial competition, which in principle will be open to new entrants from anywhere in the world. This also means that the world will no longer be held hostage by a small group of petro-states, of which most are authoritarian regimes.

There is, however, an inherent risk that critical parts of the supply chains for renewable electricity, perhaps solar power and batteries in particular, can become concentrated and monopolised due to economies of scale, limited access to critical minerals or attempts by countries to gain strategic dominance. When the USA and other Western countries are currently scaling up their research, innovation and production in the field of solar panels and batteries, it is not only out of concern for global warming. It is, as much, driven by a wish to protect their own industries and national strategic interests from the growing global dominance of China.

Back to the Future

The captain of the *Yara Birkeland* was unmistakably proud as he showed us around the world's first fully electric and autonomous container ship. The eighty-metre ship can sail without a crew and therefore has no cabins on board. The traditional wheelhouse has been replaced by a provisional container equipped with large windows and instruments for manual steering and navigation. The container will be removed as soon as the ship's systems have passed all the practical tests. A thousand interconnected batteries give the ship a cruising speed of six knots, and a maximum speed of a little more than twice that. The *Yara Birkeland* is designed to sail from port to port round the clock, to dock, charge its batteries and load and unload containers without any need for human intervention. When I asked the captain what he thought about the fact that his own job was thereby made redundant, he smiled broadly: 'I will continue to be a captain, but from now on, for several ships at the same time. From a control room on land I will have responsibility for a number of fully automated ships. I will miss the sea but I am still as excited as a little boy.'

When the *Yara Birkeland* is fully operational, this vessel alone can do the work of 40,000 annual trips made by lorries in southern Norway. Transport will be free of hazardous air emissions, eliminate the risk of traffic accidents and produce no road abrasion, noise or traffic congestion on the motorways. The sea lanes need not be constructed or maintained – and, in winter nations such as Norway, they need neither be ploughed and salted.

Although most ships in the world fleet are still powered by heavy fuel oil and marine diesel, there are no other forms of transport that can claim greater environmental efficiency in the transport of large quantities of cargo over great distances. The greater the amount of cargo and the longer the distances, the greater the advantages of the use of ships. To transport two tons of cargo over a distance of one mile, a ship would typically emit roughly half as much carbon dioxide as a train, one fifth as much as a lorry, and merely a fiftieth compared to an aeroplane.[59] Many countries are therefore actively pursuing a policy to achieve a modal shift in the transport of goods from land to sea.

Since more than eighty per cent of all international trade of goods is transported by ship, the total emissions are nonetheless substantial. International maritime transport is responsible for just under three per cent of global greenhouse gas emissions, approximately the same amount as Germany as a whole, the world's third-largest economy.[60] Therefore, great efforts are underway to make shipping a zero-emission industry, as it already was for thousands of years, up until the introduction of the steam engine in the late nineteenth century. Shipping must go back to the future.

Extensive research is being done on the development of more energy-efficient engines, propellor designs, hull types and coatings that minimise water friction. Research is also being carried out on different types of devices that can exploit the energy of wind, waves and currents, such as kite-like sails and on-deck Flettner rotors. Artificial intelligence and software advancements are emerging that will better optimise sailing by adapting the ship's speed, ballast and route to real-time information about wind, waves and currents. Also under development are 'digital twins' that improve the efficiency of maintenance, and computerised solutions that reduce the time spent queuing, unloading, loading and bunkering in ports, and thereby the energy consumption. There are also companies exploring online solutions that will better exploit the existing fleet's capacity, a kind of 'Uber of the Seas', offering a more efficient streamlining of supply and demand for shipping.

All of these innovations are important but, all the same, only supplementary measures. The goal is to find solutions that can eliminate all harmful emissions to air, as well as to sea and land. This will most likely require a number of solutions because it is unlikely that a single solution for all types of ships and transport can be found.

The *Yara Birkeland* illustrates that it is fully possible to build emission-free vessels, but use of fuel-cell batteries is still limited to short-distance transport between ports that are equipped to provide charging. A large container vessel sailing from Shanghai to Rotterdam, for example, would require several hundred thousand modern EV batteries, so for the time being this is not a viable alternative. Some have suggested exploring the use of nuclear power, which is already the most common energy source for large aircraft carriers and strategic submarines. In a world characterised by peace, collaboration and mutual trust between benevolent actors, this may have been an attractive option, since a lump of radioactive material the size of a golf ball can fuel a large ship for several decades. But given the state of the world today, few are willing to assume the environmental and security challenges related to equipping the more than 70,000 vessels of the international fleet with radioactive material. It makes little sense to solve one existential threat by creating two new.

To the extent that nuclear solutions are discussed in the maritime sector today, they seem to be rather focused on concepts for power barges that can generate electricity for ports and the production of low- and zero-emission fuels for ship.

The uncertainty that continues to prevail regarding regulations, profitability and availability means that many shipping companies continue to invest in ships with fossil fuel systems. Still, today, more than two thirds of the vessels in order for delivery until the end of the 2020s are with mono-fossil-fuel engines. These ships will presumably be in operation beyond the middle of the century. However, an increasing number of shipping companies are ordering vessels with engines that can later be converted to zero-emission alternatives and 'optionality' has become an industry buzzword. Others have chosen to invest in liquefied natural gas (LNG), which can reduce carbon dioxide emissions by one third and eliminate virtually all emissions of sulphur, nitrogen and soot particle emissions. On the other hand, LNG emits large quantities of methane, which is a far more potent greenhouse gas than carbon dioxide. Many stakeholders therefore consider LNG to be a costly and transitional solution that risks delaying the progress towards the goal of an entire international fleet of zero-emission vessels.

Other types of fuel must be utilised in order to eliminate the harmful greenhouse gas emissions. Solutions based on ammonia, hydrogen and methanol appear particularly promising. They are all considered 'zero-emission solutions' provided that they are also produced and distributed in an emission-free manner. The zero-emission requirement must apply from 'well to wake'. Production of these fuel substances is highly energy-intensive and it would make no sense to base that production on electricity from coal-fired power plants. Along the same lines, distribution of emission-free fuel over great distances by diesel-fuelled tank lorries would be counterproductive.

Efforts are therefore being made in a number of places to locate production facilities for renewable energy close to ports, ideally supplied with electricity from nearby offshore wind parks or solar farms. In Egypt, for example, the production of ammonia for ships is now underway, right beside the new port being built by the Suez Canal, using electricity from the site's newly built solar farms.

Green Corridors

Shipping companies will not be particularly interested in investing in emission-free fuel if they cannot trust that sufficient quantities of the fuel will be available in ports of call, and at prices that are competitive compared to fossil fuels. The same holds true for ships as for electric cars: if the power source is costly and charging stations few and far between, the cars will not hold much appeal. Large-scale investments in infrastructure on land must therefore be made to provide bunker facilities and sufficient supply of different kinds of zero-emission fuels. Neither individual shipping companies nor the shipping industry as a whole can address this challenge alone. National and local governments must contribute, and, since shipping is international, this calls for extensive cooperation across national borders.

Several countries and ports have already begun collaborating on 'green corridors', designated zones where ships are guaranteed access to zero-emission fuel in the ports of call. Joint projects to this end have already been introduced in different parts of the world, such as between the ports of Los Angeles and Shanghai, and between Australia and Japan. There have also been launched larger, industry-wide projects where the International Chamber of Shipping (ICS) is developing concepts to serve as blue prints for green corridors together with a number of governments, including Canada, Greece and the United Arab Emirates. These undertakings to decarbonise shipping and establishing green corridors are not only important in and of themselves. They are also important because the international merchant fleet is estimated to move half of the low- and zero-emission fuels needed by other industries and users in different parts of the world.

Zero-emission fuels for ships can nonetheless not be put to use until seafarers, dockworkers, safety inspectors and the people who build and maintain the vessels have been trained to handle them safely. Hydrogen, ammonia and methanol all have characteristics that distinguish them from the heavy fuel oil and marine diesel used today. They are either extremely toxic, very explosive or highly flammable – or all three. The transition to 'green shipping' will therefore impose new requirements for the training and competency of seafarers, dockworkers and others employed in the maritime industries. Millions of maritime workers will need to be retrained or upskilled. Unfortunately, so far, only modest progress has been made in establishing new regulatory schemes and certification requirements, and developing the necessary curriculums for maritime schools and training centres.

For the shipping industry, as is the case for so many other sectors, it is people, profit and politics, not technology, which represent the most important hurdles in the work of cutting greenhouse gas emissions.

Cruise Control

Although merchant shipping serves as the backbone of the global economy, the direct value added of marine tourism, in the form of cruises, recreational fishing, leisure boats, sightseeing and diving is greater than that of commercial fishing, aquaculture and shipping combined.[61] The cruise industry in particular has experienced powerful growth over the past few decades.

After having been virtually shut down altogether throughout the Covid-19 pandemic, cruise companies now anticipate that in the latter half of the 2020s they will welcome approximately thirty million holidaymakers on board annually. This represents almost sixty per cent growth over the previous ten years. In some of the smaller, more luxurious and niche segments, the number of passengers for the same period has quadrupled.[62] The industry has no shortage of plans for the future. I have spoken with high-ranking representatives of Asian shipyards that envision the construction of large artificial supply bases that will be stationed in the Arctic Ocean to serve as hubs for long-distance cruise vessels.

Although the more exclusive segments are experiencing the most rapid growth, mass tourism today constitutes the backbone of the cruise industry. The largest of these propellor-driven satellite towns have ten decks and capacity for up to ten thousand passengers and crew.[63] Many are so large that they are divided into different 'districts' and 'neighbourhoods', each with their own unique features. On board, passengers can enjoy stage shows, cinema, casinos, swimming pools, training centres, spas, beauty salons, restaurants

8. BLUE GROWTH FOR A GREEN FUTURE

and endless shopping arcades. Ports of call offer sightseeing, guided tours, beaches, bicycle rides, cultural experiences and an abundant selection of other activities. Adverts promise all-inclusive hospitality, entertainment and activity programmes, and the thrill of awakening in a new location almost every day. 'Sustainability' has become a key buzzword in advertising, and cruise companies are actively displaying and promoting their efforts to reduce air emissions and environmental footprints. Many of these companies are also making substantial progress in their environmental performance.

But cruise operators are more reticent about the extensive energy consumption involved in operating the ships, up until now fuelled exclusively by marine diesel and heavy fuel oil. On the coast of Norway alone, the greenhouse gas emissions from these vessels have tripled since the first half of the 2010s.[64] I recently attended an international conference, at which the director of one of the world's largest cruise companies claimed that even if all the available surfaces on their ships were to be covered with state-of-the-art solar panels, the electricity generated by these would not be enough to power even the lifts on board. The operators also speak seldom and in subdued tones about the emissions to air and environmental impact of transporting thousands of passengers to and from the ships before and after voyages.

Nor do the tourist brochures explain that these ships, like all other vessels, are permitted by law to dump chemicals, wash water and sewage into the ocean, as long as they are a few kilometres away from shore. The distance depends on the region, but the rules are in any case unsatisfactory from an environmental perspective. The discharge of large cruise ships carrying thousands of passengers is clearly greater than that of large cargo ships which typically have a crew of only fifteen to twenty on board. The negative impact of the inadequate regulatory schemes is correspondingly far greater in the case of former. Off the west coast of Canada alone, cruise ships annually expel more than thirty billion litres of sewage and polluted waste water on voyages to and from Alaska.[65]

For the passengers it is of course not particularly amusing to spend many days of their holiday on board, staring across the endless surface of the ocean. The cruise ships therefore predominantly remain close to land, navigating between islands and into fjords, before calling at the ports of famous and idyllic locations. But the cruise operators do not often make mention of how the noise from these sailing cities disturbs underwater wildlife, or how the waves and propeller-induced currents destroy coral reefs, mangrove forests, Arctic coastal regions and fragile marine ecosystems. They rarely speak about how the giant ships degrade the views of iconic urban landscapes of

Venice and Cozumel, or how they navigate through narrow World Heritage fjords on the west coast of Norway.

The ad campaigns also do not show how swarms of tourists from large cruise ships overwhelm the local daily life of small coastal communities or how the tourists are transported in huge buses down narrow roads through fragile natural surroundings. Little is said about how these tourists spend pretty modest amounts of money in the places they visit, since they eat, drink, sleep and are entertained on board. Neither is much mentioned about the strain put on local hospitals and health-care services by cruise passengers who in the course of the cruise have become too sick to be treated on board. Given the size of the ships and the number of passengers, riding roughshod over the natural gems, cultural sites and local communities they visit is difficult to avoid.

Though some destinations benefit from the call of cruise ships as the only income opportunity for their population, little attention is devoted to how the cruise operators actively pit different destinations against one another to avoid paying their way. In the Caribbean, where around forty per cent of the world's cruise ships in operation can be found at any given time, several large industry actors are known for threatening to divert to other destinations if they would be obliged to pay local taxes or duties.

No responsible, sustainable enterprise can base their business model on degradation of nature and consumption of global commons or public amenities free of charge. The cruise ships travel the ocean for free, they disrupt and pollute their surroundings, their ownership is often registered in tax havens and they profit from duty-free schemes on all onboard sales. The cruise operators offer experiences based on nature, cultural legacy and local communities, often with little regard for the consequences of their activities for the environment or local communities.

It is therefore high time that we switch on the cruise control for this industry. I do not begrudge anyone a nice holiday or lovely travel experiences, but far stricter international regulations must be imposed on this segment of the tourist industry. There is an urgent need for stricter regulations that will limit the operations of the 'sea giants' and ensure more sustainable conduct, economically, culturally and environmentally. This would require the introduction of taxes, levies and systems to ensure that the cruise companies' activities are mutually beneficial for society at large, and for the local communities affected. I also believe that these companies would be well advised to engage more extensively with sustainability initiatives for the communities they impact.

8. BLUE GROWTH FOR A GREEN FUTURE

Cobalt Crisis

Throughout the ages and across civilisations, gold has enjoyed a unique position, as a value standard, a subject of myths and a symbol of status and achievement. Today, however, there are far less glamorous substances, called 'critical minerals' that are every bit as attractive and valuable as gold. These are commonly defined as 'mineral resources that are essential to the economy and whose supply may be disrupted'.[66] This means that the 'criticality' and importance of the different substances encompassed by this term can change over time. In Roman times, for example, sodium chloride, table salt, was such a valuable commodity that it was referred to as 'white gold'. Large quantities of the critical minerals are found in the seabed, which presents us with great opportunities – and even greater dilemmas.

The powerful rise in the demand for these substances is driven by the technological development and the heightened focus on climate, the environment and sustainability. Minerals, and a subset of these called rare earth elements (REE), are found in all the electronic gadgets many of us surround ourselves with in modern daily life, such as smartphones, computers, pulse watches, waffle irons and refrigerators. An iPhone contains several dozen different types of minerals. Nickel, dysprosium and neodymium enable your phone to vibrate when you receive a message. Potassium, indium oxide and tin are used for the screen's touch functions and tungsten for electric connections. Minerals are also used in data chips, semiconductors and all the key electronic components for aeroplanes, ships, automobiles, submarines, radars, robots, satellites, base stations, X-ray machines, camera lenses and industrial production equipment.

The growing demand of recent years is however also driven by the 'green transition' from fossil fuel to renewable energy forms. Electric vehicles typically contain six times more minerals than fossil fuel-powered cars, and wind farms require nine times as much minerals as conventional gas-fired power plants.[67] Tellurium and silicon are important solar cell panel components, while wind turbines contain large quantities of neodymium, copper and aluminium. Electric car batteries typically contain several dozen kilos of lithium, cobalt and nickel, along with manganese, graphite and aluminium.[68]

The International Energy Agency (IEA) estimates that reaching 'net-zero globally by 2050 would require six times more mineral input in 2040 than today'.[69] The World Bank projects that the production of lithium must increase tenfold, cobalt sixfold and graphite production must quadruple in the period leading up to the middle of this century, solely to cover the growing demand from low-emission technologies.[70] The World Bank also stresses that the mining of such important minerals already represents more than one tenth of the world's total energy consumption.

All extraction of critical minerals is currently done from mines on land, and most of these are located in Asia and Africa. The mines typically occupy large natural areas, and their operations often disturb and destroy animal life, ecosystems and biodiversity. The extraction of these valuable substances also consumes large quantities of freshwater. For the time being, there is no effective and reliable global system in place to track where the minerals are produced, or if they are processed in an environmentally sustainable or socially responsible manner. It is therefore very probable that the electric car, the smartphone or computer you use contains minerals that are produced under environmental and safety standards, working conditions and with wage levels that would not be tolerated in any modern, industrialised country.

For example, mining operations in the Democratic Republic of Congo, in central Africa, are responsible for one half of the world's production of cobalt. The mining industry there experienced powerful growth when the mobile telephone came onto the market in the 1990s, since the mines in Brazil and Australia could no longer meet the global demand for cobalt used in mobile telephone batteries. Although many of the mines in DR Congo have modern industrial operations, there are also hundreds of mines in which the work is done manually and outside the realm of any control by the authorities. Many of these unauthorised mines are run by criminal bands and local militias, who rake in lucrative profits on the sale of the valuable minerals. The mines are often found in impoverished districts where there are few other employment opportunities.

In many of these mines the working conditions are horrendous. Tens of thousands of people working for a pittance spend more than twelve hours a day, seven days a week, carrying out this heavy work in narrow, dusty and dark mining tunnels. Some of the tunnels are so small that only children can wriggle through the narrow passages deep in the earth. In these mining communities there are few who are aware of, and even fewer who care about, modern-day working environment regulations or that child labour is prohibited under the UN Human Rights Conventions. The work is gruelling and dangerous, and every year thousands of the workers in these mines are killed, injured or permanently disabled.

When Congolese physician Denis Mukwege accepted the Nobel Peace Prize in 2018, he used the occasion to spotlight the conditions in the mines in DR Congo, stating that '[w]hen you drive your electric car, when you use your smartphone or admire your jewellery, take a minute to reflect on the human cost of manufacturing these objects'.[71]

8. BLUE GROWTH FOR A GREEN FUTURE

Even if one were to succeed in cleaning up these objectionable conditions, making mining operations more water- and energy-efficient, ensuring traceability and transparency and recycling more of these minerals, the growth in global demand will probably outpace the supply extracted from existing terrestrial mines. The IEA warns that 'today's mineral supply and investment plans fall short of what is needed to transform the energy sector, raising the risk of delayed or more expensive energy transition'.[72]

The Allure of the Deep Sea

Many politicians, companies, investors and others are therefore intrigued by the potential opportunities offered by the large deposits of critical minerals that can be found on the ocean floor, commonly referred to as 'seabed minerals'. Today, the extraction of these minerals is carried out only in some national waters and on a very modest scale. No permits have been issued to date for extraction in the vast ocean depths in the areas beyond national jurisdictions, outside of countries' own continental shelves. There is widespread concern about the impact extraction will have on deep-sea life and marine ecosystems, and there are still difficult technical and practical challenges to be solved. Even on the part of large-scale industrial actors, uncertainty prevails about whether it will be possible to perform deep-sea mining in a profitable manner.

Seabed minerals is the common term for three types of geological deposits that are classified either as nodules, manganese crusts or sulphides. The nodules are small, loose stones, around the size and shape of a potato, and are found on oceanic plateaux all over the earth. These were first discovered during a research expedition in the Arctic Ocean in 1868, and for the next century, the nodules were first and foremost viewed as geological curiosities. When scientists began taking a serious interest in the nodules, they struggled to comprehend why they were not covered up by sediments on the ocean floor. It was long believed that the zones where they were found were wholly without marine life, because there was scarcely any sign of life on the large underwater plateaux. We now know that the scientists' amazement about the former was due to a misunderstanding about the latter. A teeming population of micro-organisms and tiny crustaceans lives on each of these small stones, and the cleaning activity of this population ensures that the nodules are not covered by sediments from the ocean floor.

Now we also know that the nodules are formed through a process similar to that of the creation of pearls: a tiny kernel, whether it is a shark's tooth or a pebble, attracts sediments made up of tiny grains of minerals. These settle layer by layer in a process that can take millions of years. When the nodules are cut in half, the cross-section reveals rings much like those found in a tree trunk. Each of the rings represents a new layer, and the growth of the nodule occurs through the accumulation of these layers. A practised geologist's eye can quickly determine which types of minerals are found in the nodule, since different minerals create different structures on the surface of these small stones. Each of them usually contains several types of minerals, and for that reason some people like to call them 'battery packages'.

Dense concentrations and large clusters of the nodules can be found on the ocean floor. The largest and thus far best explored area is the Clarion-Clipperton Zone, known as the CCZ, which is located at 3,000- to 6,000-metre depths in the ocean midway between California and Hawaii. This is the area where scientists discovered the 'dark oxygen' that we discussed in Chapter 2. Here more cobalt, manganese and nickel deposits have been mapped than in all known reserves on land.[73] These deposits can replace more than ten years of the world's land-based production of cobalt, twenty years of nickel and thirty years of copper.[74] It is also estimated that one tenth of this region alone contains enough of the minerals required to replace all the world's petrol and diesel-powered vehicles with electric counterparts.[75]

While the nodules can be literally picked up from the large oceanic plateaux, critical minerals are also found in the solid geological structures. Some have been formed by ocean currents that have carried small fragments of these valuable substances, depositing them layer by layer around mountains and rock formations on the ocean floor. Such deposits, called manganese crusts, can be found on Tropic Seamount, for example, a mountaintop located a kilometre beneath the surface and 500 kilometres off the north-west coast of Africa. Here the ocean currents have for millions of years transported sediments that have settled into a ten-centimetre-thick, asphalt-like carpet of valuable minerals. It is estimated that this underwater peak alone contains enough cobalt for more than 250 million electric vehicles and enough tellurium for solar panels providing half of Great Britain's electric power requirements.[76] It is likely that Tropic Seamount is just one of around 30,000 large mountain tops found beneath the surface of the ocean.

There are also large concentrations of seabed minerals in the areas where the intercontinental tectonic plates meet. Here there are long, deep cracks in the seabed, and when the water flows several thousand metres downward through these cracks it eventually hits the red-hot magma beneath the seafloor. The intense heat and pressure convert the water to steam, which reaches temperatures of several hundred degrees. The steam is then pushed upward to the ocean once more. As it travels upward from the subfloor region, the steam also carries tiny particles of minerals that are deposited on the seabed as the steam is cooled by the cold water of the ocean. Where the steam is expelled, characteristic 'black smokers', or sulphides, are eventually formed. Such minerals can be found, for instance, along the mid-Atlantic ridge in the Norwegian Sea.

These 'black smokers' can grow several centimetres every year and reach heights of several dozen metres. On the basis of specific features, the black smokers are, like volcanoes, categorised as either 'active', 'dormant' or 'extinct'. Around the active black smokers, the hot steam also contains large quantities of sulphur from the subfloor region. This triggers a reaction known as chemosynthesis, which involves chemical processes similar to those that provided the basis for the emergence of life on earth almost four billion years ago. Around these vents, a teeming community will often be found, containing everything from microbes to starfish, crayfish and fish. Many of these creatures have genetic characteristics that are not found anywhere else on the planet.

Tiptoeing Across the Ocean Floor

Seabed mineral sites are often located far from land and at great depths, sometimes several thousand metres below the water surface. There are for this reason large technical, practical and cost-related challenges involved in the extraction of these minerals for industrial purposes.

But the greatest challenge and uncertainty pertains to the consequences for the marine environment. Scientists still know little about the life and connections of these ultra-deep ecosystems. They do, however, know more than enough to confirm that extraction of seabed minerals will potentially have serious consequences for some of the earth's most extraordinary and fragile ecosystems. These ocean depths are the habitats of unique and highly specialised species, which have lengthy reproductive cycles and are extremely sensitive to external influences.

It can also be stated categorically that all seabed mineral exploration and extraction activities will seriously disturb these fragile ecosystems. It is not possible to tiptoe around in the depths, and there are no gentle methods for extracting these minerals. The nodules, which lie scattered in the sand on the ocean floor, must be gathered using shovels, pumps or steel brushes. Large machines have been produced for this purpose that resemble the combine harvesters found in wheat fields all over the world. The newest models of these seabed harvesters measure almost 100 square metres, the size of a spacious apartment. To harvest the substances from the manganese crusts and sulphides, machines must be used that can drill and dig in the geological structures. It doesn't take much imagination to envision the impact of this on marine ecosystems and life forms that for thousands of years have lived undisturbed, in darkness and utter isolation.

Whether they are picked, drilled or excavated, the geological material must be brought to the surface from depths of several kilometres. This requires suction devices and flexible pipe systems that are connected to large ships. Once the material is onboard the vessel, it must be sorted, preserving the valuable substances, while small stones, clay and other bottom sludge are dumped back into the ocean. Even with today's relatively modest level of activity, we can see how this dumping produces large sediment clouds of contaminated and often toxic mud that is transported for miles by the waves and ocean currents. It is almost impossible to believe that the consequences for the marine ecosystems will be restricted to the areas where industrial activity is taking place.

Green Dilemmas

There are few other topics that provoke stronger feelings on the part of marine biologists and environmental organisations than the extraction of seabed minerals. The majority are deeply concerned about the impact on marine wildlife. A number of countries, including Germany, Canada, Brazil, Chile, Mexico, Spain, UK and France, have united behind a demand for a precautionary waiting period or an international moratorium on such activities until we have acquired a far better understanding of the field. The Norwegian government, however, has taken a different approach. Despite massive protests, Norway is one of the first countries in the world to open for the commercial exploration and extraction of seabed minerals on large parts of its own continental shelf.

Personally, I am very sceptical about seabed mining. In general, policy decisions and regulatory design in all areas must be data-driven and fact-based, and on the topic of seabed mining we are still faced with a large deficit of both aspects. The deep-sea floor is the only area on our planet which is still largely untouched by man, and I do not think we should permit such activities without first acquiring much more, literally in-depth knowledge about the impact on underwater life. At the same time, I am mindful of the distinction between seabed mining in national and international waters, respectively. Although I am in the group of people who are quite scared of the prospect of seabed mining in any part of the deep sea, I also think it is crucial to respect the UN Convention on the Law of the Sea (UNCLOS) and the right of states to decide how to use resources within their own jurisdictions. We have to expect that this is done in a responsible and precautionary manner; if not, they will themselves probably bear most of the consequences. But I think, for seabed mining in areas beyond national jurisdiction, the global consequences will be larger, and the dimensions of questions related to fairness, equity and responsibility entirely different.

Simultaneously, I think that the subject of seabed mining confronts us with dilemmas that are far larger and more difficult than those suggested by many scientists and environmental activists. On the one hand, such activity will be disturbing, harmful and perhaps destructive for many of the marine life forms and ecosystems in the ocean depths. We do not currently know the full range of these consequences, either for life in the ocean or life on land. On the other hand, the seabed contains large reserves of minerals upon which we will probably be dependent in our endeavours to combat global warming and secure a sustainable development of prosperity for the world population. We can therefore not draw a conclusion solely based on the potential consequences for marine life. We must assume a far broader and more holistic perspective, and weigh considerations for the latter against the consequences of more terrestrial mining. Only then can we find reasonable answers to the questions of if, where, when and to what extent the extraction of seabed minerals should be permitted.

The first priority in addressing these dilemmas must be to reduce the use of such substances where we can. We must follow the guiding rule of the three Rs: reduce, reuse and recycle. It is also possible that innovation and new technology will to a certain extent free us from our reliance on these substances, such as through the development of synthetic alternatives.

But we must at the same time develop a more realistic attitude about whether this approach is sufficient in terms of meeting future needs. I sometimes talk to people who advocate 'urban mining', a catchphrase for the reuse of such substances, but I have yet to see reliable data on how much can actually be recovered in this way. It is also risky to base conclusions on vague ideas about the potential contribution of future innovations and technological breakthroughs. We will then find ourselves easily blinded in the same way as when we pin our hopes on the emergence of groundbreaking technology that will successfully remove large quantities of greenhouse gases from the atmosphere. Hope can provide encouragement and inspiration, but it should never constitute the basis of policy and strategies.

So, if recycling, 'urban mining' and synthetic alternatives should turn out to be insufficient, what should we do then? Shall we then increase the extraction from mines on land? What additional strain will this put on local communities, human beings and the environment? Will it be enough, or will we risk obstructing or delaying the transition to a zero-emission society? How shall we weigh considerations for marine species and ocean ecosystems against human beings and the environment on land? And at which point will the consequences for marine life be so extensive that they also affect life on land?

The short answer to these questions is that today we know a great deal about how mining operations on land affect human beings and the environment. We also know a lot, though not nearly enough, about the negative consequences of seabed mining. At the same time, we know far too little about the consequences of *not* harvesting the seabed minerals. Basically, we need far more hard facts and documented findings in order to calibrate the consequences and weigh the different considerations against each other.

The issues relating to these types of dilemmas extend far beyond the scope of discussions about seabed minerals. There are many other areas in which conflicting considerations must be factored into assessments. For example, if all the ships in the world merchant fleet were to go green today, the electricity consumed by the production of zero-emission fuel for these vessels alone would be three times the amount of the total energy production in the European Union.[77] In that case, there would not be enough supply for other purposes and electricity prices would soar for everyone.

Either way, we should have far more deliberate, informed and thorough discussions about many of the difficult *dilemmas* inherent to the green transition – not solely about the *solutions*.

Yet there are few fields in which the dilemmas and potential consequences are as great as for the extraction of seabed minerals. There are also few other areas in which the conditions for the green transition are so tightly bound up with national strategic interests, security policy and rivalries between the major powers.

In a world of growing geopolitical tensions, we are facing new and tougher competition over the ocean's resources and trade routes.

'We risk a competition for the seabed and rights of transit as fierce as the European competition for Africa in the nineteenth century.'
Professor Eliot A. Cohen,
Johns Hopkins University

9.
THE NEW BATTLE OVER THE OCEAN

About the ocean as a driver and domain of international tensions and conflicts, about current and emerging maritime hot-spots and flash-points, and the growing competition over marine resources and control over maritime trade routes. About the strategic considerations of the USA, China and Russia, and why many countries are expanding their military fleets and reinforcing their coastal defences. How the tense international situation of today is challenging the authority of the United Nations and the Constitution of the Ocean, the global regulatory system defining nation states' rights, obligations and activities at sea.

Beach Holiday

I was in my early twenties and had just attended a student congress in Chicago. A reckless withdrawal from my savings account financed a few lazy days in the sunny Bahamas before my return to a wintery Norway. From where I was lying lazily in the shade of a beach parasol, I could observe the cruise ships, luxury yachts and colourful small crafts streaming in and out of the bustling Nassau port. All of a sudden, the vessels deferentially slid aside to make way for a grey giant gliding into the port. Its dark silhouette conveyed authority and power.

Early the next morning I was among the first in the queue for the 'open ship' tour of the 250-metre Tarawa-class amphibious assault ship USS *Nassau*. The US war machine could carry more than 3,000 Navy and Marine Corps personnel, and was equipped with missile batteries, sophisticated anti-aircraft guns, fighter jets and heavily armed attack helicopters.[1] The ship had a hospital with 300 beds and the engine room housed the power station that produced 140,000 horsepower for the giant's propulsion. It was a fascinating and thought-provoking guided tour on board one of the USA's most important symbols and means of global power.

Forty years later, the US Navy warships have become even larger and more powerful. On a summer day in 2023, I stood on my terrace at home watching the world's largest aircraft carrier, USS *Gerald R. Ford*, glide into the Oslo Fjord. The US armed forces and the superpower's global position still rely heavily on its ability to exercise military power on, from and across the ocean. It is also the US Navy that since the Second World War has enabled the country to behave as a 'world police force' and guarantor of the UN-led, rules-based world order.

But in the current geopolitical situation the USA's position, the global order and the United Nations' role are all under increasing pressure. The tensions between the major powers are intensifying, trust is withering and rivalry is increasing in international relations.

This, as always, exacerbates the tensions at sea.

An Intensified Maritime Dynamic

All the most powerful states, and many of the less powerful, are thus upscaling their naval capacity and reinforcing their coastal defences. They are doing so to defend their own waters, safeguard important trade routes and assert rights on fishing and the extraction of energy and minerals in international waters. Many are doing so in response to the military build-up of other states, some to strengthen their international position or impose their will on other countries, and still others to mark their territory in the human race's largest public commons. Regardless of the motives, they are all responding to robust, historical lessons: power at sea means power in the world, and an absence of power at sea is a recipe for national impotence and insecurity.

A number of factors specific to the heightened tensions of our times underscore the ocean's significance and the maritime dynamic in international relations. First among these is that the ocean defines key historical lessons and present-day geopolitical conditions underpinning the growing rivalry between the USA and China, and the increasingly isolated position of Russia. In a practical sense, the USA is an island state, surrounded by vast ocean expanses, and self-sufficient in terms of key resources. This has for generations nurtured a belief in large parts of the American population that the country can 'retreat from the world and into itself'.

China also has a long coastline, but just off its coast lie island states, all of which are US allies. Historically, as we have seen in Chapter 4, it was the

9. THE NEW BATTLE OVER THE OCEAN

absence of a strong coastal defence and a powerful navy that in the beginning of the nineteenth century laid open the Middle Kingdom to attacks from the sea and the subsequent Century of Humiliation. These lessons are still etched into the memories of China's political leadership and military planners. Moreover, modern-day China is totally dependent upon sea lanes for its massive exports and the import of large quantities of food, energy and other critical resources to keep the wheels of society turning. This makes the country, and ultimately the political leadership, vulnerable to any circumstances that might threaten trade in its coastal regions, across the ocean regions of the world, and in strategic bottlenecks such as the Suez Canal, the Bab el-Mandeb Strait, the Strait of Hormuz and the Strait of Malacca.

Russia on its part has few powerful friends beyond China and can with relative facility be denied access to the ocean in the south, in the west and, to a certain extent, in the east. As a result, throughout all of modern history the Kremlin has suffered from an oppressive 'encirclement syndrome' and a sense of strategic claustrophobia. In the current situation Russia's northern ocean regions have become even more important in mitigating these concerns.

Second, the development of precise, inexpensive and long-range drones is in the process of changing the conditions of maritime conflicts. From the invasion in the beginning of 2022 to the summer of 2024, Ukraine, a country without a navy of any significance, was able to sink a third of the Russian fleet in the Black Sea by using rapid seagoing drones packed with explosives, aerial drones and missiles strikes from shore. In the Red Sea, the Iran-backed Houthi militia employs air drones, cruise missiles and ballistic missiles in attacks on military and merchant ships. Drones for some thousands of dollars are used in attacks on vessels worth billions. It is also probable that the Houthi militia is being supplied with modified Chinese missiles by way of Iran. In the Red Sea, many of the attacks are launched from lorries on land, and the choice of target is made using information found on the internet, on sites such as MarineTraffic. When small, mobile groups on land can attack civilian and military ships in international waters, this means that important ocean regions and trade routes can no longer be protected at sea. This creates a need for new naval strategies and doctrines to ensure control in these regions. It also raises difficult legal dilemmas because attacks on such mobile, land-based weapons platforms, which can easily be in the hands of non-state actors, would violate these countries' sovereignty and can be construed as acts of war according to international law.

The third factor adding tension to maritime dynamics is the gradual shift of the world's economic, political and military centre of gravity from the north-western to the south-eastern part of the globe. This heightens the role of the ocean in international relations, particularly because the topographic features of the eastern part of the southern hemisphere are completely different: 'Europe is a landscape; East Asia is a seascape. Therein lies a crucial difference between the twentieth and twenty-first centuries,' Robert Kaplan writes.[2] While the largest territorial conflicts of the last century were about dry land on the European continent, it is the ocean regions between East and South Asian countries that are the source of the most dangerous tensions in our times. In Southeast Asia, countries surround the ocean, not the other way around, and this produces other types of issues and frictions between those countries. This is, not least, made manifest in the South China Sea – currently one of the world's most tension-fraught ocean regions.

Fourth, as discussed earlier, the ocean's role in economic growth, food production, energy supply and combating global warming is becoming far more important. Already today, the 'blue economy' is estimated to have an annual turnover of three to six trillion US dollars and it is expected to grow more rapidly than the land-based economy in the coming decades.[3]

Finally, and intrinsically related to the latter, the natural resources of the ocean itself have become even more critical economic and strategic topics. As we have seen in Chapter 8, by producing food and electricity from territorial waters off their own coastlines, countries can become more self-reliant, and thereby enhance the resilience of the supply of vital goods in an increasingly unpredictable and unstable world. Also, as discussed, global fish stocks are under increasing pressure due to overfishing, poor management and climate change. 'Rising ocean temperatures alone are expected to push nearly one in four local fish populations to cross an international boundary in the coming decade, reshuffling access to this critical resource and incentivizing risky illegal fishing and labor abuse in the sector. It is not hard to imagine how, in this context, a fish-related fight could spiral,' according to Sarah Glaser and Tim Gallaudet of the WWF US.[4]

The potential for international conflicts over access to seabed minerals is probably even larger and more dangerous. China and a small number of companies today dominate the global market for the extraction and refinement of critical minerals, and the IEA warns that 'these high levels of supply concentration represent a risk for the speed of energy transitions, as it makes supply chains and routes more vulnerable to disruption, whether from extreme weather, trade disputes or geopolitics'.[5] In December 2023, the Biden administration unilaterally extended the US continental shelf by one million square kilometres, an area twice the size of California, and declared that these seabeds are now to be considered American.[6] The US press release states that the extension is in accordance with the rules of

9. THE NEW BATTLE OVER THE OCEAN

the UN Convention on the Law of the Sea (UNCLOS), although the latter was never signed by the USA. The press release also mentions that the newly claimed areas have abundant resources, such as 'for example coral and crabs', but few entertain doubts about whether the USA is far more interested in the large mineral occurrences that may be found here. The US claim immediately triggered protests from China and Russia, both of whom stated that the USA cannot invoke rights under a convention it has never ratified.[7] It was in fact precisely the discussion about seabed minerals which led the USA to refrain from signing the UNCLOS. With the recent declaration of a unilateral expansion of the continental shelf, there is a risk that the USA may have added fuel to tensions in other parts of the world, setting a dangerous precedent for other countries to follow.

Seabed minerals are also from time to time a topic in Russian propaganda, such as the claim that President Vladimir Putin holds a doctorate in strategic planning and use of Russian mineral resources.[8] Although most foreign experts question the veracity of this claim, it is striking that Russia has chosen to highlight this subject in particular when outlining President Putin's academic achievements.

There is, however, no reason to doubt the Chinese authorities' expertise in the field of critical minerals, be they on the seabed or on land. To secure its own supply and strengthen the country's strategic influence, China has since the early 2000s been systematically buying up mines in Africa, Asia and Latin America. Today China stands for more than ninety per cent of the global production of a number of critical metals. China can effectively paralyse most of the world's high-tech industries by cutting off access to these materials, were they so inclined. They demonstrated as much in 2010, when they forced Japanese companies to their knees following a conflict over territorial claims on a small group of islands in the East China Sea.

China's virtual global monopoly on the production of a number of critical metals and rare earth elements represents an economic and strategic position of power, which is often underestimated and overlooked. That position is growing stronger due to the world's increasing dependence on these resources. As the terrestrial mines of today are gradually depleted, it becomes more difficult and expensive to extract new resources on land. This makes the seabed minerals all the more attractive and strategically significant, for China and for the rest of the world.

It is therefore no coincidence that Chinese strategy documents make reference to how 'the traditional mentality that land outweighs sea must be abandoned and great importance has to be attached to managing the seas and oceans and defending maritime rights and interests'.[9]

THE OCEAN

Whisky War

We cannot take for granted that the UN-led world order in general, and the UNCLOS in particular, will withstand the pressure of the rapid, large-scale geopolitical changes we are witnessing at this time. A long and quickly growing list of ocean-related conflicts and disputes is intensifying the pressures on the international system.

In the Black Sea, because of the war in Ukraine, merchant vessels, many carrying wheat to the Middle East and Africa, face the threat of mines and Russian attacks. In the Red Sea, civilian ships are attacked, sunk and hijacked by the Yemeni Houthi militia and in the Persian Gulf there is ongoing risk of attacks orchestrated by Iran. In these regions little or no respect is paid to international rules regarding merchant vessels' right to 'freedom of navigation and innocent passage'.

In the East China Sea, there is a disagreement between Japan, South Korea and North Korea regarding ownership of a number of small reefs and rocks, and China has constant confrontations with Japan about a group of uninhabited islands called Diaoyudao by the Chinese and Senkaku by the Japanese. Another group of rocks, called Tokodo by South Korea and Takeshima by Japan, have been a source of main political disputes between the two countries for more than seventy years.[10] The issue shows no sign of being resolved, and South Korea has pointedly named its only aircraft carrier.*Tokodo*. In the Mediterranean Sea, Greek and Turkish warships are on patrol to mark their territory in the conflict over the continental shelf surrounding Cyprus. In the Caribbean Sea, the USA and Haiti continue to discuss who is the rightful owner of the tiny uninhabited Navassa Island. In the Indian Ocean, Great Britain, Mauritius and the Maldives were having discussions for more than half a century about the ownership of the Chagos Archipelago, before the dispute was settled through negotiations in the summer of 2024.

In the North Atlantic, Ireland, Great Britain, Denmark and Iceland still disagree about who owns Rockall, a tiny islet jutting out of the ocean. In the Arctic Ocean, Canada, the United States, Russia, Denmark and Norway have submitted overlapping claims on extended continental shelves. For more than fifty years Denmark and Canada both claimed the ownership of Hans Island, a 1.3 km^2 uninhabited limestone island north-west of Greenland, before the dispute was settled by an agreement in 2022 which divided the tiny island between them in a rather amicable and equitable manner.[11] While historically significant to both the Greenland Inughuit and the Canadian Inuit people, tensions never escalated between the two countries. For decades, while Denmark and Canada were still unable to reach a solution to the dispute, 'ownership' of the island was shared in a way that Danish Arctic patrol vessels dropped by and left a Danish flag with a bottle of Danish schnapps, which then later in the year was traded for a Canada flag and Canadian whisky, a practice which became known as the 'Whisky War'.

There is, however, no friendly exchange of schnapps and whisky on the contested islands in the South China Sea. On the contrary, here there is a real risk of real war. In this region, China is in conflict with the Philippines, Malaysia, Brunei and Vietnam regarding ownership of a number of tiny, uninhabited islands, reefs and sandbanks, such as the Spratly Islands, Scarborough Shoal and Mischief Reef. In a number of these locations China has already settled in, building airstrips and naval port facilities. The Chinese navy and coast guard are known for behaving aggressively and menacingly in these waters and dangerous situations often arise in meeting with merchant ships, fishing boats or other states' coast guards and military vessels. There have also been a number of close calls in this area when Chinese naval vessels and US warships have almost collided.

Dry Land and Wet Rights

With a few important and spectacular exceptions, first and foremost related to China's construction of military bases in the South China Sea, these conflicts are not first and foremost about the small islands, islets, rocks, reefs and sandbanks. The strategic interests and large values lie in the ocean regions surrounding them. These countries' interests in barren, inhospitable and apparently wholly insignificant locations stem from the fact that the UN Convention on the Law of the Sea (UNCLOS) defines countries' rights based on their geography: the coastline's trajectory, the islands that are theirs and the nature of the underwater topography.[12]

In short, it is the 'dry' land of each country that defines the 'wet' rights they can claim. On this basis, the UNCLOS grants four main types of rights to coastal states.

The first is related to the so-called 'baseline', which is defined as the line that can be drawn between the outermost points of the coast along the mainland and around any islands. Within the baseline zone, often referred to as 'internal waters', the coastal state has full sovereignty. All the country's laws and regulations are in force here, so the Philippines, for example, can deny Chinese ships entry or passage closer to the shoreline.

Beyond this, there is an area of twelve nautical miles (approximately twenty-two kilometres) outside the baseline that is called 'territorial waters'.

This is also considered the coastal state's own territory, but here the state is bound to grant the vessels of other countries 'innocent passage'. This means that Norway, for example, must allow Russian naval vessels embarking from the Kola Peninsula passage close to the mainland as they head out into the North Atlantic. Based on practical considerations, an agreement was reached that a coastal state's territory includes the airspace above territorial waters. In this way, the UNCLOS defines the area of a country's sovereignty both at sea and in the air.

In several places in the world, the twelve-nautical-mile boundary extends into the territories of other states. There are therefore special rules for international straits, such as the Straits of Malacca and Gibraltar. In these cases, the surrounding states do not have the right to deny other countries' merchant or military vessels passage.

Outside territorial waters, in an area extending 200 nautical miles (approximately 370 kilometres) from the baseline, lies the 'exclusive economic zone'. Here the coastal state has sovereign rights on all the natural resources found underwater and on the seabed. The surface of this zone is nonetheless considered international waters. Other countries must have the permission of the coastal state to carry out fishing, crabbing, aquaculture, explorations for seabed minerals or oil drilling in these waters. But, as long as no attempt is made to acquire resources from these areas, other countries may lay cables and pipelines here, and ships and submarines have right of passage through coastal states' economic zones.

The definition of the exclusive economic zones provides uniquely favourable benefits for coastal states with little land. For example, Fiji has exclusive rights on an ocean region fifty times larger than its own land area – which can perhaps explain why Fiji was the first country to sign and ratify the UNCLOS in 1982. Jamaica can claim rights on an ocean region twenty-four times larger than the island itself, and Norway has jurisdiction in an ocean region six times larger than its land area. But no country beats Cook Island, the tiny island state in the Pacific Ocean, where the ocean represents 99.99 per cent of the area under the country's jurisdiction.

Finally, a coastal state has special rights related to its 'continental shelf', which is the natural submerged prolongation of the state's land territory to the outer edge of the continental margin. A map will often depict the ocean floor as sloping gradually downward to depths of a few hundred metres as we move away from the coastline, until the floor abruptly drops off to several thousand metres. In cases where the continental shelf extends beyond the exclusive economic zone, the coastal state itself must apply to the UN Commission on the Limits of the Continental Shelf (CLCS) to claim the

9. THE NEW BATTLE OVER THE OCEAN

rights. The seat of the commission is in New York City and is made up of twenty or so experts in the fields of geology, geophysics and hydrology, who are responsible for assessing the scientific and technical documentation a state submits in support of its claim. If the commission is convinced, it may determine that the coastal state's continental shelf extends up to 150 nautical miles (approximately 278 kilometres) beyond the exclusive economic zone.

On the continental shelf outside the economic zone, the coastal state has exclusive rights on all the resources on and beneath the ocean floor but not to the living resources in the water column, such as fish and krill. In these zones, the coastal state must nonetheless pay a fee to the United Nations if they should choose to harvest the seabed resources, and the income from this is distributed to a number of recipients, which include developing states.

The coastal states' exclusive economic zones today encompass about one third of the world's total oceanic regions. The remaining two thirds, which cover half of the earth's surface, shall, according to the UNCLOS be open and available to all countries. These regions are referred to as simply the High Seas, or Areas Beyond National Jurisdictions. Here all countries, also those that do not have coastal borders, such as Switzerland, Afghanistan, Rwanda and Bolivia, have the same rights and duties as the coastal states. Under the rules of the UNCLOS, the ocean is the public commons and humanity's common heritage.

Because both international tensions and the strategic significance of the ocean are on the rise, the conflicts surrounding overlapping claims to continental shelves and areas around islands, islets, rocks and reefs have in recent years proliferated in number and magnitude. The majority of the cases are brought before the United Nations, but not all the rulings are accepted or respected. China, for example, refuses to accept the 2016 ruling of the Permanent Court of Arbitration stipulating that the waters around Mischief Reef and Second Thomas Shoal belong to the Philippines' exclusive economic zone.[13] China therefore continues impassively constructing naval bases, exploring the seabed and exercising force in these waters, based on what they claim to be the country's 'historical rights'.

Management of these numerous and emerging conflicts at sea in this way also function as a general litmus test for the UN-led, rules-based world order. If the member states do not respect the rules regulating the ocean, an erosion of respect for the rules on land may inevitably follow.

We will return to these issues in Chapter 10 and elaborate on how the UNCLOS, and international law and order in general, is being challenged and eroded, and how other bodies of international cooperation are acquiring greater significance.

A Unique Island State

In the coming decades, it will be the balance of power between the USA and China that will define the most important terms for the global legal system in general and, more specifically, for the rules of play at sea. Since the early 2010s, successive US administrations have described China as the USA's greatest 'strategic competitor'.[14]

The relationship between the two countries is characterised by growing rivalries over economic power, technological domination and diplomatic influence. This relationship is already on the verge of sliding into another cold war, which in the worst-case scenario could culminate in a military confrontation. Nonetheless, as we will discuss later, in a number of important areas it will be in the enlightened self-interests of both countries to collaborate rather than challenge one another. This is the case first and foremost with an eye to avoiding conflicts that could spin out of control and lead to a mutually destructive nuclear war. Since the economies, supply chains and financial markets of the two countries are so tightly intertwined, both also have a great deal to lose from trade wars and technological disconnects. Finally, important common challenges, such as global warming, the loss of nature and biodiversity and the destruction of marine wildlife can only be resolved through international collaboration, in which the USA and China must play key roles.

Despite China's strong growth and the USA's many problems, the latter remains the dominant world superpower. The country still has greater economic, military and diplomatic power and influence than China. The USA accounts for more than one third of the world's total military spending and has a military budget larger than the next nine countries combined.[15] The USA also has significant strategic advantages in several areas where China is exposed and vulnerable.

Unlike China, the USA has sufficient resources within its own borders to supply its population with fresh water, food, energy and raw materials indefinitely.[16] Americans are therefore less dependent on other countries, and on maritime supply routes, to keep society up and running. The large ocean expanses surrounding the USA also make the country in many ways a large island state, not solely geographically and economically speaking, but also in a cultural sense. Non-reliance on other states and ideological differences with and geographic distances to 'the rest of the world' have historically contributed to upholding the US population's belief in the country's exceptionalism – the idea of its unique position, character and international responsibility.

9. THE NEW BATTLE OVER THE OCEAN

The ocean also gives the USA specific military advantages. The country has unhindered access to the Pacific Ocean in the west, the Atlantic Ocean in the east and the Caribbean Sea in the south. The coastline is almost twice as long as its land borders with Canada and Mexico, two neighbouring states that regardless do not pose any military threat.[17] American military planners therefore need not consider the strategic risk of the US Navy being landlocked. Like the island state of Great Britain, and unlike all other major powers throughout history, the USA has never experienced being invaded or ruled by others.

At the same time, the vast ocean regions represent some of the most important strategic risk factors for the USA. In particular, the risk of a surprise nuclear missile attack launched by submarines hiding in the depths of the Pacific or Atlantic Ocean ranks high on the list. But surface vessels can also constitute a significant threat. Modern warships are weapons platforms with immense capacity and firepower. They can be rapidly deployed to remote locations and navigate close to US territorial waters. The USA received an unpleasant reminder of this when for the first time Chinese and Russian warships made an appearance off the coast of Alaska in August 2023, in what was viewed as a carefully choreographed 'visit'.

This is why, as mentioned previously, the USA's overall military strategy is based on two doctrines related to the ocean: the first is the Monroe Doctrine's 200-year-old imperative regarding the importance of maintaining control of the region in and around the Caribbean Sea. This dictated President John F. Kennedy's willingness to risk an all-out nuclear war in order to keep the Soviet Union out of this region during the Cuban Missile Crisis in 1962 (which in Cuba is called 'Crisis de Octubre').

The second is the doctrine of 'forward-deployed defence', a corollary to the approach of the Monroe Doctrine, according to which the USA attempts to keep threats and conflicts as far as possible from its own shores and preferably on the adversary's side of the ocean.

US defence and military power are therefore first and foremost built around the US Navy. The doctrine entails that the USA must at all times have a large, long-range and powerful fleet and numerous supply and maintenance bases all over the world.

90,000 Tons of Diplomacy

Although China today has more surface vessels than the USA, it is still the latter who can mobilise the most military power at sea. US naval vessels are consistently larger and more advanced than the Chinese. The USA also has more nuclear-powered submarines than all the other countries in the world combined. With more than one hundred naval bases – owned, leased or borrowed in seventy countries throughout the world – the US fleet can always be deployed on short notice in regions experiencing conflicts or tensions.

The USA currently has eleven large nuclear-powered aircraft carriers. Though some of them are at any given time out of operation due to maintenance or retrofitting, most are deployed in strategically significant maritime regions. The total deck space of these carriers is more than double the space of all other aircraft carriers in the world combined.[18] With the exception of long-range nuclear missiles, it is only aircraft carrier groups that have the power and capacity required to project definitive military might in remote world regions. In the US Navy, aircraft carriers are often referred to as '90,000 tons of diplomacy'. That may sound impressive, but it is actually an understatement. The aircraft carriers are the USA's most heavy-weight and long-range combat groups. Each of these vessels has typically a crew of 5,000 and more than sixty fighter jets, surveillance planes and helicopters. An aircraft carrier never sails alone; it is usually escorted by one cruiser, several destroyers, frigates and supply vessels, and a number of submarines for protection.

Each of these carrier strike groups represents a formidable capacity, and in wars or conflicts will often serve as forward-deployed bases for operations on land. They are also actively used to send signals of military strength and political resolve, such as when US aircraft carriers regularly sail through the narrow strait separating Taiwan from mainland China.

For its own part, China has a large fleet of surface vessels, and currently two, soon to be three, operative aircraft carriers, as well as a small number of nuclear-powered submarines. China has one official foreign naval base, in Djibouti, by the entrance to the Red Sea, though Western analysts believe that two ports in Cambodia and Tajikistan are also serving this purpose.

But, even though China does not yet have the capacity to challenge the US Navy on a global scale, the country already has sufficient firepower to contend with the US fleet in and around its own region.

Never Again!

We have already addressed how Western countries attacked, occupied and dethroned the millennia-old Chinese Empire through military superiority at sea. The Century of Humiliation was a watershed era of historical trauma that left deep and permanent scars in China's national consciousness and strategic ideology. Never again will the country leave itself open to an attack

9. THE NEW BATTLE OVER THE OCEAN

from the sea. Never again will foreign powers be permitted to usurp control of the maritime regions surrounding China.

Since the birth of the People's Republic in 1949, military ambitions at sea have therefore expanded apace with the country's economic growth and rising prosperity. Jiang Shigong describes how the build-up of China's naval forces essentially mirrors the three main phases of the country's modern history: China 'woke up' under Mao Zedong, 'became rich' under Deng Xiaoping and 'became strong' under Xi Jinping.[19]

From 1950 to the 1980s, although the impoverished country had its hands full with the management of social and economic challenges at home, it built a 'brown-water fleet' that operated along the coast and inland on the large rivers. Under Mao the Chinese built many navy vessels, but they were simple, small and lightly armed. The objective was to secure control over their own territories and ward off attacks by foreign powers. In this period, the Chinese navy had an extremely limited range and only modest capacity for operations beyond its own coastal waters.

When economic growth began to take off during the 1990s and into the new millennium, the navy was upgraded into a 'green-water fleet' to ensure control and influence in the country's nearby ocean regions. China became a regional power with capacity for military engagement in the maritime areas extending from the Strait of Malacca in the south to Japan and Korea in the north.

In the early 2010s, China began to venture outside of its own region, first by participating in operations targeting the pirates who terrorised the waters off the coast of Somalia. At the time China invested heavily in the construction of a 'blue-water fleet' of long-range navy vessels. When in 2015 China for the first time released an official document about its military strategy, it highlighted the need for a strong defence of its own maritime regions. The document clearly stated that the Chinese navy would 'gradually shift' its emphasis on operations in maritime regions farther away from China.[20] Here it was acknowledged in black and white that the country's security and its geopolitical influence relied upon its ability to exercise military power in both nearby and more remote maritime regions. It later emerged that China's strategic planners now view it as most probable that any future military conflicts would take place within the maritime domain.[21]

Swift and decisive action followed the announcement of the strategy. In 2016, China launched its first aircraft carrier, *Liaoning*, symbolically the same year that the World Bank predicted that China was on the verge of eclipsing the USA as the world's largest economy measured in purchasing power. With *Liaoning*, the Chinese navy was for the first time in modern history able to project global military force. A mere few months later, the country's second aircraft carrier, *Shangdong*, was completed. In early May 2024, China launched its third aircraft carrier, *Fujian*, named after the province on the coastline facing Taiwan. The vessel is larger and more advanced than *Liaoning* and *Shandong* and has an electromagnetic catapult system with capability to launch fighter jets with a longer range and higher munition loads. For the time being, only one of the US aircraft carriers is equipped with corresponding systems, which means that, when *Fujian* is fully operational, subsequent to systems tests and trial voyages, China will in this area have advanced technology matching that of the USA.

But also other parts of the Chinese fleet, which goes under the official name the People's Liberation Army Navy (PLAN), have been built at an impressive pace. In the period 2014–2018 alone, the number of naval ships, submarines, amphibious assault ships and support vessels launched by China surpassed that of the combined fleets of Great Britain, Germany, Spain, India and Taiwan.[22]

Earlier in the book we have seen how control and disruption of supply chains can have widespread economic and strategic consequences. This holds no less true when it comes to the expansion of the world's fleets of merchant and military vessels. Today close to ninety per cent of global shipbuilding capacity is in South Korea, Japan and China, of which the latter accounts for almost half.[23] Also, Chinese yards are getting bigger, better and more efficient. They often outperform US and other Western shipyards not only on price but also on quality and delivery time.

The US Navy today has close to 400 combat vessels serviced by four major American shipyards, and the Chinese navy has 450 vessels serviced by some twenty-five domestic yards. While these American yards are working at full capacity to maintaining and expanding the US military fleet, it is estimated that less than ten per cent of the Chinese shipyard capacity goes to servicing the PLAN. As it would easily take a decade or so to expand yard capacity, building dry docks, installing cranes and machines and training people, this means that the USA has little or no capacity to ramp up production of combat vessels in the short term.

Should China, on the other hand, wish to double the size of its navy, it could probably do so in a handful of years.

The build-up of its naval fleet and shipyard capacity is not merely a reflection of China's economic growth and altered international position. A strong navy and a large domestic shipyard capacity is also a condition for advancing and securing the country's interests in an ocean that is becoming more important both strategically and economically.

9. THE NEW BATTLE OVER THE OCEAN

While the USA is largely self-sufficient, China relies on the import of large quantities of food, energy and strategic raw materials to keep society running. While home to a fifth of the global population, China has less than a tenth of the world's arable land and only six per cent of the world's total freshwater resources.[24] Former Chinese premier Wen Jiabao has warned that water shortages threaten 'the very survival of the Chinese nation'.[25] The Chinese economy is also far more dependent on export to the world market, in which almost everything is shipped by sea. While the foreign trade of goods constitutes only one fifth of the USA's gross domestic product, it makes up more than one third of China's.[26]

It is therefore not empty rhetoric when China's official defence strategy contends that 'the traditional mentality that land outweighs sea must be abandoned'. The statement is rather a reflection of the ocean's existential significance for the country's social and economic development, political stability and national security. For China it is becoming increasingly important to protect its own ocean regions, secure maritime trade routes, harvest marine resources and develop a sustainable 'blue economy'.

A Great Wall in Reverse

China's location and geography also give the country strategic parameters wholly different from those of the USA. China shares land borders with neighbouring countries in the north and west and borders the ocean in the south and east. The country's coastline, the home of seven of the world's ten busiest container ports, is somewhat shorter than USA's.[27] But, unlike the USA, China has no coastlines facing vast, open ocean waters. The Chinese military planners must therefore factor in the possibility of a blockaded navy in the event of an international conflict.

While Americans can look eastward to the vast blue expanses of the Atlantic Ocean, the Chinese have an eastward view of the Korean peninsula and a chain of islands – which are not their own. Furthest north lie the Russian Kuril Islands, and to the south are Japan, Taiwan, the Philippines, Brunei and Malaysia. With the exception of Russia, none of these countries are China's allies, and they all have close ties with the USA.

This chain of islands, sometimes referred to as the First Island Chain, constitutes in a strategic sense a 'Great Wall in reverse'.[28] In the event of a military conflict, it could become a barrier restricting China's access to the high seas and serve as a base for enemy attacks on mainland China. Further out at sea, we find the Second Island Chain, which roughly speaking extends from Japan to Guam, Micronesia, Palau and further south to New Guinea.[29]

The First and Second Island Chains represent China's greatest military weakness and source of concern. These island chains, and the ocean regions in between, are of equal, if not greater, strategic importance today as during the early nineteenth century, when they served as bases for the attacks that ushered in the Century of Humiliation. At that time, China was an introverted, poor, agrarian society, and the majority of the population lived in villages in the interior of the vast empire. Since then, a widespread migration to the coast has taken place, especially to the large cities. Today, two thirds of the country's population live by the sea. Also, China now has more than fifty cities with populations exceeding two million, and the majority of the largest are found on or near the coast, such as Guangzhou, Shanghai, Beijing, Hangzhou and Hong Kong.

It is also in the coastal belt that we find the engines of the country's economic growth: all the most important centres of industry, trade and finance are located here. Along this coastline, economic value creation per inhabitant is twice that of the rest of the country.[30] In the ocean off the coast are ships transporting more than ninety per cent of China's imported and exported goods. Here there are also abundant fishery resources and rich deposits of oil, gas and minerals. On the seabed are fibre-optic cables, the digital umbilical cord connecting China to the rest of the world.

While all of the islands of the First and Second Island Chains are of strategic importance, none of the others can compete with Taiwan in a military, political and historical sense. Taiwan is China's painful phantom limb.

A Sealed Fate?

With its beautiful, palm-tree-lined beach promenades, its reputable university and broad avenues flanked by classic townhouses, for many foreign visitors Xiamen may be somewhat reminiscent of Barcelona. This old 'treaty port' of Amoy, formerly used by several Western navies and companies during the Century of Humiliation, is also every bit as popular among the Chinese as the Spanish city is with Europeans.

I have been there several times on business, but on this particular day, I was enjoying a cappuccino at a small outdoor café, while taking in the bustling activity on the beach. There was little to suggest that the glittering waves in the background were rolling in from one of the most tension-fraught maritime regions in the world. There was nothing about the swimming children or their holiday-photo-snapping parents to suggest that the sea off the coast here is the site of intensifying rivalries between mainland China and Taiwan, and between the world's two major powers. There was little indication that

9. THE NEW BATTLE OVER THE OCEAN

the young couples strolling flirtatiously along the quay gave any thought to how the worst-case scenario of the events unfolding here could be a conflict between superpowers and, potentially, an all-out nuclear war.

From the beachfront café in Xiamen, I had a clear view of the tiny group of islands called Kinmen, which with its some 100,000 inhabitants is a Taiwanese outpost. The main island itself is located a short distance farther out to sea, separated from the Chinese mainland by a narrow strait – and an ocean of history, emotions and expectations. My friendly local host told me that, the day before, a US aircraft carrier had sailed through the Taiwan Strait, clearly visible from the mainland. It was a demonstration and warning of the USA's military power and presence.

In terms of military strategy, Taiwan holds the same position in Chinese thinking as Cuba holds for the USA: both are islands with 'alien' ideologies and political systems, they have 'hostile' allies, and they are located uncomfortably close to their own coastlines. But, in sharp contrast to the relationship between the USA and Cuba, Taiwan occupies a unique and prominent position in China's history and national psyche. The island was ceded from the Chinese Empire when it was occupied and annexed by Japan in 1895, and it remained in Japanese hands until the capitulation in 1945. When China was invited to take part in the establishment of the United Nations that same year, it was Chiang Kai-shek who took the seat at the table, as discussed earlier. He had at this point been the country's formal leader for almost two decades, but was heavily undermined by the ongoing brutal civil war between his own Western-supported nationalist forces and Mao's communist army of farmers. Eventually, Mao's army prevailed on the mainland, and in 1949 Chiang and his loyal supporters fled across the narrow strait to continue fighting from Taiwan.

Since this time, the governments in both Beijing and Taipei have claimed to be the sole legitimate representative of the Chinese people. This is reflected by their formal names: while the official name of mainland China is the People's Republic of China, Taiwan's official name is the Republic of China.

Taiwan's position in the dark and menacing shadow of powerful mainland China is nonetheless full of striking paradoxes. Although only a few countries formally recognise Taiwan, almost every country in the world accepts the island's passport for international travel. And, even though China does not accept that any state have diplomatic ties with Taiwan, Beijing has silently accepted that almost every country trades with the island. China, like the rest of the world, is almost wholly dependent upon Taiwan Semiconductor Manufacturing Company (TSMC), the dominant company of the island's highly specialised and internationally leading community in the production of advanced semiconductors. In the foreseeable future, neither China nor any other country will be in a position to manage without these components, which are used in all modern electronic equipment. But Taiwan is also heavily dependent upon economic ties with China. China is Taiwan's most important trade partner, and a number of Taiwanese companies have established factories on the Chinese mainland, where wage levels are lower. However, recent years have seen a number of Taiwanese companies leaving China and a decline in the share of Taiwan's exports that go to the mainland.[31]

Already in 2005 the Anti-Secession Law was passed as a domestic law in China, which states that the authorities are bound to employ all necessary measures, also 'non-peaceful', in the event Taiwan were to declare independence.[32] But, since President Xi came to power in 2013, the diplomatic, economic and military pressure has increased. For all practical purposes, Xi has promised his people that Taiwan will be 'reunited' with the mainland. The reunification is one of his highest political priorities and an integral element of his 'great rejuvenation of the Chinese nation'.[33] He declared in a speech made in November 2016 that 'China's 1.3 billion people could not accept independence for Taiwan' and that his own regime 'will be overthrown' if he fails to resolve this.[34] These vows were repeated and underscored during the National Congress of the Chinese Communist Party in October 2022.

For the Chinese president, Taiwan is a *casus belli*. Should Taipei declare independence, this would constitute a justification for war.

A Deliberate Uncertainty

Due to the enduring threat from the mainland, Taiwan has over the years been transformed into a virtual fortress. On the shores along the western side of the island, facing the Chinese mainland, miles of barbed wired fences, minefields, tank chicanes and bunkers have been constructed as a defence against invading troops. Taiwan also has a missile defence system, air force and navy, the size of which, in an international context, is hugely disproportionate to the country's population and economy. Yet the chances are minuscule that the country's less than twenty-four million inhabitants will manage to hold their own in the face of a large-scale Chinese attack without military assistance from the USA, for all intents and purposes their only ally.

The USA has not had bases in Taiwan since 1979, when they switched sides and established diplomatic ties with China. But the USA was not willing, then as now, to risk putting all of its eggs in one basket. The US Congress therefore simultaneously approved the Taiwan Relations Act, as we discussed in Chapter 5. Since then the USA has supplied Taiwan with huge quantities of arms, and in recent years also increased its military presence on and around the island.

The USA has, however, no formal agreement stipulating an obligation to defend Taiwan. It therefore caused a stir when President Biden on several occasions stated that the USA would defend Taiwan in the event that China attacked the island, as if it were a clear policy.[35] Other members of his administration have, in the aftermath of these remarks, tried to explain with varying degrees of success that the US policy has not changed following Biden's remarks. Since 1979, the US policy has been to act with what they call 'strategic ambiguity', which means that the USA demonstrates a political willingness and military capacity to defend Taiwan but without obligations that are carved in stone. This strategic ambiguity is understood as a deliberate attempt to create uncertainty in both Beijing and Taipei about whether the USA would intervene in a war. This serves the American's goal of dual deterrence: the threat of US intervention prevents China from invading, and the fear of US abandonment prevents Taiwan from sparking a war by declaring independence.

There are nonetheless many who would hold that the credibility of the US strategy of ambiguity is diminishing in step with the shifts in the military and diplomatic power balance between the USA and China. China's desire to gain control over Taiwan is not solely politically and historically motivated, but also militarily. By conquering Taiwan, the Chinese can assume the role of gatekeeper at the entrance to the Philippine Sea and gain better control over the fishery resources, seabed minerals, shipping lanes and the subsea cables in this maritime region. They would also be able to move their seabed listening posts farther into the Pacific Ocean, and thereby secure a longer warning period in the event of approaching submarines. This would force the USA to move its fleets farther east in the Pacific Ocean in the event of a military confrontation.

Because of China's economic and military growth, Taiwan's inhabitants live with enduring and growing uncertainty about their own future. The geopolitical and economic forces of gravity seem to be pulling Taiwan slowly but surely into China's solid grip. On the shores of Xiamen, the carefree laughter is falling silent and more Taiwanese citizens go to bed at night with growing sense of uneasiness about what tomorrow will bring.

An Honest Mistake

With the exception of Taiwan, there are no territorial issues that hold greater importance for the Chinese leadership than the South China

Sea. The conflicts here can be traced all the way back to the Century of Humiliation, and later to China's first participation in international talks after it replaced Taiwan as a member of the United Nations in 1971. The subject of these talks was specifically the UNCLOS. China ratified the agreement in 1996 and has since been bound by this international regulatory system. It is therefore reasonable to expect that the UNCLOS holds a particularly prominent position in China's awareness of its obligations under the international legal system.

This is indeed the case, but not in the sense one might expect.

During the negotiations on the UNCLOS, the most significant lines of conflict ran between the Western industrialised countries and the developing parts of the Global South. To secure their own factory trawlers and fishing vessels access to the coastal regions of Africa and South America, the industrialised countries wanted to restrict the size of the exclusive economic zones. For exactly the same reason, the developing countries were in favour of broad zones to retain control over fish stock and other valuable resources in the areas off their own coastlines. As it turned out, China would prove to be the country to tip the scales.

Further information has recently emerged about China's position in these talks. Ling Quing, the head of China's negotiating delegation and permanent representative to the United Nations, writes in his memoirs of how Mao, whose hands were full at this time with the domestic challenges of his own country, gave the Chinese delegation three guidelines: 'Be anti-hegemony' (in other words, anti-USA and anti-Soviet Union), 'support the Third World' (the developing countries) and 'protect national interests' (the interests of China). The delegation was also instructed to 'give something in return' out of gratitude to the developing countries for having secured a majority enabling China to take Taiwan's place in the United Nations as the sole recognised representative of the Chinese people.[36] Mao used to say that 'it is our African brothers who have carried us into the United Nations'.[37]

China therefore formed an alliance with the African and South American countries in support of ratification of 200 nautical miles as the boundary for the exclusive economic zones, as it stands today in the UNCLOS.

In retrospect, it might seem as if Mao's mandate was skewed more towards political ideology and international alliances than towards China's own national interests. Such an obvious setting aside of pragmatic self-interest in favour of a gesture of grateful recompensation has neither before nor after been characteristic of Chinese politics. It could also appear as if the Chinese had not thoroughly thought through all the consequences of their negotiating position.

9. THE NEW BATTLE OVER THE OCEAN

There is every reason to believe that China today views this as having been an honest, but costly mistake in need of rectification.

Monroe in Mandarin

It is precisely the 200-mile zone that is creating the greatest problems for China in their own territorial maritime regions and that lies behind the highly charged conflicts in the South China Sea. The issue is purely topographical: while the coastlines of Africa and South America offer unfettered access to the ocean, the South China Sea is surrounded by islands and territorial states. When the countries around the South China Sea measure out their 200-mile borders, towards the middle of this 'doughnut', several of the economic zones overlap and come into conflict with one another. The Chinese maintain that this principle leads to a division of these areas in the South China Sea which violates 'historic rights'.[38]

China therefore claims 'historic rights' to the ocean region within the 'nine-dash line', a U-shaped line on the map which China has drawn with a liberal hand, extending from the coast of its own mainland and around nine (originally eleven) select coral reefs and sandbanks in the South China Sea. What these reefs and sandbanks have in common is that they are located far from the Chinese mainland, are uninhabited and tiny – some barely visible at high tide – and, according to the UNCLOS, belong to other countries.

Nonetheless, China claims ownership of the entire maritime region within this line and full rights to the territorial waters, exclusive economic zones and continental shelves around all of these 'nine dashes'. In addition to the region's important strategic location, more abundant resources are likely to be found here, in the way of marine life, oil, gas and minerals, than in any other ocean region off the coast of China.

The claim encompasses almost eighty per cent of the entire South China Sea and is predominantly based on tradition, historical evidence and 'acquisitive prescription', a legal term for the right to ownership acquired through possession over a longer period of time. The claim is meeting, to put it mildly, with little sympathy from China's neighbours and the international community. The government in Beijing does not, however, appear bothered by the loud, angry protests voiced by the Philippines, Vietnam, Malaysia and Brunei, all of whom state that China is helping itself to islands and ocean regions that, according to the UNCLOS and decisions by the Permanent Court of Arbitration, belong to them.

China's conduct in this region cannot be described as subtle or diplomatic. On the contrary, China is employing a determined, heavy-handed approach that has serious ramifications not solely for international relations, but also for the environment, marine ecosystems and biodiversity.[39] The small sandbanks and coral reefs along the nine-dash line have been swiftly demolished and converted into ports and bases for missiles, fighter jets and navy vessels. In this way, China is not solely altering the facts on the ground; they are altering the ground itself. China is in the process of establishing a 'bastion defence system' to secure control and deny potential adversaries access to what they hold to be their own maritime region. China has already installed missiles on these bases with capacity to attack hostile marine vessels and carrier strike groups in the event of a military confrontation. US strategic planners have long since nicknamed these missiles 'carrier killers'.

China's neighbours in the South China Sea are not simply sitting on their hands, passively observing this development. They have all commenced an expeditious upgrade of their naval capacity and coastguards. The US Navy has stepped up the number of what they call Freedom of Navigation Operations (FONOP), and has reopened its naval base in Subic Bay in the Philippines. Countries such as France, Great Britain and Australia now more often also send naval vessels into the region as a demonstration of military strength and presence.

Without referencing the South China Sea by name, few international observers have any doubt about what President Xi Jinping is alluding to when he states that 'we cannot lose even one inch of the territory left behind by our ancestors'.[40] But the South China Sea is not solely a disputed ocean region, it is also China's 'exposed underbelly', of key significance to the country's territorial defence and the security of critical supply lines. The Chinese are therefore about to carry out what amounts to a *de facto* annexation of a maritime region that is larger than the Mediterranean Sea.

The approach is for all practical purposes a 'Monroe in Mandarin': a revised, Chinese version of the US doctrine that for two hundred years has been one of the main pillars of the US defence strategy.

The Tyranny of Distance

The tensions in the South China Sea could, in the worst case, culminate in a direct military confrontation between China and the USA. It is not inconceivable that China would emerge triumphant if such a duel were to take place. When the commander of the US fleet in Southeast Asia testified before Congress in the autumn of 2018, he stated that 'China is now capable of controlling the South China Sea in all scenarios short of war with the United States'.[41] Six years later, Mark Milley, former US chairman

9. THE NEW BATTLE OVER THE OCEAN

of the Joint Chiefs of Staff, wrote that 'Chinese hypersonic missiles could sink US aircraft carriers before they make it out of Pearl Harbor. Beijing is already deploying AI-powered surveillance and electronic warfare systems that could give it a defensive advantage over the United States in the entire Indo-Pacific'.[42]

Also, while the Chinese can stroll down the shores of the South China Sea, the USA must contend with what military planners often refer to as the 'tyranny of distance'. The distance from this tension-riddled maritime region to the US coast is equivalent to almost one third of the distance around the earth. The USA is therefore wholly reliant on having bases in this part of the world where they can refuel, carry out maintenance, effectuate crew changes and uphold combat readiness. In the Indo-Pacific, the US fleet today operates from bases in Japan, Singapore, Guam and Bahrain, and from Camp Justice on the British-leased coral island Diego Garcia in the Indian Ocean. Three of the eleven US carrier strike groups are usually out on missions in the Southeast Asian waters.

Its impressive firepower and advanced defence systems notwithstanding, the US carrier strike groups would be vulnerable in a military confrontation in the South China Sea. It is strategically challenging, tactically risky and logistically complex for any country to engage in combat far from its own shores. The difficulties are compounded in a maritime region where China has not only large naval presence, but also has a militia of tens of thousands of fishing boats and 'little blue men' that can be or are already armed.[43] China has also invested considerable resources in building bases and defence fortifications on the mainland, islands and coral reefs. In this 'theatre of operations', China enjoys military advantages stemming from shorter supply lines, simpler logistics, more soldiers on the ground, greater dispersal of force and greater latitude in choosing where and how they will engage in battle. It is also always more difficult to defeat an enemy who is entrenched. Finally, as a US naval officer laconically remarked when we were discussing the subject, 'you cannot sink a reef'.

While China is both aided and trapped by geography – they have no choice but to remain where they are – the USA must at all times actively choose its military presence and priorities in the South China Sea and other parts of the world. Given the challenging political, economic and social problems the USA is contending with on the home front, it is difficult to imagine that it will in the long term prioritise allocation of resources on a scale that would offset China's rapidly growing military force in the Indo-Pacific region. This would entail even tougher budgetary decisions than those the USA is currently facing, since the cost increase of naval vessels, fighter jets, missiles, weaponry, radars and other defence materials tends to exceed that of other goods and services. The interaction between the tyranny of distance and the dictates of financial obligations is growing decidedly more intense.

The USA's strategic dilemma is not only a military and national affair. If the USA withdraws from the South China Sea this would amount to a surrender of its role as the guarantor and enforcer of the UNCLOS and the UN-led, rules-based world order that it spearheaded after the Second World War.

This would mark the end of Pax Americana of our times and 500 years of Western global dominance. At the same time, it would inaugurate a new chapter in the history of the world. With a global distribution of power in which China, again, dominates its own region, more than half of the world population will be living under what will effectively amount to a Pax Sinica.

A Gold-Decorated Rolls-Royce

When I travelled in the autumn of 2019 to the idyllic paradise island of Koh Rong off the southern coast of Cambodia for a few days of sun, beach and diving, the traffic on the hot, shimmering and spotty asphalted road from the airport to the coastal city of Sihanoukville resembled a slowly flowing, organic mass of honking cars, buzzing mopeds and diesel-choking trucks. It was manoeuvred with great creativity, zigzagging often on the very edge of the roadway and sometimes on the opposite side to avoid large and small potholes, daring pedestrians, and unfortunate fellow travellers with engine failures. On both sides of the road, there was noisy and dusty construction activity. The roadside was sprinkled with simple stalls and small umbrella tables selling food and drinks, toys, cigarettes, plastic sandals, kitchen utensils, mobile phones, and other electronic gadgets. Parked between the stalls and tables were hundreds of 'clothing shops', bicycles with a colourful assortment of goods hanging off long poles.

The hectic activity is part of an 'Indo-Chinese Corridor', which branches out from the city of Kunming in South China into Vietnam, Thailand, Laos, and Cambodia. Cambodia is one of Southeast Asia's poorest countries, and the authoritarian rulers have welcomed Chinese capital with open arms.

Once I arrived in Sihanoukville, I had to wait for a few hours at the port before the hotel's boat was to take me further out to the island itself. In a modest, low building in the shadow of the large warehouses on the dock, I was welcomed by a smiling young Khmer woman who served tea, cookies, and encouraging updates about the expected arrival of the boat. It was just her and me in the small waiting room, and, as the hours went by, the conversation became more personal and trusting. She told me about herself

9. THE NEW BATTLE OVER THE OCEAN

and her upbringing, and how she earns far better in her hotel job than her husband, who has followed in his father's footsteps as a local fisherman. After years of saving, they had recently bought a plot of land to realise the dream of their own house. But the plot was located in an area hours away from the city, family and her workplace. 'We can't afford to buy anything closer,' she told me, 'because the prices here have risen much faster than we can manage to save.' Now they were also uncertain whether they could manage to save enough money to build the new house. The young couple has no chance of competing with Chinese capital, which in just a few years has changed both the city and their future.

Since they got married, Sihanoukville and the surrounding area have been completely transformed. A large new international airport is under construction and the existing one is being expanded and upgraded so that Sihanoukville can become 'a new Southeast Asian traffic hub like Kuala Lumpur, Singapore, and Bangkok', as boasted in the glossy presentation from the Chinese-controlled development company. A four-lane highway to the capital, Phnom Penh, which is two hundred kilometres further north, is also being built. Meanwhile, construction is underway for a number of skyscrapers in and around the city, which currently has a population of far less than 200,000 people. Chinese road signs and billboards are beginning to replace signs in Khmer and English. Everything is happening under Chinese auspices, with hotels, shopping malls, and casinos popping up in large numbers. The casinos are accessible only to Chinese and other foreigners since such gambling is prohibited for local residents.

'They don't just come with money; they bring workers, machines, and materials, and they buy up all the shops, homes, large plots, and small pieces of land. This pushes prices up and the locals out,' said the manager of a local business I met later, explaining: 'Rent has multiplied, and with a wage level equivalent to 100–150 US dollars a month, the local population can no longer afford to live in their own city. When the Chinese first came, they used local workers for some tasks, but now they do almost everything themselves. The Chinese practically run their own economic and social ecosystem.' Maybe it's just as well, according to another local guy I met. He was, to put it mildly, not very enthusiastic about what is happening and had no desire to work again for the Chinese developers. 'They treat us like slaves,' he softly confided to me.

The gigantic development, which is the largest in Cambodia's history, is taking place effectively under full Chinese control. The key to it all is the Dara Sakor Investment Zone, which encompasses one fifth of Cambodia's total coastline. Here, the Chinese front company has a leasing agreement with a duration of ninety-nine years, which practically gives them nearly unlimited rights over the area.[44]

Few believe that this is primarily about casinos, high rises and shopping centres. It is the port, the airport, and the transport corridor northward through Cambodia and Laos that hold strategic interest for China. In a military conflict situation, the two airports in the city could be used as bases for large Chinese aircraft. Sihanoukville also has Cambodia's only deep-water port, and, even before it is fully developed, it can accommodate submarines and large Chinese naval vessels. From here, the Chinese fleet can effectively operate throughout the Gulf of Thailand and close to the coastlines of Thailand, Malaysia and Vietnam. The sailing distance from the port in Sihanoukville to Ho Chi Minh City, formerly Saigon, in Vietnam is just under 500 kilometres, and it is less than 1,000 kilometres across the bay to Bangkok. From here, it is also a short distance to the Strait of Malacca and the eastern parts of the South China Sea.

With the strategic foothold in Sihanoukville and the supply routes through Laos and Cambodia, China has also laid a 'strategic loop' around Vietnam, just as they have done around India as we shall see later. For the Vietnamese, this is very uncomfortable, as not only do they have the Chinese as neighbours to the north, but they also face them with an 'economic corridor' to the west, with strategic bases to the south and large naval forces off the coasts to both the south and east. This is unlikely to strengthen the Vietnamese motivation for an armed conflict with China over the disputed reefs and sandbanks off their own coast. However, the Americans are also sufficiently concerned about what is happening around Sihanoukville that they in 2020 imposed sanctions against the Chinese company behind most of the development. The formal justification was that the development has caused massive and irreparable damage to coastal areas and marine ecosystems. This is likely entirely correct, but the real justification is probably much closer to what the Chinese themselves claim: that the US intensely dislikes China's increasing presence in this area.[45]

There is little reason to believe that all the details in the agreements between Cambodia and China are publicly known. Here, as in many other places, there are claims that large sums of Chinese money end up lining the pockets of powerful intermediaries and corrupt ministers. Perhaps a little hint about the answer lies in the almost Hollywood-like scene that unfolded on my way back to the airport a week later: suddenly, in the opposite direction, appeared a shiny, light blue, and richly gold-decorated Rolls-Royce surrounded by motorcycles on the sides and escort vehicles in front and behind. The blaring, intimidating convoy pushed its way through the dense, dusty chaos of cars, pedestrians and bicycles. It was a brief glimpse of vulgar wealth and power in shameless display.

Shortly before, the smiling young Khmer woman had waved me off. She is not among those who will reap large sums of money from the Indo-Chinese corridor, nor will she likely ever drive in a shiny gold-decorated Rolls-Royce. But China's large investments, which is part of a massive programme to 'promote peace, cooperation and trade from China's east coast to Europe's west coast' will nonetheless shape her daily life – and mould her dreams, hopes, and concerns for the future.

Malacca Dilemma

Even if China were to succeed in acquiring greater control over the South China Sea and the Gulf of Thailand, the country's economy would still be vulnerable to the problems of strategic bottlenecks in international trade. One of these is the narrow and heavily trafficked Strait of Malacca.

More than one fifth of world trade is transported through the 900-kilometre-long strait, which at its narrowest point has a width of less than three kilometres.[46] Around 100,000 vessels travel through the strait annually – five times as many as through the Suez Canal – making it the world's most heavily trafficked sea route. This number is expected to increase significantly in the years ahead due to growth in Southeast and East Asian economies.[47] Almost all trade between Europe in the west and China, Taiwan, Vietnam, the Philippines, Indonesia, Japan and South Korea in the east passes through here. Approximately four fifths of China's exported goods, and an equally large share of its oil and gas import, is transported by ship through this strait.[48]

The Strait of Malacca has historically been one of the most infamous stretches of water in the world, a notorious site of brutal attacks, hijackings and pirate raids. The tsunami in 2004, however, not only killed more than 200,000 people, it also wiped out important infrastructure and bases along the coast used to launch these attacks. To further improve security, there has been a pronounced increase in the number of coordinated operations headed by the three countries surrounding the Strait of Malacca: Indonesia, Malaysia and Singapore. This has helped mitigate the risk of attacks, which in recent years have been few and far between. Yet many seafarers still feel a sense of unease when sailing through the long, narrow strait.

But seafarers must be on the alert for dangers other than pirates in this strait. Apart from large merchant vessels, the heavy traffic of local ferries, small cargo ships, traditional junks and colourful longtail-boats shooting back and forth across the strait produce dangerous situations every day. The risk of serious accidents is also heightened when visibility is diminished to less than a ship's length due to heavy smoke from the extensive controlled jungle fires in Sumatra, started by peasants and plantation owners to clear new tracts of land.

For all East and Southeast Asian countries, the Strait of Malacca presents persistent concerns related to both trade and supply security, and for China the strait is the source of a chronic strategic headache.

In the Strait of Malacca, the waves from the South China Sea and the eastern parts of the Indian Ocean meet. The strait also forms the border between the territories of the Chinese and Indian naval fleets, two countries that combined make up one third of the world's population. India is in all likelihood the only country which at some point down the road may be in a position to challenge China's regional hegemony. China today outshines India economically, militarily and in terms of geopolitical influence. But India is experiencing consistently high economic growth and, while the population of China is decreasing, India's continues to grow. India has already passed China as the world's most populous country and in a few decades may also boast a larger economy.[49] This makes India China's largest regional rival. There are already tensions between the two countries which at times have even culminated in violent confrontations and soldiers killed in the border regions in the Himalayas.

China knows that, for India, the USA and a number of other states, blocking this critically significant strait would be a simple task, should it prove necessary due to the escalation of a conflict with China.

Living with the strategic vulnerability related to this narrow and chaotically trafficked bottleneck, which President Hu Jintao in 2003 referred to as their 'Malacca dilemma', is an ongoing source of strain for China.[50] Since then, the dilemma has been exacerbated in step with the growth of the Chinese economy, the increasing reliance on seaborne trade and the country's expanding geopolitical power. Simultaneously, the leaders in Beijing know that, if they attempted to seize military control over the Strait of Malacca, the potential for an international large-scale conflict would be even greater than if they were to attack Taiwan.

A Strategic Loop

China has therefore formed agreements with Myanmar, Bangladesh and Pakistan on land-based 'economic corridors', like the one we just discussed in Cambodia, to secure alternative transport routes to the global market.

9. THE NEW BATTLE OVER THE OCEAN

Under these agreements, China provides loans for the building of roads, railways, dams, power plants, cables, water supply, pipelines and ports. Construction is predominantly carried out by a Chinese labour force. China supplies project managers, labourers, machinery and materials. In return, China has been granted rights on transport through the economic corridors and access to ports in the Indian Ocean.

China is already able to ship oil, gas, coal, iron ore, food and other critical imported goods on roads and railways through these corridors from the Indian Ocean and northward to its own territory. From there, the goods are forwarded to the important industrial regions and populous cities further east in China. And vice versa: China can dispatch own export goods headed for the global market from ports on both sides of the Indian subcontinent.

In this way, China has not only secured alternative routes to the global market and alleviated, if not resolved, its Malacca Dilemma. It has also created a loop of strategic footholds around India, its largest regional rival.

Dystopian Hellscape

The economic corridor through Pakistan runs from the port city of Gwadar on the Indian Ocean and into Xinjiang, China's largest and westernmost province. In this part of China, the Turkic ethnic group the Uyghur make up almost half the population. This ethnic group has for many years been subjected to oppression, persecution and harassment on the part of the authorities, and militant Uyghur separatists have orchestrated a number of attacks and terrorist operations.

This situation has gone from bad to worse since the province became a transit region for the important transport corridor from the Indian Ocean. The central authorities in Beijing have an extremely sensitive radar for any sign of separatism and potential threats to national unity – and to the flow of goods through the region. The security regime has therefore been significantly tightened over the years. A number of international organisations have accused the Chinese government of implementing policies designed to eradicate the Uyghur culture, language, way of life, religion and identity. A number of independent sources estimate that, as of today, at least one million Uyghurs are being detained in so-called re-education camps. There are reports of mass surveillance and systematic violations, because of which an entire ethnic group lives in fear for their lives and future. Amnesty International has accused the Chinese authorities of having 'created a dystopian hellscape on a staggering scale'.[51]

Xinjiang has no coastline, and large parts of the sparsely populated province consist of dry plains, arid valleys and high mountains. This is also where we find the point in the world located farthest from the ocean. The distance from the Dzoosotoyn Elisen Desert to the nearest coast is more than 2,600 kilometres.[52]

Yet both the province's future and the Uyghur people's tragic fate are inextricably tied to the ocean and China's 'Malacca Dilemma'.

New Silk Roads

The economic corridors extending through Cambodia and surrounding India form the pillars of the large-scale, prestige project the Belt and Road Initiative (BRI). President Xi Jinping launched the project in the summer of 2013 as an investment in the country's own economic growth and to promote a 'vision of a global community of a shared future'.[53]

The BRI is often referred to as one of the largest and most ambitious infrastructure projects in the history of the world, through which China is leveraging a broad range of its economic, industrial and diplomatic resources for strategic purposes. The official ambition is to build a trade network and infrastructure connecting China to Europe and Africa, based on the template of the old Silk Roads. On the ten-year anniversary in the autumn of 2023, 150 countries had signed partnership agreements under the BRI.[54] *The Belt* currently includes six land-based Eurasian economic corridors, predominantly roads and railways, all of which meet in the north-western Xinjiang province. *The Road* is a network of ports extending from the south-eastern Fujian province and along the important sea routes in the South China Sea, the Indian Ocean and the Mediterranean Sea.[55] As we shall see later, in 2017 a third leg was also added, *The Ice Silk Road*, a development of infrastructure along the sea route between Asia and Europe through the Arctic.

The BRI currently passes through more than seventy countries, representing one third of the world's gross domestic product and two thirds of the world population.[56] The majority of the investments are in Southeast Asian, Central Asian and East African countries. Many of these countries are poor, under-developed or facing difficult financial challenges. For these countries, the Chinese investments provide much welcome economic stimuli and opportunities to build important infrastructure which they could otherwise not afford.

The projects are financed through loan agreements that are often shrouded in secrecy. Often the loan amounts, interest rates, payment and maturity terms are wholly or partially unknown even to the parliaments of the receiving states. Still, with time it has come to light that the BRI loans have some

9. THE NEW BATTLE OVER THE OCEAN

typical characteristics: the size of the loans are often disproportionately large in relation to the receiving country's financial capability, and the interest rates are typically below normal market rates. In the aftermath of the formation of several of these agreements, a number of complaints have emerged about bribes and corruption. In some countries, criminal charges have been brought against government ministers and high-ranking officials.

The agreements secure right of use and special privileges for China and often contain clauses giving China exclusive rights on takeover of the infrastructure for fifty to one hundred years if a country defaults on their loan. These countries are thereby at risk of being pulled into an economic and political relation of dependency, which can subsequently force them into a strategic partnership with China. In the West there are therefore a number of critics who refer to the BRI as 'debt trap diplomacy' and a 'Trojan horse' enabling China to gain access to the recipient countries' decision-making processes and to influence their strategic priorities.[57] Several analysts have also expressed concern about the actual debt to China potentially being far greater than the amount listed in the official budgets of the recipient countries, and that mounting debt-related strain can lead to political instability and social unrest in some of these countries.[58]

The BRI project has become one of China's most important instruments for enhancing its international standing and expanding its global presence and influence. A number of analysts and commentators are, however, asking increasingly critical questions about this massive, complex project. Some believe that the different parts of the project are not as holistically managed and strategically coordinated as the government in Beijing claims.[59] Others hold that the air is about to go out of the BRI balloon and that the ambitions are being downscaled in keeping with China's weaker economic growth. Still others hold that the Chinese economy is so reliant on the infrastructure under construction through the BRI that the different projects are in themselves at least as important as the overall strategic considerations.

Regardless, the Chinese leadership continues to announce ever more ambitious international plans. During the UN General Assembly of 2021 President Xi launched the 'Global Development Initiative', through which he promised that China would contribute sizeable resources in support of the UN Sustainable Development Goals and to ensure 'stronger, greener and healthier global development'.[60] The declared objective of this initiative is to bolster the vision of 'a global community of a shared future', especially by way of contributions towards sustainable development in the Global South. Less than two years later, President Xi launched the Global Security Initiative and Global Civilization Initiative. The objective of the first is an international security collaboration based on 'non-intervention' in other countries' internal affairs, while the purpose of the second is to 'advocate respect for the diversity of civilizations'.

Many interpret the latter as a rejection of what China perceives as the West's definition of 'universal values' and an attempt to establish a world order dominated by China's ideology and interests.[61]

A String of Pearls

Half the world's container traffic and two thirds of the oil transported by sea pass through maritime routes in the Indian Ocean. The ports surrounding this large, 'semi-closed' ocean area constitute the most important regional hubs in this part of the world.[62] The routes through the Indian Ocean also connect Asia to Europe, Africa and the Middle East. The trade routes passing through this maritime region bring vital supplies to China and export goods to the world market from the country's gigantic industrial machine. Management of the country's 'Malacca Dilemma' is necessary, but not sufficient, to secure China's major supply lines.

China has therefore worked systematically towards gaining access to strategic ports in and around the Indian Ocean. By the mid-2000s, the West began referring to this network of ports as the Chinese 'String of Pearls'.[63] The work on this endeavour took off in a big way after the launch of the BRI in 2013. Through funding from the latter project, China has procured preferential rights of access or options on the long-term lease of key hubs, such as the gigantic port of Hambantota in Sri Lanka on the western side of the Strait of Malacca, the large Pakistani port of Gwadar by the entrance to the Strait of Hormuz, and Chabahar, Iran's only deepwater seaport.

At the western end of the Indian Ocean, China has formed agreements with Kenya on the construction of a railway from the port in Mombasa to the capital, Nairobi. The railway replaces the former railway built by the British Empire at the end of the nineteenth century. At that time, the British wanted to improve the supply lines to another one of its colonies, Uganda, and simultaneously establish a foothold to keep out the Germans and the French. But the railway did not solely give Uganda access to the coast; it also served to connect different tribes and groups on the large plains in between. Kenya was thereby born as the legitimate child of a stretch of railway: 'It is not uncommon for a country to create a railway, but it is uncommon for a railway to create a country,' the British colonial administrator Sir Charles Norton Edgecumbe Eliot wrote.[64]

9. THE NEW BATTLE OVER THE OCEAN

One hundred years later, it is another major global power, China, that is playing the lead role. But, while the British built the railway at their own expense, the Chinese are financing the construction through money they have lent to Kenya. China is Kenya's largest creditor, and the latter is already struggling with growing financial problems.[65] This has provoked political unrest and speculations about whether China will seize control of the port in Mombasa should Kenya default on its debt.

Further north on the African east coast, in Djibouti, lies China's only official foreign naval base. The base is strategically positioned on the western side of Bab El-Mandeb, the strait separating the Gulf of Aden and the Red Sea by the southern entrance to the Suez Canal. Djibouti, with less than one million inhabitants, is also the home of naval bases for the USA, Great Britain, France, Germany, Spain, Italy and Saudi Arabia. China's base is located just a few kilometres from the others, by the commercial port of Doraleh, which the China Merchant Port Holdings Company operates for the Djibouti authorities.

The construction of China's base was completed in record-breaking time and to the great surprise of several of the other countries with bases in this region. Only a few years passed from the time of the opening talks between the Chinese and Djibouti governments until the base was officially opened in the summer of 2017. The USA has invested a great deal of money and political capital in keeping Russia out of the region and they were apparently long unaware that China was simultaneously underway with the construction of a large naval base on the neighbouring site. Now the USA is concerned about the strategic implications of the Chinese presence. During a hearing in Congress, the head of the US forces in Africa, General Stephen Townsend, stated that in the spring of 2021 the Chinese had completed a quay facility with capacity to host aircraft carriers and that further expansions were planned. He claimed that the Chinese base in Djibouti can be a 'platform to project power across the [African] continent and its waters'.[66]

What the general neglected to mention, however, was that, formally speaking, this is China's sole overseas naval base, while the USA has thirty or more military and naval bases in Africa alone.[67]

Strategic Speculations

The US general's speculations illustrate how military planners always attempt to analyse a potential adversary on the basis of two central components: intention and capability. For the USA and other countries, it is almost impossible to verify China's actual intentions, which can of course also change with time and according to the international situation.

China's 'String of Pearls' – their foothold in ports in and around the Indian Ocean – is an example of this. China does not, like the USA, publicise its strategic intentions or military capability in this region. While the US naval bases are surrounded by tall fences, barbed wire and armed guards, China has taken possession and operative control of a growing number of commercial ports without announcing that these are 'naval bases'. Still, ports which today service container ships, cruise boats and trawlers can serve as naval bases for aircraft carriers, destroyers and nuclear submarines tomorrow.

Precisely this duality of civilian and military use, combined with how tightly the interests of Chinese private companies are interwoven with those of the Chinese government, generates uncertainty about the nature of China's actual *intentions*. President Xi has repeatedly referred to how the country must pursue a 'holistic', and not solely 'military' approach to national security.[68] But, even if China's intentions were to turn out to be the very best and entirely in keeping with its own statements about peaceful and mutually beneficial cooperation, there is no doubt that the large-scale BRI project will expand and improve China's strategic *capability*, and reduce the risk of a breakdown in China's vitally important supply lines in the event of international conflicts.

Chronic Migraine

Former Israeli prime minister Ehud Barak allegedly once said that 'in the Middle East a pessimist is simply an optimist with experience', while *New York Times* commentator Thomas Friedman has referred to the region as 'the world of dis-order'.[69] There are unfortunately few indications that such descriptions will be any less apt in the years ahead. As I am about to finish writing this book, Israeli military forces (IDF) have left most of Gaza in ruins, launched attacks on the West Bank and Beirut, killed tens of thousands, permanently injured hundreds of thousands more, and displaced millions of Palestinian and Lebanese civilians through its military response to Hamas's horrific attack on Jewish civilians and military in October 2023. Iran and the Iran-backed Hezbollah militia in Lebanon on their part are raining waves of missiles on Tel Aviv and the northern parts of Israel, and – in alleged solidarity with the Palestinian people – the Yemenite, Iran-backed Houthi militia has repeatedly attacked what they suspect to be Israel-aligned vessels in the Red Sea. Most shipping traffic between Asia and Europe therefore now circumnavigates around the tip of South Africa to avoid these waters. This increases sailing time, reduces the capacity of the international fleet, drives up freight rates and increases the inflationary pressure on the global economy.

9. THE NEW BATTLE OVER THE OCEAN

An explosive mixture of factors makes the Middle East geopolitical nitroglycerine. The region is riddled with conflict, politically fraught and socially unstable. It is the world's most important supplier of oil and gas, and the site of the narrowest choke points for two of the world's most important transport arteries: the Suez Canal and the Strait of Hormuz. Since the second half of the nineteenth century, these factors have set the stage for the Middle East's role as the most important bartering chip – and largest victim – in the gambits of the major global powers.

For alternating administrations in Washington DC, the region has been the source of a chronic strategic migraine. It was therefore a watershed moment of greater import than most seemed to realise when in 2011 Obama announced a foreign policy 'pivot to Asia'. The latter was not intended as a non-binding, vague ambition or a proposal to friends and allies in that part of the world. It was meant as a comprehensive economic, diplomatic and military effort to curb China's power by strengthening alliances and expanding trade cooperations with countries from the Indian subcontinent to the north-eastern corner of Asia.

In the USA, this foreign policy shift received strong support across the political spectrum. A political fatigue over the Middle East had long since infiltrated Washington, while a deeper recognition of the new challenges related to China's robust growth had also emerged. While the US has 'been stuck' in the Middle East, the world's demographic, economic and political centre of gravity has shifted towards the eastern and southern parts of Asia. Although US policy in almost all other areas has in recent years been characterised by intense polarisation and growing suspicion across party lines, Republicans and Democrats were at the time able to agree on at least one thing: they wanted to get out of the Middle East and shift their focus to what they perceived as the rapidly growing geopolitical threat from China.

The USA has been heavily dependent on Middle Eastern oil and gas and has for more than one hundred years invested large economic, diplomatic and military resources in its involvement, interventions and wars in this unstable world region. But in the early 2010s, the terms of US strategic and political priorities suddenly changed. Rapid technological breakthroughs opened up for the extraction of large quantities of oil and gas from geological shale deposits on the North American continent. Ten years later, the USA was not only self-sufficient in terms of oil and gas; it had also become the largest oil-producing country in the world.[70]

This also dramatically lessened US dependency on the Middle East and gave it greater freedom, both economically and strategically, in relation to the region.

THE OCEAN

As the USA started pulling out of the Middle East, China moved in. Since then, China has made large economic and diplomatic investments geared to squeeze out their geopolitical rival and secure strategic interests. China's economy remains almost entirely reliant on the Middle East's energy reserves and trade routes. Close to half of China's oil and gas import arrives by sea from this region, and ninety per cent of Iran's export of crude oil goes to China. Nearly thirty per cent of the world's seaborne oil trade is passing through the narrow Strait of Hormuz, and more than two thirds of this is headed for Asia.[71]

China is also the largest user of the Suez Canal.

Ever Given

When the 400-metre container ship *Ever Given* went aground in the Suez Canal in the spring of 2021, it was probably a random accident. The sandstorm that swept in from the Sinai Desert was so powerful that the ship was blown onto land. There it remained stranded, blocking all traffic through the canal for six days, while hundreds of large merchant vessels piled up in the Mediterranean Sea and the Red Sea, waiting to continue their voyages. In Europe and Asia, hundreds of thousands of factories, construction projects, shops and private individuals were obliged to wait indefinitely for the delivery of goods and payment. Annually, more than 20,000 pass through the Suez Canal, and Lloyd's List estimated that the stranding cost the world economy more than US $400 million – per hour.[72] The Norwegian national daily *Aftenposten* reported that onboard this ship were several dozen containers on the way from Asia to Norway, containing 'bicycles, bubble bath, tyres, wine cabinets and rice'.[73] The worldwide ripple effects of this single, random event were yet another powerful reminder of the Suez Canal's significance for the global economy.

Maritime accidents have blocked the canal on several occasions, including in 2004, 2006 and 2017. But imagine if these incidents had not been mishaps or accidents? Imagine if someone decided to block the canal by blowing up a vessel? Or if the Suez Canal were to be closed due to war or another international crisis, such as when Egypt nationalised the canal in 1956 or the eight years the canal was non-operational from the start of the Six Day War in 1967 until peace was secured in 1975? When the global economy is so sensitive to disruptions in maritime-based world trade, it is not difficult to understand why the major powers will always have a vested interest in securing strategic control over canals, straits and other exposed sea lanes.

And, above all, the *Ever Given* incident set off the alarms in Beijing.

For China, the strategic importance of the Suez Canal is conceivably a

9. THE NEW BATTLE OVER THE OCEAN

contributing factor behind the country's huge investments in hydropower projects along the Nile in Sudan and Ethiopia. These national prestige projects have an impact on the river's water levels and, by extension, the supply of freshwater to Egypt. This gives China a strategic advantage that it could leverage to expand its influence in the Suez Canal.

Limited access to freshwater constitutes a growing problem for Egypt's economic and social stability. The country's consumption of fresh water has increased dramatically due to strong population growth and a rise in the use of irrigation due to global warming. The situation is already becoming critical, and in the course of a few years, Egypt may be facing perpetual water shortage and what the United Nations defines as 'water scarcity'.[74] Since the Nile is the source of more than ninety per cent of Egypt's fresh water, the country is concerned about anything that could impact the river's watercourse. Egypt is especially nervous about the hydropower projects under construction in Sudan's and Ethiopia's respective sections of the Nile upstream.

This is above all the case for the construction of the Grand Ethiopian Renaissance Dam, which has such large reservoir capacity that Ethiopia in given situations could elect to dry out or flood parts of the river downstream. Several Egyptian politicians have therefore advocated military action. In 2013, Egyptian President Mohammad Morsi stated that 'all options are open' and that he 'was not "calling for war" but that he would not allow Egypt's water supply to be endangered'.[75] Since then the water shortfall has become even more acute. The tone is not especially reconciliatory on the part of Ethiopia either. Many of the agreements that currently regulate the river's watercourse were formed when most of the countries along the Nile were under European colonial rule. When Egyptian authorities accuse Ethiopia of violating international treaties, Ethiopia responds by stating that they aim to take to task the 'colonialist mindset' regulating the river.[76]

China plays a reserved but key role in this drama. It is China who is behind most of the financing of the Ethiopian dam and the other large, hydropower projects along the Nile. China is also the most important trade partner and largest creditor for all three countries. China has therefore secured a position which allows it to exercise strategic influence over all three.

Again, it is difficult to interpret China's intentions. But China has without doubt acquired sufficient financial and diplomatic clout to enable it to exert pressure on Egypt and thereby influence, or even dictate the terms for use of the Suez Canal – the world's most important strategic bottleneck. It would, for example, be difficult for Ethiopia and Sudan to resist were China to decide that its interests were best served by stirring up further uncertainty about Egypt's access to vital freshwater.

THE OCEAN

Hazardous Crossings

The Mediterranean Sea, the sea basin enclosed by three continents, has historically held a unique position in Europe. It was the Mediterranean Sea that first brought Europe out into the world, and the world to Europe. It was the Mediterranean Sea that set important terms for the cultural, political, religious and economic development of Europe. Today it also represents the continent's porous southern border, constantly challenged by the contrasts and chronic conflicts between the coastal states. The Mediterranean Sea is Europe's 'exposed underbelly', as the Caribbean Sea is for the USA and the South China Sea is for China.

I was CEO of the Norwegian Shipowners' Association when messages began pouring in from Norwegian ships that had picked up refugees in the Mediterranean Sea throughout the autumn of 2014. Almost a quarter of a million refugees attempted to cross the Mediterranean that year in inflatable rubber dinghies and other unseaworthy boats to escape the civil wars in Syria, Afghanistan, Libya and Somalia, the Islamic State's reign of terror in Iraq and escalating conflicts and destitution in Africa. The following year, more than one million refugees made the hazardous crossing to Europe, while thousands of children, women and men drowned in the attempt. After a few preliminary, half-hearted attempts to mobilise a joint effort to rescue these desperate human beings, the EU member states gradually pivoted their focus to preventing them from coming ashore instead. The political, social and economic costs were considered too high, and disputes erupted between the EU member states regarding the distribution of refugees.

While the eyes of Europe remain trained on the flow of refugees attempting to cross the sea, China is systematically endeavouring to acquire strategic footholds around the Mediterranean Sea. Again, the BRI project is being used for large-scale mobilisation of economic, diplomatic and strategic resources. The Chinese have procured ownership, preferential rights for access and options for long-term leases on a series of key ports around the entire Mediterranean basin.

China's 'strings of pearls' are not restricted to the Indian Ocean.

Strategic Footholds

Farthest east in the Mediterranean, the Shanghai International Port Group is responsible for the operation and construction of Haifa, Israel's largest port. The agreement formalising the group's involvement was signed in the summer of 2019, has a duration of twenty-five years and includes

9. THE NEW BATTLE OVER THE OCEAN

the development of water supplies and other infrastructure in Israel. The following year, the Chinese entered Beirut to assist with reconstruction of the port that was demolished by an explosion in August 2020. The China Harbor Engineering Company has also been contracted to upgrade and expand the port in Tripoli.

The port expansion in Tripoli is a national Lebanese project, but Beijing's interest can also stem from an interest in reducing China's dependence on the Suez Canal. They can achieve this by building an economic corridor of roads and railways between the Mediterranean Sea and the Persian Gulf. It is noteworthy in this context that in 2019 Lebanon and China formed an agreement, according to which Chinese companies will operate the railway between Tripoli and Beirut, and the affiliated line to Syria's capital Damascus.

In Syria, China has secured strategic footholds in the ports in Tartus and Latakia. The brutal civil war has left Syria's cities and economy in ruins. President Al-Assad is *persona non grata* just about everywhere in the world, but in China he has found a partner with deep pockets and few objections to his ruthless style of governance. Western military planners are concerned that these two ports may serve as bases for Chinese warships.[77] Even a modest number of Yuan-class submarines based here would represent a threat to naval forces in large parts of the Mediterranean Sea. In a situation of conflict, these submarines would also be able to lay mines in the northern entrance to the Suez Canal. This would block international traffic through the canal, while China could then utilise a land corridor between the Mediterranean Sea and the Persian Gulf.

Scenarios of this nature will potentially alter the strategic balance in both the Mediterranean Sea and the Indian Ocean. A mere suspicion about the possible presence of such submarines and the uncertainty regarding China's intentions could lead to a redeployment of US and other Western naval fleets within the two ocean regions.

Tartus in Syria is also Russia's only naval base outside its own borders (with the exception of Sevastopol on the occupied Crimean peninsula). Moreover, the ever-closer relationship between China and Russia, strengthened by the war in Ukraine, provides new opportunities for developing the collaboration between the two countries' naval fleets.

Enfant Terrible

Under the umbrella of the BRI project, Chinese companies have bought their way into the Port of Ambarli in Turkey. The port is located on the outskirts of Istanbul, by the entrance to the Bosporus Strait through which around 50,000 ships pass annually. Here the Chinese have ringside seats to the narrow and strategically important passage marking the border between Asia and Europe and separating the Sea of Marmara and the Black Sea.

The bilateral relations between China and Turkey have been strengthened substantially in recent years. Under President Recep Tayyip Erdoğan, Turkey has taken strides towards a more authoritarian, illiberal rule, and the president is actively cultivating his personal relationship with Chinese president Xi Jinping. At the same time, Erdoğan has for years behaved like an *enfant terrible* in NATO, on several occasions putting Turkey's reputation on the line. He won neither greater respect nor more friends in Western countries when he purchased anti-aircraft missiles from Russia or when he held Sweden's and Finland's NATO applications hostage as a means of forcing through his own interests.

Encumbered by large economic and domestic challenges and an increasingly strained relationship with his NATO allies, Erdoğan has embraced China's offer of expanded trade deals and large-scale investments under the BRI. Trade between the two countries has increased more than tenfold in the past twenty years and more than one thousand Chinese companies have investments in Turkey.[78] This has already given rise to a startling and embarrassing about-face in the country's official stance on China's treatment of the Uyghur people. In 2009, Turkey had been one of the first countries to criticise in cutting turns of phrase China's treatment of this Muslim ethnic group of Turkish origin. In recent years, however, Erdoğan, himself a Muslim, has extradited oppositional Uyghurs to China, denied leading representatives of the ethnic group entry to Turkey and signed agreements with Beijing involving assurances that Turkey would not condone criticism of China's persecution of this people.[79]

Blue Homeland

Turkey's collaboration with China also caters to Erdoğan's Ottoman-inspired, megalomanic dreams of restoring Turkey's former stature as a powerful contender in the central intersection of Asia and Europe. In the eastern Mediterranean, naval vessels are patrolling to back up Turkish claims for an exclusive economic zone which conflicts with competing claims made by Greece and the Republic of Cyprus. Part of what is referred to by Erdoğan's government as '*Mavi Vatan*', 'Blue Homeland', the seabed in these areas holds large deposits of natural gas, setting the stage for increased tensions between the three countries. The governments of Greece and Cyprus both argue that the Turkish claims are in violation of key provisions of the UNCLOS, at treaty which has never been ratified by Turkey.

9. THE NEW BATTLE OVER THE OCEAN

Erdoğan has also launched a proposal for the construction of a forty-five-kilometre canal between the Sea of Marmara and the Black Sea, just west of the Bosporus Strait.[80] The government's official substantiation for the so-called Istanbul Canal is that it will improve the parameters for shipping traffic, mitigate the risk of an accident blocking the Bosporus, and lead to an increase in trade between the countries around the Mediterranean Sea and the Black Sea. However plausible this explanation, it neglects to mention that the canal will produce wholly new conditions for the strategic balance of power in the region. Fearing the regional and geopolitical consequences, in 2021 more than one hundred retired Turkish admirals took the astonishing step of signing and publishing a declaration in which they warned the Erdoğan government about the risks of building the new canal.[81]

The admirals' concerns centred on the regulation of traffic through the Bosporus Strait by the Montreux Convention of 1936, an international agreement signed by all member states of the United Nations. The convention guarantees merchant ships free passage but imposes clear restrictions on military vessels. The consequences of these restrictions were brought to bear when Russia annexed the Crimean peninsula in 2014 and invaded Ukraine in 2022. The convention prevented the USA and other NATO countries from sending aircraft carrier groups and other naval vessels into the Black Sea for the purpose of containing Russia's operational mobility.

The construction of the Istanbul Canal could give Turkey new and powerful strategic advantages in the region. Since the Montreux Convention regulates solely the Bosporus Strait, it is not given that Turkey will accept that it also applies to the new canal. On the contrary, a parallel canal will enable President Erdoğan to grant passage at whim to naval vessels that are currently not permitted to sail though the Bosporus. The president can thereby assume the role of doorman and gatekeeper for naval and commercial traffic between the Mediterranean Sea and the Black Sea. This could open up a new realm of possibilities for foreign policy manipulations and ultimatums.

But Turkey is in need of funding. The government is strapped for cash, the country's economy is in a terrible state and Erdoğan's good friend Putin cannot make any significant contributions to the projected US $20 billion cost of the canal. This opens the door for China, which can offer capital, machinery, engineers and labour – and which is seeking a stronger strategic foothold in the Mediterranean.

THE OCEAN

A Greek Tragedy

For China, the collapse of the Greek economy following the financial crisis of 2008 proved to be a gift of strategic proportions. In 2016, China was able to take over operations and acquire at a bargain-basement price a controlling stake in Piraeus Port, which is not only the largest port in the Mediterranean but also the most important of NATO's south flank. The name Piraeus means 'chokepoint on the passage' and the Chinese newspaper *Global Times* did not hold back when it commented on the port's strategic significance:

> Bordering the Mediterranean Sea on the south and the Balkan Peninsula on the north, it is one of the closest Mediterranean ports on the European continent to the Suez Canal-Gibraltar main shipping routes and has railway connection to the hinterland of Central and Eastern Europe. Its strategic location makes it a key port of the Mediterranean.[82]

The newspaper, often considered the Chinese government's mouthpiece, on this occasion also described China's acquisition of the port as 'a fine example' of the philosophy behind the BRI project.

The Chinese have for many years attempted to secure strategic footholds further west in the Mediterranean. In March of 2019, Italy was formally included in the BRI project through an international agreement opening for collaboration with Chinese companies on the development and operation of the ports in Trieste and Genoa. However, due to growing scepticism about China's underlying intentions, Italy withdrew from the agreement when it was to be renewed in 2024.

Similar scepticism prevails in other parts of Europe. When, in the end of October 2022, German Chancellor Olaf Scholz greenlit the China Ocean Shipping Company's (COSCO) acquisition of a 24.9 per cent ownership stake in Hamburg Port, it caused quite a stir. This is the largest port in Germany and the second-largest container port in Europe. A barrage of criticism was unleashed, from members of his own party, political opponents, security policy experts and the European Union, expressing concerns that this would give China control over critically significant infrastructure.

This is, however, merely one of many examples of how China is systematically attempting to buy its way into the ports of Europe. Chinese companies already have significant ownership shares in a dozen large and important ports, which in addition to those mentioned above include Rotterdam, Antwerp, Le Havre, Nantes, Zeebrugge, Valencia, Thessaloniki, Marseille and Bilbao.

9. THE NEW BATTLE OVER THE OCEAN

Although the member countries of the European Union constitute only five per cent of the world population, combined they make up the world's third-largest economy, after the USA and China. The maritime routes to the European market will therefore continue to hold vital importance for China into the foreseeable future.

Despite large investments in the 'strings of pearls' and economic corridors through the Mediterranean and the Indian Ocean, China recognises that the sea lanes to Europe will still be vulnerable and at risk in the event of a war, conflicts and random accidents. This is above all the case for strategic choke points such as the Strait of Malacca, Bab El-Mandeb, the Strait of Hormuz, the Suez Canal, and the narrow Strait of Gibraltar forming the western outlet of the Mediterranean Ocean.

In 2017, China therefore included a third area of investment in the BRI project: 'The Ice Silk Road', the ambition of which is to build ports, communication systems and other infrastructure in the coastal regions from the Bering Strait in the east to the Yamal Peninsula and Kirkenes in the west. This route, better known as the Northeast Passage, or the Northern Sea Route, reduces the sailing distance between China and Europe by one third.

With this infrastructure in place, China will also be closer and have easier access to the vast natural resources found in and around the Arctic Ocean.

The Ice Silk Road

The first time I heard about this project was in a conversation with a high-ranking Chinese official in the winter of 2017. We have been acquaintances for many years and always have conversations that are more interesting and candid than any discussion I might have with other Chinese representatives. We were sitting in his office, surrounded by photographs of him in the company of both Chinese and US presidents, when I asked why China had recently bought up the large Russian facility for liquefied natural gas (LNG) on the Yamal Peninsula, located in the Kara Sea in the north-western part of Siberia. I also asked why China had financed the acquisition through the BRI fund, since at that time it was far beyond the scope of the declared ambitions for this project. My Chinese companion replied that, given 'how the world looks today', it is important for China to reduce its dependence on the Strait of Malacca and the eternally conflict-ridden Middle East. China is therefore seeking to diversify its access to oil and gas and develop alternative trade routes to Europe. He explained use of the BRI fund by stating that there was already a lot of money there, before shrewdly interjecting, 'Who knows, perhaps a third leg can be added to the BRI, an Ice Silk Road?'

Less than six months later, in July 2017, China's President Xi and Russia's President Putin announced that the two countries had plans for an Arctic collaboration. The gas facility in Yamal would be the first project of the Ice Silk Road, the new area of investment under the BRI umbrella. In early December of the same year, President Putin was in attendance when the vessel carrying the first shipment of LNG from the Yamal facility sailed away. It was headed for China, a confirmation of the close strategic collaboration with Russia's powerful neighbour in the East.

Through the Ice Silk Road, the interests of an increasingly powerful and self-assertive China and an increasingly pressured and isolated Russia became more tightly interwoven. It is in principle a partnership of stakeholders from two decidedly different weight classes: the Chinese economy is eight times larger than the Russian and the population ten times greater. While China previously represented fifteen per cent of Russia's foreign trade, this has doubled due to Western sanctions following the Russian invasion of Ukraine in 2022. Conversely, in 2024 Russia can claim only a modest four per cent of China's foreign trade.[83]

But, in the north, Russia has something to offer that China craves: access to natural resources and maritime routes in the Arctic Ocean. More than half the coastline around the Arctic Ocean belongs to Russia. While Russia has a long Arctic coastline and little money, China has no Arctic coastline and a lot of capital. This convergence of interests with resources provides ideal conditions for trade-offs, collaboration and the formation of strategic partnerships between the two countries.

The investment in the Ice Silk Road illustrates how China's interest in the Arctic region has grown in keeping with the country's emergence as an economic and geopolitical superpower. In 2010, China's State Oceanic Administration wrote that '[the Arctic] is the "inherited wealth of all humankind… The Arctic Ocean is not the backyard of any country and is not the 'private property' of the Arctic Ocean littoral states."'[84] The statement was predominantly interpreted as a reference to the fact that ninety per cent of the known oil and gas reserves, and a large portion of the seabed minerals, in the Arctic Ocean are located within the exclusive economic zone of the five Arctic coastal states: Russia, the USA, Canada, Denmark and Norway. In 2021 the Chinese government greenlit the building of the country's first heavy, polar ice breaker, designed to service Arctic shipping routes.

China also wants more unimpeded access to the Arctic Ocean to facilitate its climate and environmental research. China already has a permanent research station in Ny Ålesund in Svalbard, the Yellow River Station, and regularly sends research vessels on expeditions in this region. The Arctic

9. THE NEW BATTLE OVER THE OCEAN

region is critical to the large weather systems that dominate the Northern Hemisphere, and in no other place on the planet is global warming occurring so rapidly and with such huge consequences. The development in the Arctic is therefore an important key to understanding climate change and its ramifications for both China in particular and the world in general.

For China, the Arctic Ocean is so important that it seamlessly alternates between two arguments that can be interpreted as inherently contradictory: on the one hand, China claims that this ocean is 'the inherited wealth of humankind'. On the other hand, while granted observer status in the Arctic Council in 2013, it simultaneously claims special rights as a 'near-Arctic state'.[85]

A New Ocean

China's investment in the Arctic has not gone unnoticed by the USA or other countries. Here, at the top of the planet, three continents and three issues at the top of the global agenda meet: climate change, economic interests and geopolitical power struggles.

As we have seen in Chapter 7, the atmosphere above the northern regions is now warming more and faster than in any other part of the world. In the past thirty years, two thirds of the Arctic summer ice has disappeared. Our children, and perhaps we ourselves, will be the first generation in the modern history of humankind to experience the opening up of a 'new' ocean.

The accelerated melting of ice alters the landscape, seascape and ecosystems, and threatens traditional ways of life throughout the vast northern regions. When the ice retreats, the abundant underlying natural resources are also exposed. It is estimated that the region contains more than one tenth of the world's unexplored oil reserves and one third of the untapped natural gas, and that more than eighty per cent of these resources are found beneath the ocean floor.[86] In this region are some of the world's largest deposits of critical minerals, such as platinum, palladium, cobalt, tungsten, nickel and copper. The Arctic Ocean is also increasingly important for fishery activities, since fish stocks are migrating into the region to escape rising water temperatures in the tropical regions, as we have previously discussed.

As global warming causes the permafrost to thaw, railways, roads, power cables and pipelines are becoming more vulnerable. At the same time, the ice is melting on major rivers, opening up broad water routes allowing cargo ships to transport natural resources out to the coast. The largest rivers, the Yenisey and the Lena, both have discharge equal to that of the Mississippi, and the Ob River is the size of the Ganges River of India. The melting of ice also opens access for fishing vessels and cruise ships, enabling travel further north in the Arctic Ocean for longer periods during the year, while a short cut through the Northeast Passage will reduce the sailing distance for ships travelling between Europe and East Asia by one third or more.

For this reason, there is a growing focus on the strategic consequences of global warming and the environmental changes in the Arctic on the part of the major geopolitical powers, along with a number of mid-size and small countries. In 2021, the USA announced its new strategic concept for the Arctic, which by some was lauded as 'the USA is back in the Arctic'.[87] A number of additional states have applied for observer status in the Arctic Council and, in the course of a few years, several dozen countries have appointed their own 'Arctic ambassadors' to protect and promote their interests in the region. Simultaneously, both Russia and NATO have significantly stepped up their military activities and Russia is building ports, naval bases and airstrips at full tilt.

Preserving low tension in the high north, as it were, would appear to be an increasingly challenging endeavour. The Arctic, which during the Cold War was a frozen front, is once again the site of heated and dangerous conflicts.

Pariah State

The China-Russia collaboration in the Arctic is one of several factors confirming the countries' close ties, or what Xi and Putin on a number of occasions have called their 'no-limits friendship'. The two countries have also found one another through their mutual rejection of what they view as the West's efforts to dominate the world. But, where China's international power and influence are expanding apace with its economic, diplomatic and military resources, Russia has been weakened, marginalised and isolated after its unprovoked, brutal invasion of Ukraine. While China, in the course of a few short decades, has gone from a regional to a major global power, Russia has evolved in the opposite direction.

Unlike China's president, Putin has no holistic ideology to offer for the development of his country apart from a claim to revive the dream of a Russian empire. Neither does Putin have the support of a strong, sole-ruling party. His power base is first and foremost the security and intelligence services, the military and an inner circle of trusted cronies. Even in times of peace, Russia has little to offer that the world might want, with the exception of oil, gas, grain, minerals and other natural resources. Before the invasion of Ukraine in 2022, oil and gas accounted for a full sixty per cent of Russia's export income and forty per cent of the state's annual

9. THE NEW BATTLE OVER THE OCEAN

incomes.[88] Now the country's dependence on these two commodities is probably significantly greater.

While up until a few decades ago, the Soviet Union was the only major power that could challenge the USA on a global scale, the Russia of today is a society in economic, political and moral decline. Their gross domestic product is the size of Spain's and less than that of New York City. After the invasion of Georgia in 2008, the occupation of the Crimean peninsula in 2014 and full-scale invasion of Ukraine in 2022, Russia has become a pariah state, ostracised and out of favour with Western countries and other large parts of the international community. The country's global influence now relies on the threat of its huge arsenal of nuclear missiles. Everyone must take seriously a regime in possession of 6,000 nuclear warheads, which attacks neighbouring countries, drops bombs on civilians and issues poorly veiled threats of nuclear warfare. 'The essence of Russia's foreign policy is strategic relativism: Russia cannot become stronger, so it must make others weaker,' American historian Timothy Snyder writes.[89] The Russian regime is therefore increasingly aggressive in its attempts to sow discord between countries, intimidate neighbours, undermine the UN-led rules-based world order and destroy the democratic processes of liberal societies.

Strategic Claustrophobia

The Arctic is of critical importance to Russia's economy and defence, and as the Polar ice cap melts, the significance of the region grows. The harvest of natural resources north of the Arctic Circle had until the invasion of Ukraine represented one tenth of Russia's annual economic value creation and one fifth of its export.[90] In 2007, a titanium metal flag painted in Russia's national colours was planted in the seabed of the North Pole, 4,261 metres under water. Boris Gryzlov, chairman of the State Duma at the time, hailed the mission as 'a new stage of developing Russia's polar riches', stating further that 'this is fully in line with Russia's strategic interests', and that he was 'proud our country remains the leader in conquering the Arctic'.[91] The statement was immediately dismissed by the Canadian foreign minister: 'This isn't the fifteenth century... you can't go around the world and just plant flags and say "we're claiming this territory!"'[92] Some may appreciate the small, historical irony of this statement, made by the foreign minister of a country founded by British and French colonists who planted flags, helped themselves to land and displaced and subjugated the indigenous population only a few generations ago.

But the purpose of the rules-based world order of our times is to ensure that it is right, not might, that decides territorial claims. While Russia, in violation of international law, has gone to war to take Ukraine by force, the country has simultaneously submitted claims under the UNCLOS for the extension of its continental shelf in the Arctic Ocean. Should the Russian claim be upheld, it will mean that the country's continental shelf will expand by more than one million square kilometres in one of the world's most resource-abundant regions. Canada and Denmark, both of whom have submitted partially overlapping claims, are among those contesting Russia's claim. According to the UNCLOS, such claims are to be assessed by the UN Commission on the Limits of the Continental Shelf (CLCS) on the basis of scientific criteria. Should it turn out that all three countries have legitimate claims, and that the scientific criteria alone are not sufficient to resolve the issue, the countries will be obliged to negotiate a settlement on the division of the contested regions. It is hardly a bold conjecture to assume that the negotiations will be taxing given the current tensions in the international climate.

Since the start of the 2010s, Russia has carried out extensive construction of naval bases, airstrips and military infrastructure along its 24,000-kilometre-long Arctic coastline, equivalent to more than half the distance around the equator. The aim is to develop its 'bastion defence' further to deny adversaries access and secure military control in these maritime regions. The Arctic bases are also strategically important because in other parts of the world the Russian fleet is at risk of blockade in the event of an international conflict. The total length of Russia's coastline is greater than that of the USA's and China's combined, but in a large-scale military conflict, the country can be denied access to the ocean and prevented from redeploying military fleets along its own coastline in other parts of the world.

To the west, the Russian fleet has bases in St Petersburg and Kaliningrad, but all passage out of the Baltic Sea must proceed through the narrow and strategically important Öresund strait or Greater Belt. In the south, the remaining vessels of the Black Sea Fleet, which has been decimated by the war with Ukraine, and the Russian fleet in the Caspian Sea must transit the narrow and heavily trafficked Bosporus to reach the Mediterranean. Here they can be blocked at the Strait of Gibraltar in the west and the Suez Canal in the east. Russia's Pacific Fleet, based in Vladivostok by the Sea of Japan, can be prevented from sailing south through the strait between Japan and South Korea. To the north on the Pacific coast, the Russian warships can be blocked in the slightly less narrow passage between Japan and Sakhalin Island.

The Northern Fleet, headquartered in Murmansk on the Kola Peninsula close to the Norwegian border, is therefore Russia's largest, most important and most flexible naval military force. The Northern Fleet is also a part of and protects the strategic 'triad' of nuclear weapons that can be launched from land, sea and the air. Combined, the Northern Fleet and the bases on the Kola Peninsula constitute the most critical hub of Russian military power.

But the Northern Fleet can also be at risk of blockade. When it sails east towards the Pacific Ocean, it must pass through the shallow and narrow Bering Strait. The strait is so narrow that Sarah Palin, former Republican vice-presidential candidate, and governor of Alaska, erroneously claimed that she could see Russia from her kitchen window. Still, only eighty kilometres separate the US and Russian shores. When the Northern Fleet sails west and out into the North Atlantic, it must pass between Svalbard and the Norwegian mainland. A bit further south, it encounters NATO's most important line of defence at sea, the region between Greenland, Iceland and Great Britain, by the military often referred to as 'the GIUK gap'.

Russia is thus still burdened by a modern-day, blue version of its historically oppressive encirclement syndrome. Military planners in the USA, Europe and China all understand how this feeds Russia's sense of strategic claustrophobia – and how it can be exploited in a situation of international conflict.

Black Smoke

Even with the third-largest navy in the world, Russia is nowhere in the vicinity of mustering a real challenge to the USA's global power at sea. Russia currently has only one aircraft carrier in operation, the *Admiral Kuznetsov*, while another is under construction. The carrier, which was launched in 1985, is the Russian fleet's flagship. It has subsequently been upgraded and modernised with sophisticated weapons systems, so its fighting capability is beyond reproach. But the *Admiral Kuznetsov* has been plagued by a seemingly endless series of accidents, from fires and persistent engine failure to fighter jets that have crashed when the cables on the runway that are supposed to safely 'arrest' the aircraft have snapped. Western naval officers exhibit poorly concealed glee when the subject of the carrier comes up. A well-informed source told me ironically that 'you don't need a military radar to monitor her movements, because the black smoke from the old engines can even be seen on meteorologists' weather maps'. It was, by the way, the sister vessel of the *Admiral Kuznetsov* that was converted into a casino in Ukraine, only to be resold a few years later, after which it became China's first aircraft carrier, *Liaoning*.

After the collapse of the Soviet Union in 1991, the Russian fleet experienced an accelerated decline for almost two decades before it fell under the mandate of a broad upgrading, modernisation and strategic reprioritisation programme. The extensive and costly fleet renovation is expected to be completed by the second half of the 2020s. Russia is carrying on the tradition of the Soviet era, in the sense that a large, advanced fleet of nuclear submarines constitutes the core of the navy's capacity. However, Russia is no longer investing heavily in numerous large surface vessels. The surface vessels are fewer and smaller today, but they are also faster, more advanced and operationally more flexible.

Monster Torpedo

In addition to a traditional fleet of submarines and surface vessels, Russia claims it is underway with development of autonomous, high-speed underwater torpedo drones armed with nuclear warheads that have extended range capability. Commonly known in Western military circles by the names Poseidon, or Status-6, these twenty-four-metre-long torpedoes can carry nuclear warheads one hundred times more powerful than the Hiroshima bomb. They allegedly have a range of 10,000 kilometres, can operate at depths down to 1,000 metres and move at speeds up to 100 km/h.[93]

Digitised film clips of the Poseidon torpedo were broadcast on prime-time Russian television in May of 2022, just after the USA, Great Britain and other Western countries had redoubled their arms deliveries to Ukraine in the wake of Russia's invasion. Enthusiastic Russian news anchors reported several evenings in a row that a single Poseidon torpedo would be capable of moving swiftly and imperceptibly towards the waters off the coast of Great Britain. The detonation would produce a 500-metre-high radioactive tsunami which would immediately wipe out the British Isles in their entirety.

There is more or less a consensus among Western military experts that Russia's claims about the Poseidon's capacity are misleading and exaggerated. Nonetheless, few are in doubt about the weapon's violent and destabilising potential. While ballistic missiles can be detected immediately upon firing, it would be almost impossible to track a Poseidon speeding through the ocean depths, producing little noise and virtually no wake trail or heat signature. It would therefore also be almost impossible to intercept before powerful nuclear warheads had been detonated that were capable of destroying large cities such as London, Washington, New York and Los Angeles – or Beijing and Shanghai.

Russia has also already introduced a new type of long-range strategic submarine, Belgorod, which can function as a launcher for these torpedoes.[94] It is estimated that these virtually noiseless submarines are more than 180 metres long, eighteen metres wide and can carry six Poseidon torpedoes. This implies that a fully loaded Belgorod alone can wipe out all of the largest cities in the USA. The submarines can also carry mini submarines designed to sabotage seafloor pipelines and fibre-optic cables.

The Poseidon torpedoes and Belgorod submarines exacerbate the USA's already pressing concern that an attack could be launched from precisely the ocean regions that have historically been perceived as forming a protective buffer zone around the USA.

Despite extensive expansion and modernisation, the Russian navy's capacity and stamina for long-term, offensive operations far from own waters remain limited. The Russian fleet of surface vessels is predominantly designed to secure strike force in the country's nearby maritime regions, first and foremost the Arctic and the Atlantic Oceans. But, with the Poseidon torpedoes and the Belgorod submarines, Russia is also building maritime capacities which uphold the threat of long-range destructive attacks.

Even a weakened and pressured Russia will have the ability to send the world up in flames.

Headbanger

The ocean regions around the Svalbard archipelago form a transition zone between the Arctic Ocean and the North Atlantic basin. This situates the island group at the intersection of heavy-weight global economic and strategic interests at the top of the Northern Hemisphere.

I have been to Svalbard several times in conjunction with my work, but on this particular occasion I went there on holiday with my family. The six of us, plus a guide, each drove snowmobiles on a several hours' excursion from the town of Longyearbyen through Spitsbergen's wild, Arctic landscape on this winter-white, sunny day. (Yes, I readily accept the objections and am not proud of the noise pollution or the greenhouse gas emissions. The next day we went on a similar excursion – this time, each of us with a dog sled.) Our hope was, as for most tourists in this region, for a polar bear sighting, and the guide therefore also had the obligatory rifle within reach in a side pocket of the snowmobile. Although we encountered no *Ursus maritimus* on this day, we had a memorable visit to Barentsburg, the small Russian mining community by Grønfjorden. When we drove between the large, exposed pipe systems, beneath the dirty conveyor belts and into the town, with its buildings of an unmistakably Eastern Bloc era, it was like entering another world. The town was, to put it undiplomatically, cold, dirty and dilapidated.

The four hundred or so Russians and Ukrainians who live and work in Barentsburg have been sent from the mainland to produce coal. The majority have traditionally been Ukrainians, though many have returned to fight in the war at home. The quantities of coal extracted are quite modest, but sufficient to support Russian claims of active mining operations. Had we driven a little further into the fjord, we would have reached another Russian mining community, Pyramiden, which lies completely untouched and just as it was on the day when it was abruptly abandoned in the autumn of 1998. Until recently, there were still coffee cups on the tables in the living rooms, cheap plastic potted plants on the window sills, and, in a small dormitory for toddlers, slippers on the floor beside the beds. A few years ago, an eccentric heavy-metal musician from Russia arrived and settled in this ghost town, where he was later joined by a handful of like-minded souls. He refers to himself as 'the world's most northerly headbanger', and I think that may be an apt description for most of his fellow inhabitants there, including those who are not musicians. For my own part, I would probably have beaten my head against the wall a great deal, had I been forced to live in Pyramiden.

But the Russian presence is about neither black rock nor heavy rock. It is about maintaining a façade of activity to secure Russia a physical foothold in this strategically important archipelago. Tourists can get a sense of this when they spot the Russian 'Consulate General', which is located on top of a small ridge on the outskirts of Barentsburg. This large, fortress-like concrete building is overloaded with antennae and surveillance cameras, and protected by tall, oversized steel fences.

It is an obvious candidate for the setting of the next James Bond film, and a completely out-of-place structure in the desolate Arctic landscape, thousands of kilometres from the nearest mainland settlement.

No Man's Land

Until the end of the First World War, Svalbard was *terra nullius*, in terms of international law, deemed a 'no man's land'. Here one finds oneself in the inhospitable north, and the name Svalbard stems from the Old Norse word *Svalbarð*, meaning 'cold shore'. For centuries, hunting and fishing have been carried out there, and in more modern times also coal mining and mineral extraction.

But despite its abundance of natural resources – and in striking contrast to the spirit of imperialism and colonial ambitions that continued to reign in Europe by the end of the nineteenth century – no country seized ownership of the island group. On the contrary, to minimise the risk of international conflicts, the major powers encouraged the tiny, neutral and unassuming country of Norway to claim sovereignty over these islands. Fearful of the

9. THE NEW BATTLE OVER THE OCEAN

prospect of shouldering excessive economic and international obligations, successive Norwegian governments showed little interest in the proposal. Norway's attitude, however, changed throughout the early years of the twentieth century, and the authorities began instead working actively to formalise such a solution. During the Paris Peace Talks in the wake of the First World War, it was therefore agreed that Norway would be granted supremacy and sovereignty over the entire island group. The decision was laid down in the international convention now known as the Svalbard Treaty. But the recognition of sovereignty included a number of conditions and reservations. The most important was that all countries that ratified the Svalbard Treaty would have equal rights to carry out research, hunting, fishing and commercial activities on and around the island group.

The Svalbard Treaty went into effect in 1925, and today more than forty states from all over the world are listed as signatories, among them all the Arctic states, including the USA and Russia, as well as China, South Korea, New Zealand, Afghanistan, South Africa, Monaco, North Korea and the Dominican Republic.

Cold Calculations

The Svalbard Treaty also prohibits the establishment of naval bases or military fortifications on the island group. When Norway joined NATO in 1949 an exception was therefore made for Svalbard.

With time, as the relationship between Russia and Norway has deteriorated in step with growing international tensions, the interpretation of this exception has become a more sensitive topic. At the same time, there has been an escalation of accusations on the part of Russia. I was in Longyearbyen to give a lecture for NATO's Parliamentary Assembly when the members visited Svalbard in 2017, the first such visit for more than ten years. The parliamentary visit provoked vehement protests from Russia, including accusations that NATO was attempting to 'militarise' Svalbard. Along the same lines, Russia has repeatedly made similar accusations when Norwegian frigates have sailed into Isfjorden and Norwegian Hercules military aircrafts have landed in Longyearbyen. Russia also claims that the civilian radar facility for satellite communication on the outskirts of Longyearbyen serves as a covert military installation. Russia has moreover, along with a number of EU member states, accused Norway of an unfair and self-serving distribution of quotas for fishing and crabbing.

Accusations of this nature from the Russian regime are a dime a dozen, and it can be tempting to dismiss them with a shrug of the shoulders. It is, however, a cause for concern that Russia on several occasions in recent years has repeated that the Svalbard Treaty is only valid as long as Norway upholds all the provisions of the treaty.[95] Taken at face value, this is a valid claim, but, given the backdrop of growing international tensions, an uncomfortable interpretation can be that the Russian regime is building up to a claim that Russia is no longer bound by the 100-year-old treaty. The next step for those in power in the Kremlin could potentially be to announce that they also do not accept Norway's sovereignty over this strategically important island group with abundant natural resources.

The tensions relating to Svalbard have been heightened following Russia's invasion of Ukraine in 2022. Both Sweden and Finland decided to join NATO as a result in early 2024. Barring Russia, all the countries around the Baltic Sea are now members of this transatlantic defence alliance. In the event of a large-scale international conflict, the Russian naval forces in the Baltic Sea basin will be surrounded and blockaded, a situation underscoring the importance of unfettered access to the Atlantic Ocean for the Russian Northern Fleet. Should the Russian regime continue to pursue its aggressive and destructive imperialistic ambitions, seizing control over Svalbard could prove tempting.

For Russia, the strategic gains of such a scenario would probably be greater, and the military and political costs lower, than an attack on the Baltic countries.

Little Boy Dreams

The escalating tension between the major powers is also reflected in the large-scale and growing military activity in the Bering Sea and the Norwegian Sea. In these regions NATO has aircrafts in the air space, vessels on the surface, submarines in the depths and listening stations on the seabed to monitor, round the clock, all traffic in and out of Russia's largest naval base, Murmansk, on the Kola Peninsula.

On a beautiful winter day of brilliant sunshine and a cloudless, blue sky, I was on board one of the Norwegian Orion surveillance aircraft on patrol above these ocean regions. The flight was a routine trip wholly without incident and the most important thing that occurred had nothing to do with geopolitics: I was invited into the cockpit to sit behind the controls for a brief half-hour (with the vigilant pilot-in-command seated beside me, ready to intervene). Little boy dreams can apparently come true, even late in life.

Recent years have seen a surge in military exercises carried out in these regions on the part of both the West and Russia. In 2018, NATO carried out its largest military exercise to date here, and the following year, Russia

9. THE NEW BATTLE OVER THE OCEAN

responded in kind with its own naval exercise of an unprecedented scale in the Norwegian Sea. Norway's then-chief of defence described the activity as 'the greatest military threat to Norway in more than forty years'.[96] Russia drilled blatantly offensive fleet formations and at one point a line of surface vessels and submarines formed in the GIUK gap between Greenland, Iceland and Great Britain. In an actual situation of conflict, such a formation would enable Russia to deny NATO naval forces entrance to the Norwegian Sea from the south.

This would leave Norway, Sweden and Finland behind the Russian lines of defence at sea.

An Oversized Embassy

South of the GIUK gap, the EU member states are endeavouring to form a more close-knit and expanded military collaboration at sea, on land and in the air. After four turbulent years with Donald Trump in the White House, many Europeans no longer trust that the United States will remain an equally stable and reliable ally.

Russia's illegal annexation of the Crimean peninsula in 2014 and full-scale attack on Ukraine in February 2022 also marked dramatic, historical turning points. When joining NATO in 2024, Sweden and Finland abandoned the policy of neutrality which both had upheld for 200 and seventy years, respectively. Denmark simultaneously joined the formal defence collaboration of the European Union, despite having remained steadfastly outside since its formation in 1992. Germany, which has maintained a low military profile since the Second World War, immediately passed a resolution for a substantial increase of its defence spending. Most of the EU coastal states also commenced modernisation and rearmament of their coastal defence and naval fleets.

The same holds true for Great Britain. Three months before Russia's attack on Ukraine, the Defence Committee of the House of Commons published a report, the conclusion of which was given in the title: 'We're going to need a bigger Navy'.[97] The committee pinpointed Russia and China as the primary threats at sea and urged the Ministry of Defence to 'be honest with the public about the deteriorating international security situation, the capabilities the Navy will need to protect Britain in this environment, and the funding required to deliver those capabilities'. None of the report's analyses and conclusions have been deemed less relevant or accurate after Russia's full-scale invasion of Ukraine.

As European countries are uniting in a more compact and vigilant protection of their local maritime regions, an aggressive and isolated Russia is leaning more heavily on China. Since the early 2010s, joint Sino-Russian fleet exercises have been carried out in the South China Sea, the East China Sea, the Gulf of Aden, the Mediterranean Sea, the Indian Ocean, the Baltic Sea and the Arctic Ocean.

China is also attempting, independently of Russia, to establish maritime footholds in the North Atlantic basin. It is well known that a Chinese 'estate agent' attempted to buy up large territories along the coasts of Iceland, Northern Norway and Svalbard. In Svalbard the purpose was allegedly 'to build a holiday village, offering hotels, vacation homes and fishing boats'.[98] It is of interest that the government-controlled newspaper *South China Morning Post* in this context reported that 'China gets chance to buy first foothold in Arctic'.[99] Neither has it escaped notice that the Chinese embassy in Reykjavik is strikingly overstaffed for the requirements of ordinary diplomatic activities. When I mentioned this to a couple of high-ranking officials at the Icelandic Foreign Ministry, they responded with an ironic smile that 'we don't actually know how many people work there, but we are quite certain that the staff outnumbers the entire Icelandic foreign service'.

Reggae and Rockets

Tensions are also rising in the waters around Latin America.

Through trade agreements and large investments financed by the BRI project, China has systematically strengthened its economic and political ties to this region. Trade between China and the Latin American countries has more than doubled in the last twenty years, and China is now the largest trading partner for the countries in South America.[100] In Peru, just north of the capital Lima, the large port of Chancay is the largest investment to date under the BRI project on this continent. When the agreement was signed in 2019, Peru's president stated that 'it will be the most important hub in South America'. The stated ambition of the project is to contribute to increasing trade between the countries on both sides of the Pacific Ocean. The Chinese company COSCO has a sixty per cent stake in the new port, and the work has mainly been carried out by the China Harbor Engineering Company and China Communications Construction Company. Chancay Port is planned to be opened by President Xi in November 2024, in connection with Peru hosting the Asia-Pacific Economic Cooperation summit.[101]

To the growing distress of the USA, China's BRI project also includes a dozen or so other states in and around the Caribbean Sea, such as Jamaica, Barbados and Grenada. When you drive north on the impressive, newly constructed motorway from the capital, Kingston, in Jamaica, you will regularly

9. THE NEW BATTLE OVER THE OCEAN

see banners and signs celebrating the collaboration with China. This island state is one of the poorest countries in the Caribbean, and the road-building agreement is as simple as it is attractive for the Jamaicans: China finances, builds and maintains the roads in exchange for the toll revenues for the first fifty years. But, China's primary interest is in neither roads nor asphalt. In Jamaica they have gained access to large supplies of bauxite, a rock used in the production of aluminium. Here, as in other places in the world, the bilateral partnership also produces a legitimate pretext for frequent arrivals of Chinese naval vessels.

In the region around the Caribbean Sea, both China and Russia are behaving like allies of Venezuela, Cuba and Nicaragua, the three countries the USA's former national security advisor John Bolton described in 2019 as the 'Troika of Tyranny'.[102]

The situation in Venezuela in particular is a source of concern for the USA. President Maduro's illegitimate and authoritarian regime has led the country into a humanitarian, economic and political crisis. China has for many years provided aid for the country, including large loans through the BRI project. The USA fears that this is a means of establishing political influence and strategic footholds in the region. Venezuela also receives economic and military aid from Russia in exchange for cheap oil, which the Russians are probably reselling at a higher price. In March 2019, a Russian aircraft carrying soldiers, weapons and military equipment landed in Caracas for the first time.[103]

However difficult it may be for Washington to witness China moving into the USA's 'own backyard' in this way, what are viewed as Russia's endeavours to gain military footholds and renewed political influence in the Western Hemisphere are no less troubling. The two-hundred-year-old Monroe Doctrine has therefore been pulled out of the US military drawers and dusted off once again.

But US concerns about the growing Chinese and Russian presence in the region are not solely related to defence of its own territory. The USA is also hypersensitive about everything that could imperil the maritime traffic through the Panama Canal, the volume of which amounts to almost 15,000 passages annually. Global warming and drought are already having an impact on the canal's capacity since the water levels are at times too low to accommodate ship crossings. In 2023 this led to a reduction in traffic of almost one third.[104] Vessels trading between the east and west coasts of the USA would then typically have to spend an additional three weeks circumnavigating the South American continent. I have been told by well-informed sources that, if the lock systems of the Miraflores Lake on the Pacific side were to be blown to pieces, the water would not reach a level permitting the reopening of the canal for several years. If this is correct, such an incident would have dramatic consequences for the US foreign trade of goods, the world economy, and the strategic flexibility of the US Navy.

The gradual accumulation of challenges related to global warming and drought, and the risk related to more dramatic conflict scenarios in this region, are considered a grave military threat by US strategic planners. Reduced capacity or closing of the Panama Canal would mean that the US Navy would not be able to transit as quickly between the east and west coast, or between any areas of conflict in the Atlantic and Pacific Oceans.

It is therefore not difficult to imagine that both China and Russia see the benefits of stoking a permanent environment of tension in this region. The Caribbean Sea is the USA's most strategically sensitive maritime region, and everything that takes place here will consume the country's political focus and military resources. This may also in turn reduce the USA's capability and willingness to engage in other parts of the world, such as in the South China Sea, the Middle East or Ukraine.

Out of Sight and Mind

Like the Arctic, large areas of the Antarctic are covered with ice. But in many other ways the northernmost and southernmost parts of the earth are literally poles apart. While the Arctic is ocean surrounded by land, the Antarctic is land (Antarctica) surrounded by a body of water known as the Southern Ocean, which constitutes about ten per cent of the world's total ocean surface. In the Arctic the ice is mainly on the ocean; in the Antarctic it is mainly on land.

While millions of people live in the Arctic region, Antarctica is the only continent without an indigenous population, and it is located far from the populated world. For most people, Antarctica and the large ocean regions surrounding the remote, ice-covered continent are out of sight and out of mind. It is located literally beneath the horizon of our attention.

But everything that occurs here is influenced by what takes place in the rest of the world. And everything that occurs in the Antarctic will sooner or later have an impact on the rest of the world.

South of sixty degrees latitude – which on the other side of the globe corresponds with the location of Stockholm and St Petersburg in relation to the North Pole – the ocean and land are regulated under international law by the Antarctic Treaty. The treaty went into force in 1961 and was originally a measure intended to mitigate the international tensions of the Cold War.

9. THE NEW BATTLE OVER THE OCEAN

Focused on peace and science, the Antarctic Treaty did not, however, contain provisions for the protection and management of marine species and ecosystems in the Southern Ocean surrounding Antarctica.

The treaty's express purpose is to preserve the southernmost part of the globe as a region for research, international cooperation and peaceful coexistence. The relatively brief text of the treaty with affiliated protocols lists explicit bans on military activity and the extraction of minerals, including oil and gas. Military aircraft and vessels can be used here for logistic purposes and civilian supply.

The treaty has been signed by fifty-two states, of which half are so-called 'consultative parties' due to substantial research activities. Seven countries – Argentina, Australia, Chile, France, New Zealand, Norway and Great Britain – have claimed territorial sovereignty in different parts of the treaty region. The majority of the other signatories, including the USA, Russia and China, have, however, never formally accepted these claims. Several of them have on the contrary proclaimed the right to submit partially overlapping territorial claims. This sets the stage for frictions and difficult conflicts at a time when geopolitical tensions are escalating. Three issues in particular represent challenges and sources of concern.

First, there is growing concern that several of the research stations are being used covertly to download satellite data for military purposes. If so, a war could initiate the steering and tracking of missiles, submarines, aircraft and naval fleets. US scientists and officials point the finger at China and Russia when I mention this in private conversations. This of course invites follow-up questions about whether the US stations are not in a position to do the same, but then I seldom receive a clear response.

Second, the ocean region around Antarctica contains an abundance of attractive natural resources such as krill and minerals. In the mid-1970s, parties to the Antarctic Treaty raised concerns about the impact on marine ecosystems of a dramatic increase in fishing for krill. This led to the adoption of the Convention on the Conservation of Antarctic Marine Living Resources (CCAMLR). However, in recent years, global warming has caused the migration of fish stocks from the warm regions around the equator to the cooler polar waters in the south and north. With growing international competition over access to natural resources, cooperation in the Southern Hemisphere can meet with new and difficult challenges.

Last but not least, the geopolitical tensions are already having a direct impact on the ongoing discussions about the enforcement and updating of the Antarctic Treaty. Since all decisions are made by consensus – in other words, unanimous agreement – a minimum of reciprocal trust, understanding and good will between the involved parties is required. The basis for fulfilment of these prerequisites is now eroding. The signatories are finding it increasingly more difficult to agree on anything, from the enforcement of existing regulations to the addition of new ones, such as new and expanded marine protected areas.

Upholding the treaty's overall intention and estimable objectives increasingly seems to be burdened with a plethora of challenges.

Russian Jacuzzi

What was probably the first and only military attack on offshore infrastructure in Europe since the Second World War took place on 26 September 2022. On this day, detonated dynamite blasted holes in the two pipelines on the seabed of the Baltic Sea that supply Germany with natural gas from Russia. The explosions, which destroyed parts of Nordstream I and II, caused large quantities of natural gas to flood to the surface. The water surface was virtually boiling with gas bubbles within a radius of several hundred metres. There was so much gas in the water that had any large vessels attempted to sail through the spill area, they could have sunk. Quick-witted commentators wasted no time in dubbing it the 'Russian jacuzzi'.

The party responsible for the explosions remains unknown at this time of writing. Although Putin's propaganda machine quickly issued routine denials, initially there were few beyond Russia who believed it could have been anyone else. Later, information emerged suggesting that the sabotage was carried out by Ukrainian actors, with or without the consent of the government in Kyiv, in order to cut off this important source of revenues for Russia.

Either way, since Russia's export was already dramatically reduced due to sanctions, Norway has become the most important supplier of gas to Europe. Because of this, if Russia should wish to paralyse Europe's energy supply, the activity on the Norwegian continental shelf and pipelines from there constitute a uniquely significant military target. Immediately after the Nordstream incident, substantial resources were therefore mobilised on the part of both Norway and NATO to intensify surveillance and protection of the 9,000-kilometre network of Norwegian gas pipelines on the North Sea ocean floor.

Earlier the same year the 900-kilometer fibre-optic cable connecting Svalbard with mainland Norway was cut, and just a few weeks after the explosions in the Baltic Sea, the international media reported that one of the two links in the fibre-optic subsea cable running north from Scotland

9. THE NEW BATTLE OVER THE OCEAN

– which is the Shetland Islands' digital connection to the rest of the world – had 'suffered a break'. The breakage occurred while repairs on the other link, which had also recently 'suffered a break' were still underway. Because of the damage to the two latter subsea cables, vital social functions such as health care, fire services, the police force, defence and other emergency preparedness services were obliged to rely on emergency backup solutions. Internet service was disabled, as were banking and transaction services, from payment terminals in stores to cash machines. Following the incident on the seabed between Scotland and Shetland, only a few hours passed before another fibre-optic cable was severed. This time, the incident occurred in Aix-en-Provence, just north of Marseille, at an onshore facility that is the hub of three subsea cables connecting Europe with Asia and the USA.

In October 2023, a fibre-optic cable and a gas pipeline between Finland and Estonia were severed, probably by the anchor of a Chinese vessel that in the preceding weeks had sailed in and out of Russian ports. Just a few days later, a fibre-optic cable between Sweden and Estonia was destroyed by what Swedish authorities described as 'external damages'. For the time being, the party responsible for these acts of sabotage remains unknown, although Finnish, Swedish and Estonian authorities have all pointed the finger at China and Russia. In response to these incidents, NATO and the three affected countries have increased their naval and aerial patrols of the Baltic Sea.

A few months later, in the winter of 2024, four cables installed on the floor of the Red Sea were severed. This affected approximately a quarter of the communications traffic between Europe, the Middle East and Asia. The cause of the breakage is still unclear, but the incident occurred only weeks after the government of Yemen had issued warnings that Houthi rebels were planning an attack targeting the cables.

Cable Maps

The destruction of offshore gas pipelines and fibre-optic cables, whether due to random mishaps or targeted sabotage, serve as powerful reminders of the vulnerability of modern-day society. The incidents also illustrate the virtually incomprehensible naivety and complacency informing the attitudes of so many democratic countries, regarding protection of critical seabed infrastructure. In open societies all over the world, both pipelines and fibre-optic cables are often meticulously drawn and documented on detailed nautical charts. A simple internet search using the term 'submarine cable map' accesses regularly updated information about location, length, landing sites and ownership of every single subsea cable running between countries and continents. The purpose of such transparency is of course to prevent ships dropping anchor or bottom-trawling fishing vessels from inadvertently damaging the cables.

But the downside is that actors with malicious intent and the requisite know-how are able to maintain an overview at all times of where and how they can effectively disrupt and paralyse energy supply and communication. The risk of such disruptive attacks is heightened because fibre-optic cables and gas pipelines span large distances in international waters, where national authorities' abilities to monitor activity and intervene are limited. It was no coincidence that the explosions of both Nordstream I and II occurred in international waters.

While gas pipelines run across the seabed between countries located in the same region, the fibre-optic cables are linked into a total length of almost one and a half million kilometres extending all over the planet. This network of more than 400 submarine cables constitutes the central nervous system of the modern-day world. More than ninety-nine per cent of all data traffic passes through these fibre-optic cables, and it is estimated that every day they transmit financial transactions amounting to US $10,000 billion.[105] The fibre-optic network undergoes constant development, and the longest cable to this day, 2Africa, became operational in 2023. It is 45,000 kilometres long, equivalent to the distance around the equator, and extends around the entire African continent, connecting Asia, Africa and Europe.[106] Antarctica is the only continent that is not yet linked to the global cable network. However, plans for installing fibre-optic cables to connect the research stations there to the rest of the world are underway.

Despite their huge capacity for data transmission, modern fibre-optic cables have a diameter not much larger than that of an ordinary garden hose. At the core of the cable are thin fibre-optic glass pipes cast in silicone gel and wrapped in layers of nylon, steel, copper and plastic to protect the cable from breakage and damage. Data is transmitted by light signals that are sent at an ultra-high frequency from powerful laser machines on land. To prevent attenuation of the light signals over large distances, so-called repeaters are installed along the cables at regular intervals.

The fibre-optic cables are laid using ships equipped with large spools, and it is only in the areas close to land that the cables are buried in trenches in the seabed. This is done using a plough that is dragged across the seabed a few metres in front of the cable. Further off the coast, cables lie in plain sight on the seabed itself, where they meander for thousands of kilometres across flat expanses, through narrow valleys and between steep mountains, weighted down with small sinkers to keep them from rising to the surface.

9. THE NEW BATTLE OVER THE OCEAN

One might think that this makes them fragile, but it turns out that fibre-optic cables on land suffer breakage or damage a hundred times more frequently.[107] The underwater cables, on the other hand, are more difficult, more time-consuming and more expensive to repair. Sometimes the cables are damaged by underwater earthquakes, volcanic eruptions or landslides, or even by shark bites. Still, it is far more common that damage is caused by human activity, such as bottom-trawling and dragging of ship anchors.

The word-wide network of undersea cables represents a critical infrastructure to the global society, and a large and growing concern to the national security of countries. The cables can be compromised and tapped for information during peacetime, and disrupted during times of crisis and wars. For instance, in the beginning of the 2010s, it was revealed that the UK signal intelligence agency (GCHQ) had for years been hoovering data from subsea cables landing on the south-east coast of England.[108] Russian surface vessels are often seen loitering suspiciously over areas with key subsea cables, and Western intelligence claims that Russia is rapidly expanding the capacity and activities of a secretive naval unit disguised as the Main Directorate of Deep-Sea Research (GUGI).[109] This unit commands a fleet of submarines equipped with remotely operated underwater vehicles designed to tap or cut deep-sea cables.

China most probably has similar capacities. Should China wish, for example, to impose an 'information blockade' on Taiwan as a precursor to a military attack, cutting or compromising cables connecting the island state to the outside world would conceivably be part of such an effort in the first phase.

But such actions are no longer the sole domain of maritime superpowers. Already, the rapid development of naval deep sea drones has made fibre-optic cables in abyssal depths viable targets even for small and medium-sized countries, and likely also so for well-funded non-state actors.

The tension of the international situation is exposing this central nervous system of the modern-day world to growing risks. The question of who makes, lays, operates and owns the cables connecting continents and countries, and where they are laid in the deep sea, is no longer merely a matter of commerce and topography. It is increasingly a matter of geopolitics and national security at the highest level.

Weapons of Mass Disturbance

The challenges posed by kinetic intrusions and disruptions of the world-wide web of subsea fibre-optic cables are part of the wider issue of cyber security.

Like all other industries, maritime operations are increasingly exposed to cyber attacks aimed at compromising their integrity, be it vessels, port operations, offshore energy production or fish farming. A report by the European Hybrid Center of Excellence, headlined 'Weapons of Mass Disturbance', outlines fifteen scenarios for maritime hybrid threats, including cyber attacks.[110] Already over the past years, a number of serious incidences have taken place. Some of the world's largest shipping companies, like Maersk, COSCO, CMA CGM and Carnival Corporation, have all been attacked by ransomware. Hackers have also, for example, been able to break into the computer system of the Port of Antwerp, Belgium, to access data on the movement, location and security details of the containers.[111]

Cyber attacks on all kind of operations and infrastructure, be it maritime or land based, may have disturbing and devastating consequences. One can only imagine what could happen if hackers were able to take control of large vessels sailing into busy ports, through narrow straits or in congested sea lanes.

Today, on average one ship is hacked every day all year round, and there is no reason to believe that this number will decline in the years ahead.[112]

Dark Fleet

It was luck alone that prevented the *Andromeda Star* from causing an environmental catastrophe off long stretches of the coastline of northern Europe. In early March 2024, the ship was headed to St Petersburg to pick up a shipment of Russian oil when it collided with another ship just north of Denmark. The *Andromeda Star* was in ballast condition when the accident occurred, otherwise 700,000 barrels of oil – more than 100 million litres – could have ended up in the ocean. The ship is one of several hundred, perhaps over a thousand, large tankers in the so-called 'dark' or 'shadow' fleet that transports Russian oil sanctioned by Western countries to the global market. The ownership of such ships is also shadowy. Little is known about the *Andromeda Star*, other than its formal owner at the time was registered under a sole proprietorship on the Goa Island in India.[113]

The 'dark fleet' is a result of the US and EU sanctions on exports of oil and liquefied natural gas (LNG) from Russia that were issued following the invasion of Ukraine in February 2022. The aim of the sanctions is to reduce Russia's revenues from this export and thereby its ability to fund the ongoing war. Yet neither the USA nor Europe have been willing to shoulder the full consequences of their own sanctions. Russia is an important oil-producing country, and fully fledged Western embargo on Russian supply to the world market could have led to politically untenable price hikes on oil and gas. A more restricted system of sanctions was thus implemented, through which Russia is allowed to sell its oil, but only at a price substantially lower than world market rates. In the fall

9. THE NEW BATTLE OVER THE OCEAN

of 2024, this price cap was at US $60 per barrel of crude oil, while otherwise oil was traded at prices approximately twenty to fifty per cent higher. While before the sanctions were introduced Russia provided almost half of the EU's oil import, today that amount is down to around five per cent.[114]

The large price gap of course provides profitable opportunities for stakeholders indifferent to the sanctions. Large import countries, such as China and India, are not participating in the sanctions and continue to purchase Russian oil as long as the price is at or below that of the world market. Since almost all export of oil is transported by ship, a two-tier system for transport of Russian oil has emerged: one 'legal' system for the transport of oil priced under US $60 per barrel and a 'shadow system' for sanctioned oil sold above this price cap. While not illegal or illicit outside of the national jurisdictions of the sanctioning countries, the dark fleet is often operating sub-standard vessels without proper insurance. To avoid detection, they frequently switch off their AIS transponders, which is the automatic tracking system providing identification and positioning information for other ships and vessel traffic services. This increases the risks of collisions and accidents, which can cause harm to seafarers, ships and the maritime environment all over the world. Already, several such serious incidents have occurred. Shipping companies and other stakeholders participating in this shadow system risk being slapped with steep fines and exclusion from the US and European markets.

According to estimates, the dark fleet comprises tankers which, combined, have the capacity to ship one fifth of the world's total crude oil trade. Recently, it has emerged that also Russian export of liquefied natural gas (LNG) is carried by the dark fleet, the latter requiring more specialised vessels and operations.[115] Since the ships risk detention should they call at foreign ports in countries taking part in the sanctions regime, the oil is often diverted onto other ships as it approaches its destination. Such ship-to-ship transfer operations can be complex and entail a greater risk of oil spills and accidents. When the ships avoid ports of call, they also dodge mandatory inspections by port states, classification societies, or representatives of the seafarer unions. This creates uncertainty regarding the technical and operative standard of the ships and the competence, safety and well-being of the crew, but the general opinion is that many of the ships are old and in substandard condition and many of the crew members lack proper training and valid certifications.

Because the opportunities for profit are substantial, an entire ecosystem of actors connected to the dark fleet has emerged. At international conferences, I have spoken with directors of large maritime insurance companies who state that they now meet people from shipping companies, brokerage firms, insurance companies and classification societies they have never heard of and from flag states they hardly knew existed. There is also a growing awareness that this shadow system has become so vast and lucrative that it would probably not disappear even if the sanctions were to be removed. This parallel 'ecosystem' of illicit shipping activities not only constitutes a grave danger to life, health and the environment. It also represents a growing challenge for upholding the UNCLOS and the global regulatory system of the UN International Maritime Organization (IMO).

It is probably only a matter of time before dark fleet vessels cause further mishaps and accidents. Unfortunately, it is also highly unlikely that the oil tankers will always be empty, as was the case for the *Andromeda Star*.

Water Is Coming

'Machine guns, fighter jets... are not our primary security concern. The single greatest threat to our very existence is climate change,' Fiji's Defence Minister Inia Seruiratu declared at a large international security summit in Singapore in June 2022.[116] In his home country, tropical cyclones and flooding in recent years have triggered huge economic problems and displaced thousands of people from their homes. The minister continued, 'Waves are crashing at our doorsteps, winds are battering our homes, we are being assaulted by this enemy from many angles.'

In recent years, a number of countries have begun including the security-related consequences of global warming and rising sea levels in their strategic-planning documents. One important reason for this it that drought, flooding, extreme weather events and rising sea levels can destroy communities, infrastructure and crops, cause social unrest and displace millions of people both internally and across state borders, not least in the low-lying delta regions of Southeast Asia

Climate change and rising sea levels also pose a threat to naval bases worldwide. A report to the US Congress states that:

> A number of coastal military installations already routinely experience high-tide flooding, and storm surge from recent hurricanes has exacerbated flooding, disrupted operations and caused extensive damage to infrastructure. Likewise, infrastructure outside of military installations, (e.g., mission critical access roads) can be impacted by sea-level rise, further impeding military operations.[117]

9. THE NEW BATTLE OVER THE OCEAN

The report states further that 'U.S. military installations on low-lying atolls in the Pacific Ocean will be negatively impacted when "mean sea level is 0.4 metres higher... the amount of sea water flooded onto the island will be of sufficient volume to make the groundwater non-potable year-round."' The US Department of Defense's strategic planning now factors in a likely sea-level rise of these proportions by the middle of this century. Climate-driven events in the years since the report was published in 2019 have underscored its assumptions and highlighted the significance of its message.

In another report from 2021, the US Department of Defense discusses how the anticipated sea-level rise will put dozens of military installations on the east coast of the USA at risk, including Norfolk, the world's largest naval base.[118] The report states emphatically that the sea level rise 'poses a risk to the Navy's ability to conduct and support operations in the Atlantic'.

Although the speed, scale and consequences will vary, strategic planners in coastal states all over the world face challenges similar to those of their US colleagues. All over the globe, global warming and sea-level rise will necessitate the reconstruction, closing or relocation of naval bases and military installations.

Water is coming.

'But I know, somehow, that only when it is dark enough, can you see the stars.'
Martin Luther King Jr (1929–1968)

10.
HOPE ON THE HORIZON?

About what must be done to halt the destructive, self-perpetuating dynamic between a warmer atmosphere and a warmer ocean. About the erosion of the United Nations' authority and legitimacy, and the growing role of G-groups. About what will be required in the manner of international cooperation, policies and regulations, from governments and the corporate world, and from citizens – like you and me – to restore the health and productivity of our most important public commons and tap the ocean's vast potential for contributing to a better future for coming generations.

Generation R

We had invited film stars and heads of state to lend their sparkle to the event taking place in a large, Orientalist-inspired conference tent, which for the occasion was set up on the beautiful sandy beach in Cascais, on the outskirts of Portugal's capital. Hundreds of students from all over the world were in attendance, all finalists in the international youth competition designed to elicit new, innovative solutions for taking care of the ocean. The winners and their proposals would be presented on the following day, during the opening of the UN Ocean Conference 2022.

One of the day's surprise celebrity guests was the UN Secretary-General António Guterres. Without a script and seated in an armchair on the stage, he softly addressed the expectant students who had gathered in a semicircle before him: 'Let me start by just following on the words of the Portuguese President [Marcelo Rebelo de Sousa] and to apologize, on behalf of my generation, to your generation, in relation to the state of the oceans, the state of biodiversity and the state of climate change.'[1] He then went on to speak about how the young people of today will be obliged to shoulder the consequences, assume responsibility for – and

10. HOPE ON THE HORIZON?

hopefully manage with greater success – the great challenges passed down by his own generation.

The young people of today are destined to become the Generation R, the 'Restoration Generation'.

A Silent Miracle Stalling

The size of the world population is today eight times what it was at the time of the Industrial Revolution, while the global economy is 250 times larger.[2] For the first time in history, the global middle class makes up half of the world population.[3] Never before have so many people lived on earth, and never before have such a large percentage lived such good lives as in our times.

The development skyrocketed after the Second World War. Famine, infant mortality, poverty and illiteracy have been reduced, while billions of people have gained access to education, higher incomes and a better standard of living. A far larger percentage of the world population now live in a country where they have a say in political decisions of importance to themselves and society. From the first half of the 1990s to the middle of the 2010s, low- and middle-income countries in East Asia and the Global South on average had a faster pace of economic growth than affluent countries in the Global North.[4] The Swedish professor Hans Rosling used to describe it as 'the silent miracle of humanity's advancement'.[5] A single statistic can sum up the amazing evolution: since the time of my childhood, the life expectancy for a newborn citizen of the world has increased by twenty-five years. This is wholly unique in the history of the human race.

Countries which for hundreds of years were under Western colonial rule have experienced renewed pride in and awareness of own national identity, culture and history. These countries have recovered the strength and self-confidence to defend own values and national interests.

However, these large strides of progress have not been to the benefit of all. In Sub-Saharan Africa the annual economic value creation per capita still today is at the same level as in 1970. Also, since the middle of the 2010s, the 'catch-up growth' of the low- and middle-income countries have subsided. Ten years on, the average country in Africa, the Middle East and South America is no closer to the per capita income level of the USA. In the same period of time, the world has become less democratic. The number of electoral democracies has declined, and according to one report the number of people with democratic rights have plummeted from 3.9 to 2.3 billion.[6]

The Covid-19 pandemic, wars, armed conflicts, growing international tensions and global economic fragmentation all serve to explain these developments. So do the subsequent inflation and rising interest rates which are weighing disproportionally on the poor countries of the Global South.

As nationalism, populism and protectionism have taken hold, both rich and poor countries have lost their appetite for trade agreements, and new barriers to trade and tariff hikes have become more prevalent. A report shows that from the mid-2010s and through the first half of the 2020s, there were introduced five times as many harmful trade intervention policy measures as liberalising ones.[7]

Last, but not least, the impacts of climate change, environmental degradation and the rich countries' reckless overconsumption of the Earth's natural resources are taking their tolls.

The Luxury Trap

The Scandinavian television series *The Luxury Trap* (*Luksusfellen*) has become popular in a number of European countries since the broadcast of the first episode some twenty years ago. In the series we meet people who have been living far beyond their means and receive help to escape financial ruin. This means they must pay off debts, reduce consumption and adapt their daily lives within the parameters of what they can actually afford. For most viewers it is almost impossible to comprehend how some people can be so clueless and irresponsible in handling their personal finances.

But that is in many ways precisely how we have organised our modern society. We live beyond our means and 'borrow' from nature and future generations. The total amount of natural resources currently consumed by the global community is equivalent to the ecosystem supply of almost two Earths, and this disproportionate overshoot continues to grow.[8] If the lifestyle of the entire world population were comparable to that of the USA and parts of Europe, four planets would be required to uphold this rate of consumption over time. At the same time, global warming is in high gear, biological diversity is being diminished and the natural environment degraded and demolished. The access to food and fresh water is under threat in large parts of the world. The ocean is heating up, and marine ecosystems is being destroyed.

This means that we are living on credit and at all times shouldering additional debt which we will pass on to our children and grandchildren – without giving much thought to how they will make good on what we have borrowed.

The consequences of global warming and the degradation of nature are becoming profoundly visible and they are emerging more quickly and with greater force than scientists have assumed so far. Tens of thousands of scientific reports conclude unanimously that climate change is human-made and the consequences are dire for all life on earth.

Simultaneously, estimates by the IMF, the World Bank and other leading financial institutions indicate that the growth of the global economy is expected to remain at around three per cent per year moving forward, more or less the same as the average growth rate since the Second World War.[9] This is, indeed, also the minimum of what most governments around the world attempt to achieve. A growth rate of this scale implies that the size of the global economy will have doubled by around 2050 and quadrupled by the beginning of the 2070s.

By implication, this means that the international community's stated goal of net-zero emissions and a sustainable balance of natural resource consumption are to be achieved while the global economy expands exponentially and the world population increases by two billion.[10]

There are some who claim that this can only be achieved through large-scale systemic changes.

Less Is More?

Since the chase for profit, the increase in material prosperity and the overconsumption of natural resources have created the existential challenges we are currently facing, an increasing number of people have begun advocating the concept of 'degrowth'. The idea is that economic activity must be curbed and that we must instead utilise and redistribute the resources and material goods that already exist. The approach can be traced back to the British economist Thomas Malthus, who more than two centuries ago wrote that, since the population was increasing more rapidly than the food supply, the world was headed for famine, war and poverty. Population growth therefore had to be curbed. The degrowth philosophy is also inspired by the Club of Rome's *Limits to Growth*, which in the early 1970s made the argument that unlimited economic growth on a planet with limited resources is not possible.

Degrowth can in principle seem like a rational idea, and the approach and intention behind it are not difficult to understand. Half of the world population already have enough, and many have more than enough material prosperity to live good lives. There is also an obvious need for redistribution in a world in which the poorest half have only ten per cent of the income and two per cent of the total wealth.[11]

However, partly because of the concept itself and partly for practical reasons, I put little stock in the idea that this is the right path. First, there is no one-to-one correlation between economic growth on the one hand and environmental degradation and overconsumption of natural resources on the other. For example, we have already addressed how containers of goods can be shipped from Asia to Europe producing far less emissions than a few decades ago and how zero-emission vessels are already in operation in coastal and short-sea shipping. This type of disconnect between economic growth and greenhouse gas emissions can also be found in the statistics. From 1990 to 2022, the world economy almost tripled in real terms, while emissions increased by 'only' fifty per cent. In the same period, the size of the US economy more than doubled, while, while there was a slight decrease in the country's total emissions. This illustrates that it is the use of fossil fuels and the destruction of nature, not economic activity as such, that are causing global warming. Also, a main driver of economic growth and a main feature of well-functioning markets is increased productivity, which entails a more efficient use of scarce resources.

It has already proven difficult to induce affluent countries to fulfil their stated commitment of transferring US $100 billion annually to the Global South, although this only constitutes 1/1,000 of global economic value creation.[12] We can, moreover, get a sense of the scale of the challenges by the fact that Africa – the continent in most need of investments for development of infrastructure, health systems, education and other basic societal functions – in the years between 1970 and 2022 lost more than two trillion US dollars through capital flight to other parts of the world. These outflows, corresponding almost to the annual gross domestic product of all sub-Saharan African countries combined, exceed annual inflows of development assistance and foreign direct investment received by African countries. According to the UN Trade and Development (UNCTAD) 'addressing the problem of capital flight and related issues such as trade mis-invoicing, money laundering, tax evasion, and theft of public assets by political and economic elites will require national and global efforts with a high level of coordination'.[13] I believe both of these examples serve to underscore the profound need for major changes – and the challenges of achieving this in the timeframe available.

And finally, there is little if any historical evidence to support the claim that unselfish attitudes can surpass the potentials of innovative companies and well-functioning markets in the efforts to secure effective utilisation of scarce resources, or in promoting development of new solutions that can benefit society at large and the natural world. In fact, the opposite is more the case. It is precisely the attitudes of human beings, rather than the market system as such, that constitute the main problem. If the basic conditions for degrowth had been in place, we would probably have already moved further and faster in that direction in order to solve the existential challenges we are facing in today's world.

10. HOPE ON THE HORIZON?

Personally, I do believe that most people are decent and well-meaning, and we can always hope that we will take better care of one another. But hope is not a strategy. It is therefore difficult to see how degrowth can be a practical solution within the time frame available. On the contrary, I am afraid that it can be a time-consuming dead end. We must focus on what we can realistically achieve before the challenges we face as a global community become insurmountable.

That said, I strongly believe that attitudes matter. I also believe that attitudes can – and must – be changed, and that consistent attention and informed discussions among politicians, business leaders and public at large serve to improve awareness, literacy and the sense of urgency about these fundamentally important topics. Moreover, as more people around the world are faced personally and close-up with the dramatic consequences of global warming and degradation of nature, this will probably also effectively focus their minds, change their attitudes and alter their priorities. It has already proven possible for the global community to come together to take swift action to address urgent, existential threats, such as combating Covid-19 or fixing the growing hole in the ozone layer in the 1970s.

But the sheer scope, scale and urgency of the challenges that we are facing now are several orders of magnitude larger, and so are the inherent, conflicting economic, political and strategic interests. Therefore, even if it is morally and ethically desirable, the fundamental shift in attitudes required by the idea of degrowth will most probably require much more time than we have at our disposal.

Importantly, I do believe it will still be possible to address most of the challenges within the parameters of the dominant economic and political systems.

But, to be clear, this also will require immediate political action and wide-ranging economic and industrial changes. It will require rewards for enterprises that benefit society at large, and sanctions for those that do not. And, above all, it will require mechanisms for the redistribution of income and wealth, as well as hope and opportunities, in a much more just and equitable manner than is the case today.

Heaven and Earth

Fossil fuel energy sources and maritime trade have been two of the most important underlying conditions for the powerful economic growth of the last two centuries. Coal, oil and gas have made it possible to concentrate human activity in cities and urban areas. They have supplied factories, companies and households with power, heat and electricity. Fossil fuel has also made it possible to transport goods and people quickly and inexpensively on land, at sea and in the air.

The ocean itself is a source of food, jobs and incomes for several hundred million human beings, and one third of the world's oil and gas supply comes from the ocean floor. But the ocean's most important contribution to global economic growth and prosperity comes from the maritime trade routes and the tens of thousands of (fossil-fuel-powered) merchant vessels that make it possible for countries and companies on different continents to participate in the international supply chains.

While fossil fuel energy sources and the ocean in these ways have contributed to driving economic growth and prosperity, it is the destructive interaction between the two that is now putting our common future at risk. The mutually reinforcing interaction between the warmer atmosphere and the warmer ocean is destroying the conditions sustaining the life and future of our planet. It is a vicious cycle which must be broken.

At the same time, the ocean is an important means and key to breaking out of this vicious cycle. As we have already seen, ocean-based measures alone can help reduce greenhouse gas emissions by more than one third. This is equivalent to four times the annual emissions of the EU member states. Nature-based marine solutions can contribute to the absorption of greenhouse gas emissions without degrading marine ecosystems. Decarbonising shipping and sustainably sourcing food, electricity, medicine and perhaps minerals from the ocean can reduce harmful air emissions, environmental impact and risks of operational disruptions due to extreme weather events.

And we hereby come full circle: to solve the climate crisis we must *both* take better care of the ocean *and* use it more. And, to take better care of the ocean, we must solve the climate crisis.

Bitter Aftertaste

Hundreds of delegates cheered and shed tears of joy in the large conference centre in Le Bourget when the Paris Agreement was ratified on 12 December 2015. Finally the international community had reached an agreement on common goals for curbing global warming: 'Holding the increase in the global average temperature to well below 2°C above pre-industrial levels and pursuing efforts to limit the temperature increase to 1.5°C above pre-industrial levels.'

It was a watershed moment, but the celebration had a slightly bitter aftertaste. The Paris Agreement is based on a more flexible approach and

10. HOPE ON THE HORIZON?

other types of obligations than traditional agreements formed under the auspices of the United Nations. A fundamental principle of the agreement is that of 'nationally determined contributions', according to which each country reports its ambitions, which are expected to grow gradually. The agreement contains no mechanisms for forced implementation or sanctions. According to the agreement, it is up to each of the 196 ratifying countries to decide how much, how and when they will reduce their own greenhouse gas emissions. Neither are there any formal consequences if they should fail to fulfil their commitments. With the exception of potential critique, in the manner of 'name and shame', the cost of promising the moon and not following through is small or negligible. This undermines the power of the agreement and the probability of meeting the goals.

The unique design of the Paris Agreement could also unintentionally hold the potential of undermining the authority of the United Nations in other areas. If the member states are not willing to make ambitious, binding commitments in an area that is of such existential significance for the entire global community, why should they do so in other spheres? But this flexibility was probably what was required to land the agreement.

Although it is still technically and theoretically possible, it is hardly probable that the world will succeed in meeting the 1.5°C goal of the Paris Agreement. To do so, global emissions must be net-zero by the middle of this century. The world today is not even close to being on track for 2.0°C, much less 1.5°C. An estimate from COP28, the Climate Summit in Dubai held in Dubai in December 2023, showed that, even if all countries were to deliver on their self-declared commitments, we would still be headed for a warming of 3.0°C or more.

When the Paris Agreement was adopted, it was lauded as a milestone in the fight against climate change. And indeed it was, at the time. But today it is abundantly clear that the agreement does not guarantee sufficient and timely curbing of global warming.

It is important to understand two facts about the Paris Agreement's target goals of limiting global warming to 2.0°C, or preferably 1.5°C. Firstly, the goals pertain to global *warming*, not to *emissions*. Therefore, it is science that tells us what levels of emissions will be consistent with the goals. Secondly, the goals are not a political ambition or a social and economic choice. It is what the science tells us the earth's ecosystems are able to absorb. 'It is actually a planetary boundary,' Swedish scientist Johan Rokstrøm explains.[14] If global warming exceeds these limits, we will probably transgress what science calls 'tipping points', large-scale events which fundamentally and irrevocably alter nature and the ecosystems.

Since the start of the Industrial Revolution, humanity has emitted some 2,400 gigatonnes (billion tons) of carbon dioxide to the atmosphere. To limit global warming to 1.5°C, the 'carbon-budget' estimated by scientists by the end of 2023 would allow for emitting only an additional maximum of 250 gigatonnes.[15] As the global level of emissions is currently close to forty gigatonnes per year, this means that we are about to surpass 1.5°C already in this decade. In 2024, for the first time, temperatures exceeded 1.5°C for a full year, though observations over several years are required to establish if we have passed this goal of the Paris Agreement.

Without a radical change of pace and more ambitious, mutual international commitments, we are headed for a difficult future in a different world. The decisions made and actions taken – or not – in this decade will decide the future for centuries to come.

One for the Team

There is currently a lot of talk about the Paris Agreement, green solutions and blue innovations. There is also a lot of talk about the policies, technologies, investments and regulations that will be necessary to bring about the green transition. But there is less talk about people and emotions, and the impact of this large-scale transition on all of us and on our daily lives. There is little talk of the social, practical and psychological challenges of persuading a sufficient number of people to acknowledge that we cannot continue business as usual, while simultaneously retaining the hope of a better life beyond the transition needed.

The green transition is a large-scale societal changeover that will also bring about the redistribution of wealth, incomes and costs between groups, regions and generations. It will involve investments and sacrifices today for a better world tomorrow. It will create new and exciting opportunities for some and heavy burdens for others. It is therefore crucial to ensure that these changes take place in a way that is perceived as inclusive, reasonable and just by the majority. Without universal, popular support there will be no green transition.

We will get nowhere if we do not address this in a sincere, inclusive and empathetic manner. We must recognise the anxieties and feelings of powerlessness experienced by the individual in meeting with the far-reaching and profound adjustments we are facing. We must understand how it affects the experience of job security, and the coping mechanisms and self-confidence of large population groups. We cannot expect families who are already struggling to make ends meet to 'take one for the team'. We cannot ask poor people in Bangladesh, Sudan or Mali to sacrifice opportunities for higher incomes and better lives because the wealthy parts of the world are consuming the earth's entire climate budget. Neither can we expect factory workers in France, lorry drivers in the USA or petroleum engineers in Norway to celebrate a policy that will secure the needs of future generations if this entails a risk of losing their own jobs and homes.

Emotions are drivers of human actions, and we will not succeed unless we understand that the green transition both *creates* and *requires* feelings. There will be no transition unless enough of us understand rationally and comprehend emotionally that the status quo is not a viable option. That we quite literally are standing on a burning platform, or a boiling planet. Only then can we mobilise the psychological, social and political forces required to overcome fear, uncertainty, resistance and objections. It may be helpful to remember that the word 'emotions' was introduced into common vernacular in the late sixteenth century, in a context of social unrest and rebellion in France. The word derives from the Latin *emovere*: 'movere' signifies 'movement', and the prefix 'e' means 'away from'. An angry French population wanted to 'move away' from the status quo.

There is, therefore, perhaps an ambivalent hope to be found in the fact that many more of us are now starting to experience first-hand and close up the dramatic consequences of global warming and the destruction of the natural world. There is also a paradoxical hope to be found in how young people all over the world are becoming increasingly worried about the future and that 'climate anxiety' is putting down roots. In a large international survey of young people, almost half of the respondents stated that such worries affected them in their daily lives, and three quarters responded that they were frightened about the future. Two thirds also stated that governments and politicians had betrayed young people and coming generations. 'It's different for young people – for us, the destruction of the planet is personal,' one sixteen-year-old replied.[16]

But we can't simply 'move away' from, and it is not enough to say simply that the transition will be 'green' and 'sustainable'. We must also agree upon where we are going. The forces for change must be given a common focus and direction. A sense of commitment, aspirations and legitimate hopes must be instilled in the general population.

And there is more than sufficient cause for hope. The green transition holds great potential for growth and prosperity both in poor countries in the Global South and in affluent countries of the Global North. Green solutions for energy, transport and food production will not solely promote sustainable production. They also hold the promise of economic growth and job creation for the poor in the Global South, not least for women and young people. They can inspire new industrial adventures in wealthy countries and ignite hopes for the future in all of us.

But none of this will happen by default; it will only happen by deliberate design and dedicated action.

And it will only happen if we are able to apply a comprehensive, holistic approach and mobilise across national, political, industrial and socio-economic divides. And that, in turn, will require cooperation and collaboration between hole groups of stakeholders.

It Takes Three to Tango

Blessed through generations with generous natural resources like fish, timber, hydropower, oil and gas, and having developed global competitiveness in industries like shipping, aquaculture and finance, Norway has become one of the most affluent countries in the world. This has created the basis for developing a generous welfare system with general benefits such as free education, free health care, short working hours, long vacations, paid sick leave and twelve-month paid maternity leave which can be shared between both parents. As a result, my home country performs above average not only on economic parameters, but also in international comparisons of citizens' well-being and quality of life. There is, however, more to the story.

There are rich countries in the world that have not translated their wealth into the same general level of well-being and quality of life for their citizens. Also, there are countries in the world with larger natural resources and more human capital that have not been able to develop a corresponding level of progress and prosperity.

I believe that an important explanation for Norway's economic growth and living standard is not first and foremost the natural resources and industrial value creation in and of itself. I believe it lies elsewhere, that it is a result of deliberate political decisions aimed at a just and equitable distribution of income, wealth and opportunities, and a long tradition of social dialogue and tripartite cooperation between government, businesses and labour unions. The latter provides the basis for holistic approaches, multiple perspectives and ideas, inclusion, collaboration and consensus building, all contributing to social stability, predictable policies and industrial competitiveness. In his book *Small States in World Markets*, Peter Katzenstein shows how small countries like Norway have been able to successfully navigate and thrive in the global economy because they can adapt more quickly and efficiently to

10. HOPE ON THE HORIZON?

changing global trends. He argues that one of the reasons for this adaptability is the cooperative and consultative style of governance between business associations, labour unions and government.[17]

Through my years at the helm of the Norwegian Shipowners's Association, I witnessed first-hand how the tripartite cooperation remains a key factor in maintaining and developing the Norwegian maritime cluster, one of the largest, most complete and most advanced in the world. I am therefore also, by extension, convinced that this kind of cooperation and collective approach must form the basis for the way forward in dealing with the most pressing challenges, and reaping the huge opportunities, ahead.

This kind of approach is, for example, critical to succeeding in the large-scale, rapid changes needed to decarbonise international shipping. This transition will not happen at sufficient speed and scale without engaging the entire 'industrial eco-system' of the maritime industry. Not only regulators, shipping companies, research institutions, classification societies, yards, banks, insurance companies and maritime schools, but also seafarers, dockworkers, their trade unions and other groups of maritime employees and stakeholders must be actively included in these undertakings. Their competence, insights and commitments are urgently needed to develop, coordinate and implement the solutions necessary for the green transition, and to re-skill, attract and retain able workers in all parts of the maritime supply chain.

This is why the heads of the UN International Maritime Organization (IMO), the UN International Labour Organization (ILO), the United Nations Global Compact (UNGC), the International Transport Workers' Federation (ITF) and the International Chamber of Shipping (ICS) in the autumn of 2021 launched a joint project called the Maritime Just Transition Task Force.[18] A first of its kind for any industry, the project 'seeks to strengthen and coordinate collaboration between governments, industry, workers, academia – and their representatives – towards a safe, equitable and human-centred approach to the transition towards a decarbonised shipping industry'.

In a wider sense, the need for a human-centred, collective approach through social dialogues and tripartite cooperation also speaks to the importance of having leaders in all parts of society who invite inclusion, dialogue and cooperation, who act in the true interest of society at large, and who behave in a manner that builds trust and cohesion across national, social and ideological divides.

This applies, not least, to the political domain. Political leaders who are pitting countries or groups of citizens against each other, are not only eroding the social fibres and mutual trust so vital to a peaceful and prosperous development of their own country. They are also posing a threat to the international collaboration needed in facing the defining challenges of our time, and in reaping the huge potential opportunities of the green transition to the benefit of current and future generations.

Laziness, Cowardice and Greed

For better or worse, humanity has created this challenging situation ourselves, and it is up to us to do something about it. It is still fully possible to rein in global warming, protect ecosystems, restore the balance of nature and secure good lives for a growing world population. We already have everything we need. We have the knowledge, the technology, the resources and the money. We hold the future in our own hands, and there is still hope on the horizon.

But we must do it right, we must do it together – and we must do it now.

In my opinion, this all starts with leadership and the expectations we must be able to have of leaders in politics, business or other social domains. No reasonably informed, intelligent and sensible human being can continue to claim ignorance about climate change and environmental destruction. The scientific documentation is overwhelming, the conclusions are compelling and the consequences devastating.

It is of course easy to sympathise with leaders, given the difficult and complex issues they are grappling with in the realms of politics, business and society in general. When leaders fail to address dilemmas of existential importance for society and future generations, a benevolent interpretation might be that they are postponing the most difficult decisions in the hope that future innovations and technological advancements will solve most of the problems we are facing. Many exciting developments are indeed taking place within research, development and innovation that give cause for hope and optimism.

Two examples from my own experience as a maritime executive illustrate how quickly change can take place. In the final days of negotiations before the Paris Agreement was adopted in 2015, I was host of a seminar organised by the Norwegian Shipowners' Association. Invited participants included chief executives from the maritime sector, ministers and prominent representatives of the United Nations. The venue was the top floor of the Centre Pompidou in Paris and the topic was how shipping, as one of the few industries exempted by the Paris Agreement, should nonetheless work to reduce emissions in keeping with the goals of the agreement. It was striking how almost everyone agreed that it was virtually impossible to imagine how intercontinental, deep-sea shipping could be emissions-free unless the

10. HOPE ON THE HORIZON?

industry reverted to the use of sailing vessels. The persistent refrain was that a choice had to be made between greenhouse gas emissions and economic growth: either accept that shipping emissions are a necessary condition for international trade and global economic growth or reduce the number of voyages and thereby accept economic decline. It would, in short, not be possible for us to have our cake and eat it too. Today, less than ten years later, there is a broad consensus that this is a false dilemma. There is overall acceptance in the industry and on the part of governments and regulatory authorities that it is both necessary – and possible – to reduce emissions from shipping even though international trade continues to grow.

The second example pertains to political and industrial attitudes, and the question of which measures must be implemented to achieve such reductions in emissions. When the Paris Agreement was adopted, as director of the Norwegian Shipowners' Association I was also among the first to propose imposing a carbon levy on shipping that would make the use of fossil fuels more expensive and zero-emission alternatives more commercially attractive. The proposal was virtually rejected outright by most members of the international shipping community, and those of us advocating such a solution often met with fierce criticism from other industry actors. The mere notion that the industry itself should propose a global levy that might cut profits did not sit well, to put it mildly. Today a global levy for ships, through 'an annual greenhouse gas (GHG) emission fee', is nevertheless the official position and preferred solution of the International Chamber of Shipping (ICS), the organisation representing eighty per cent of the world merchant fleet.[19]

These are just two of several examples illustrating that there is cause to hope that a great deal can change and be achieved in a short period of time. But again, hope is not a strategy. When our common future is at stake, it is not reassuring to have leaders who are banking on the idea that 'something will happen' in the future. Determination, bold decisions and swift actions are required.

I think that there are three other, perhaps equally plausible explanations for inadequate leadership in the context of the green transition: laziness, cowardice and greed. Laziness in the sense that leaders find it more comfortable and easier to proceed as they have always done, trusting that 'somebody else' will take responsibility for doing what must be done. Cowardice in that leaders don't have the courage to stand up for what they know deep down to be right, because the consequences for their own lives and careers might prove inconvenient. Or, perhaps, worst of all, greed, driven by ambitions of political positions, career advancement or the prospect of economic gains.

Narrow, personal self-interest put first at the expense of the futures of their own children and grandchildren.

I believe we must be able to expect more from the leaders of society. We must also become more consistent about holding leaders accountable for how they handle the existential challenges of our era. Democratically elected politicians will then of course be confronted with a conundrum, as conceded by Jean-Claude Juncker, former president of the EU Commission: 'We all know what to do, but we don't know how to get re-elected once we have done it.'[20] Democracy as a form of government is also being put to the test because the changeover must happen quickly. As Cameron Abadi, deputy editor of *Foreign Policy*, has put it, 'Elected officials work through compromise, but a warming planet waits for no one.'[21]

But it is precisely in this type of situation that our political leaders must rise to the occasion. It is now that they must demonstrate that they are capable of mobilising support for what their country and the world actually need, not what they believe voters want to hear. This will, of course, also require a minimum level of knowledge and acceptance on the part of the electorate. Perhaps the politicians of today can find inspiration in the words of US President Franklin D. Roosevelt, spoken almost a century ago: 'The greatest duty of a statesman is to educate.'[22]

Emissions and Inflation

More often than not, political party platforms, speeches and panel debates tend to contain declarations about how considerations for the environment, nature and sustainability must form the basis and framework for all policy. But, when all is said and done, as a rule, a lot more has been said than done. We would have been in a far better position had politicians been as committed to tackling greenhouse gas emissions as central bankers are to managing inflation. And, maybe they would, if they were obliged to factor into the public budgets the true costs of climate change and degradation of nature, and the economic benefits of doing something about it. Barring the human suffering and existential threats, recent research findings indicate, for example, that the macroeconomic impact of global warming could be six times larger than previously documented, and that a further temperature increase of 1°C could reduce global economic value creation by more than ten per cent. The 'social cost' of carbon, the broader impact on society as a whole, has been estimated to more than one thousand US dollars per ton emission.[23] A study on the impact of wind and solar parks installed in the USA in the years 2019 to 2022 found that the reductions of harmful emissions to air of sulfur and nitrogen oxides by displacing fossil fuel power plants provided an estimated US $250 billion of climate and health benefits to the USA for those three years alone.[24]

The list of necessary political reforms and actions is long, and most politicians are familiar with the items on that list. Also, no political promises or reforms have value if they are not made high priority in practice, through funding and resources. They must 'plan the dive, and dive the plan', as some of us like to say. The annual budgets are often a good litmus test for assessing the actual commitment of governments. If the initiatives promised are not backed by funding, they are literally of little worth. Promises are neither credible if they are not underpinned by adequate legislation, or if the public authorities do not take the lead by adapting public enterprises, procurement and licensing policies accordingly.

The public sector should also utilise more actively the interactions and interdependence of regulatory schemes and market mechanisms. The Norwegian container-deposit scheme can serve as an example. In Norway when you buy a bottle of soda, juice or beer, shops are required to charge a nominal deposit for the bottle itself, whether it is made of plastic, metal or glass. When the empty bottle is later returned to any shop (all shops are bound to accept returned empties), the deposit is refunded. This creates a monetary incentive for picking up and returning empty bottles that for one reason or another have ended up on the street or in the ocean. The system is simple and the results speaks for themselves: while the share of returned plastic bottles in the world as a whole is fifty per cent, in Norway it is above ninety per cent.[25] When we also know that, on a global scale, 500 billion plastic bottles are sold – almost one million every minute – we gain a grasp of the potential benefits inherent to the introduction of such schemes.

Through a container-deposit scheme 'trash becomes cash', and in this way the market can work in the service of the environment.

Missing the Mark

Although the government and public authorities can achieve a great deal in their own right, in most countries it is the private business sector that owns most of the properties, buildings, factories and other physical inventory required for production, sales and logistics.

For shipping companies, the commissioning of new vessels usually represents their largest investment. Each ship can typically cost tens or hundreds of millions of dollars, and the time-horizon for returns on these investments is several decades.

Since the Second World War, all ships in the international merchant fleet have operated on heavy fuel oil and marine diesel, and the risk associated with investments has almost exclusively been related to market conditions and regulations. But the green transition introduces additional risks associated with the choice of fuel and technology, such as uncertainty about where the ships will be able to bunker and at what cost, and a potential shortage of qualified maritime workers.

Faced with the green transition and regulatory requirements to decarbonise shipping, shipowners must now choose whether to invest in new ships fuelled by ammonia, methanol, hydrogen or other types of emissions-free alternatives. Should they miss the mark when making this choice, it can in the worst-case scenario put the entire company in jeopardy. Many shipowners are therefore postponing the acquisition of new vessels to give themselves more time to acquire the knowledge and information required to mitigate the risk. Or, they continue to rely on the familiar, well-proven concept. Today still, as discussed in Chapter 8, more than two thirds of the new ships on order are designed to be equipped with mono-fossil-fuel engines.

The same risks and uncertainties hold true for shipyards, equipment manufacturers and ports. There are almost 5,000 seaports in the world and it will take time to convert these to electric operations, equip them with tankers and facilities for zero-emission fuel and train dock workers in new tasks and procedures. Potential bottlenecks and delays in the production of new ships, fuels and equipment must also be factored into assessments. There is therefore growing competition between stakeholders endeavouring to secure long-term agreements for access to zero-emission fuel. One of the world's largest shipping companies, the Danish Maersk, has already purchased several million tons of green methanol from China to be delivered over the course of coming decades.[26]

Shipping companies and seaports will be vitally important. But these maritime actors will also play a critical energy changeover role in the economy in general. The merchant fleet will, as we have seen, carry half of the low- and zero-emission fuel needed in other industries around the world. The ports will serve as energy forwarding hubs for seaborne transport of zero-emission fuel between countries and continents, and as hubs for distribution of imports of such fuel to domestic land-based activities.

It Takes a Village

No individual public or commercial actor or sector can handle all of this alone. The African proverb 'it takes a village to raise a child' also holds true for the green transitioning of the maritime industries: it requires a multi-stakeholder, industrial-ecosystem approach. Coordination and collaboration are required, internationally, between the public and private sectors, between the industry and trade unions, and along and across supply chains. Policy and regulations must contribute to mitigating the risk and increasing the expected yield on the large investments to be made in the private sector. Seaports and other onshore infrastructure must be adapted. It is often claimed that some eighty per cent of investments in the maritime industry need to be made in infrastructure, energy and fuel production. The green transition of the blue economy will take place on land.

But, since these processes will of necessity take time, policy and regulations must simultaneously be designed in such a way as to provide for more sustainable use of the ships, facilities and other equipment already in use. This is not solely because changeover takes time for technical and practical reasons. It is also because companies would prefer not to discard production equipment that can still generate profit. A ship will typically have an expected lifetime of at least thirty years, a fish farm thirty to forty years, an oil platform forty to fifty years and an office building on land can have a lifetime of fifty to one hundred years or more. It is in a practical sense only possible to replace a minor increment at a time and this limits how quickly it is possible to achieve large-scale industrial changes that promote greater sustainability. It is therefore all the more important that shipping and other maritime industries are incentivised to actively utilise technology and innovations that can reduce harmful emissions and increase energy efficiency on already existing equipment. These could typically include digital twins, artificial intelligence, advanced steering systems and new types of anti-fouling coating on vessels.

In the end, most commercial decisions are based on 'profit and regulations', as a shipping director candidly stated at an international conference. The majority of companies are simply aiming to maximise profit within the boundaries of existing regulations. This means that if the green transition is to be commercially attractive, policies and regulations must be designed to make it more profitable and viable, and less financially risky, for companies to conduct their business in a sustainable manner – and more costly and risky to fail to do so. Also, mainstream finance, not just philanthropic donations and government-supported schemes, must be mobilised to contribute the large amount of funding needed for the green transition. Some hold that 'if it is not investable and bankable, it is not doable'. In short, markets must reward the companies that try to be on the right side of the future and banish the others to the scrap heap of history.

Green business must be good business, and grey business must be poor business.

Poor Master, Good Servant

We will not be able to move the needle on the green transition in any significant manner until mainstream commercial and financial markets start to channel their resources and attention in the right direction. The market makes for a poor master but it can be a good servant.

Only when profitable practice for companies and investors benefits society at large, and vice versa, will our efforts to combat global warming, preserve nature and restore biodiversity truly gain momentum. This must be done by incorporating considerations for society, the environment and nature into the market economy through company balance sheets and calculations. We must literally induce companies to appreciate the financial value of their surroundings, what economists call 'internalisation of externalities'. The price mechanisms of the market can hereby work wonders.

The by far single most important means to achieve this will be to put a realistic price tag on the true environmental and social costs of greenhouse gas emissions, often referred to as a 'carbon tax'. Such schemes are already in operation in several places of the world by way of emission trading schemes (ETS) or different kinds of taxes and levies. In Europe, for example, there are more than twenty countries that already have introduced some kind of carbon tax, ranging from less than one US dollar to more than one hundred US dollars per metric ton of carbon emissions.[27] Such schemes serve to reduce overall energy consumption, improve energy efficiency and alter relative costs in favour of low- and zero-emission fuel.

This is also the philosophy behind the carbon levy proposed for the shipping industry mentioned earlier (there are mere formalities explaining the use of 'levy' or 'fee' instead of 'tax'; countries like the USA do not fancy an international system 'taxing' their own shipping companies. Also, the levy is intended to be charged on the bunker oil, as a proxy for carbon dioxide emissions). But, to serve the purpose, a carbon levy must be sufficiently high to alter behaviour and drive action in the right direction and at sufficient pace. That will not happen with the levels currently discussed for such a levy. The needle will not be moved, and certainly not fast enough, by a carbon levy of a handful of US dollars or even a handful of ten US dollars per ton bunker, unless revenues are used to compensate 'early movers', or actors expect this to be merely a precursor for much higher levels later on. To have any significant impact, a levy must be an order of magnitude higher, in the range of several hundred US dollars per ton bunker.

This is because market prices today, without such a levy, are in the hundreds of US dollars per ton bunker, and daily, weekly and monthly variations can already be in the tens or even hundreds of dollars. So, a levy of say five or fifteen US dollars on top of a 600 US dollar bunker price will hardly make a difference for commercial decisions. But a 300 US dollar levy will, and a higher levy even more so.

This will, of course, push up the costs of seaborne transportation, which will not please cargo owners, companies, consumers or countries that are heavily dependent on shipping. But this is what it will take to decarbonise shipping at the scale and speed required to align with the goals of the Paris Agreement. And then, since shipping is such a cost-effective mode of transportation, the mark-up of prices on goods for industries and consumers will often be pretty modest. As we have discussed previously, when you for example purchase a pair of jeans or running shoes for US $150 in a fancy Manhattan boutique today, the overseas transport cost from China or India is typically a mere fifty cents. Even a doubling of this cost would add only half a US dollar to the selling price, if the entire increase were to be passed on to the consumer. For cars and heavier consumer items, such as washing machines and refrigerators, and goods like grain, iron ore, oil and gas, the cost increases will be more significant. This again raises the more fundamental question of 'who is going to pay?' for the green transition in shipping, and in other industries.

Either way, this is what it will take, and we should expect our governments to muster the will and political courage to agree on a levy, at sufficient levels, when negotiating rules for the decarbonisation of shipping under the auspices of the UN IMO.

Breaking Bread

'Playing the market' by way of a carbon tax will take us a long way in terms of mobilising the full potential of the insights, investments and innovations of shipping companies and other business sectors. This is important not solely because most of society's assets and activities are found in the private sector. The latter also brings to the table a broad range of expertise, an immense capacity and extensive practical experience in all the fields of significance to the green transition. Since companies know that profitability and commercial viability hinge on adaptation, innovation and the ability to change, they are usually in possession of a natural curiosity, a thirst for new knowledge and a tendency to question established truths.

Yet there are no legitimate answers to the large sustainability challenges if these companies wait passively for governments to take action, and if they are not also aware of their own, individual responsibility. This awareness appears to have matured over time. Previously, ethical discussions in the corporate world mainly revolved around what was *un*ethical, in the sense of what was not morally acceptable or legally permitted. Now, the focus of 'business ethics' is gradually shifting towards the imperative of how companies can actively commit and contribute to society as a whole.

In market economies, companies live and die in a symbiotic relation to the rest of society, and, in the long run, neither can function without the other. Society depends on companies to create economic growth, jobs and prosperity. Companies, for their own part, need policies, regulations and institutions that ensure legal safeguards, regulatory predictability and level competitive playing fields. They need an educated labour force, well-functioning infrastructure, reliable legal systems, orderly social conditions and an economy that generates sufficient demand.

A sustainable society is therefore a prerequisite for the viability and long-term profitability of companies. For this reason, companies with a long-term perspective have an enlightened self-interest in being on 'the right side of the future'. Responsible business leaders also understand that companies of today need formal, political and social 'licences to operate'. They know that there is no right way to do wrong things, and that they must do good to do well.

Perhaps business executives should also keep in mind that the word *company* has the same origin as the word *companion*. Both of these words derive from the Latin *panis*, meaning 'someone with whom you break bread'.

Dirty Business

In my opinion, the idea that companies must contribute to economic, social and environmental sustainability should be a given. Companies possess the knowledge, the means and the opportunities, and with this follows responsibility. This responsibility lies, not least, on the shoulders of the hundred or so companies who produce more than two thirds of the world's greenhouse gas emissions.[28] If the twenty-five largest oil companies continue business as usual, they can, all on their own, torpedo the world's chances of meeting the goals of the Paris Agreement.[29]

One third of all oil and gas production currently comes from the ocean, and the oil companies are an issue apart when it comes to discussions about the role of business in the green transition. There are many who would hold

10. HOPE ON THE HORIZON?

that these companies bear a disproportionate share of the responsibility for the climate crisis and that they have systematically suppressed information and misled the public about the consequences of burning fossil fuel. A number of companies continue to conduct extensive lobbying campaigns in support of activities that turn a profit on destruction of the planet. The oil and gas companies represent powerful economic and political interests and can as a group hardly be described as ardent advocates for the green transition. Many people consider oil and gas to be such a dirty business that they refuse to include these companies in discussions about the road forward.

At the same time, several of these companies are in the process of becoming important stakeholders in the development of renewable energy, such as offshore wind and solar power, even though for the majority most of their activities remain based on fossil-fuel production. For example, German RWE, whose former business model was largely based on the sale of electricity from coal- and gas-fired power plants, has now become one of the world's largest companies in the renewable energy sector. The French TotalEnergies, a major oil company, also has large renewable energy investments, although the latter represent only a very modest share of its balance sheet. The Danish Örsted is a former oil company that has transitioned to the production of renewable energy only. Texas is the largest oil-producing state in the USA, but it is also the largest in both solar energy and wind power. Norway is one of the world's largest producers of oil and gas, but all domestic energy production comes from hydropower, and almost all new cars sold are electric vehicles.

Large oil and gas companies possess technical expertise and innovative capacity which can make decisive contributions to the green transition, like developing and deploying offshore installations to harness electricity from wind, solar and ocean kinetic energy. They will also most likely be important players in producing and distributing the new low- and zero-emission fuels. Some of them are, moreover, also increasingly looking into the opportunities for using existing pipelines and building new infrastructure for large-scale carbon dioxide capture and storage (CCS) by injecting greenhouse gases in geologic formations under the seabed. While often presented as an important contribution to curbing global warming, the latter is also criticised by some as first and foremost a 'strategy for industry to stall action on climate and delay the phaseout of fossil fuels'.[30] Either way, such projects are still few and at an early stage, and there are questions raised about their feasibility and concerns about their potential adverse impact on ecosystems and the climate.

I understand the reasoning of those who want to exclude oil and gas companies from discussions about the green transition and related processes, but I have after some doubt and deliberation drawn a different conclusion. I believe we need all the expertise, insight and investments that the business world can mobilise if we are to succeed in the green transition. In a number of fields, oil and gas companies are uniquely qualified to offer valuable contributions. Therefore, I also believe that we should engage in a dialogue with these companies, as long as we are clear and consistent about the 'rules of engagement' and what we want to achieve.

Legal Risks

Beyond earnings, profits and market shares, there are few things people of the corporate world are more concerned about than risk, reputation and personal accountability. There are therefore few subjects that stir up greater discomfort than the prospect of high-profile lawsuits. Legal proceedings are costly and time-consuming, and often lead to media coverage that can undermine the reputation of the company and management. In criminal cases, both executives and board members risk exorbitant fines and prison sentences.

If mining companies, oil and gas companies, and fossil-fuel energy producers were to be held legally and financially liable for the actual consequences of their activities for global warming, environmental degradation, public health and the living conditions of future generations, we can only imagine the magnitude of the risk. Several thousand such cases are already being tried in a number of countries.[31] It is no longer widely accepted that society and the environment shall simply 'absorb' the damaging consequences of the fossil fuel industry's activities. Local governments, environmental organisations and activists in several countries have already filed environmental justice lawsuits and as of today around half of these cases have been upheld in favour of the plaintiff. In April 2024, for example, the complaint of a group of elderly Swiss women was upheld when they brought charges against Switzerland in the European Court of Human Rights the Hague.[32] The court ruled in favour of their claim that the Swiss authorities are not doing enough to protect the country's population from the hazardous consequences of the climate crisis, and that this is in violation of their right to life and health. These cases have, however, not yet been sufficient to sway the policies of governments or the business practice of fossil-fuel companies in any significant manner.

But now larger and more powerful forces are being mobilised than ever before. In the autumn of 2023 the state of California filed a lawsuit against five of the world's largest oil companies, holding them accountable for climate and environmental damages, forest fires and damage to the health of the state's forty million residents.[33] The lawsuit charges the companies of having carried out 'a public campaign aimed at deceiving consumers and the public', and of having 'conspired to conceal and misrepresent the known dangers of burning fossil fuels'. The lawsuit is important, not only because California is the world's fifth-largest economy, but also because the state is known for taking the lead on a number of climate and environmental issues. California is also a large-scale producer of oil and gas, so here the state government is challenging powerful corporations, strong capital interests and important constituencies within its own borders.[34]

The outcomes of tort law as practised in the USA can sometimes be quite extraordinary in terms of payout, and if Big Oil were to be held accountable for its practices, the amount of monetary compensation awarded would be astronomical. While US oil companies are already legally obliged to include the risk of lawsuits in disclosing 'material' risks, it seems reasonable to assume that, moving forward, such legal aspects will be assigned even more weight in their risk-reward calculations for current operations and new investment projects.

Moreover, in May 2024, the International Tribunal for the Law of the Sea ruled in favour of nine small island states who filed a petition demanding that the international community do more to stop global warming.[35] The milestone verdict stipulated that the obligations of the UN Convention on the Law of the Sea (UNCLOS) pertaining to prevention of marine pollution must be expanded to include greenhouse gas emissions that cause damage to life and marine ecosystems. The verdict thereby also explicitly addressed the ocean-climate nexus, establishing a legal link between the ocean and global warming.

As the number of such lawsuits is rapidly increasing all over the world, it seems reasonable to expect that governments and companies will be forced to 'internalise externalities' to a far greater extent than they do today.

A Sea-Blue Signature

It is no coincidence that the United Nation's trademark colour is blue – 'sea blue'. The ocean covers most of the planet and it is the human race's most important global commons.

Encouragingly, there is growing awareness and recognition of the vital importance of the ocean in dealing with humanity's most pressing challenges. In 2017, the UN General Assembly proclaimed the years 2021 to 2030 to be the UN Decade of Ocean Science for Sustainable Development, aimed 'to stimulate ocean science and knowledge generation to reverse the decline of the state of the ocean system and catalyse new opportunities for sustainable development of this massive marine ecosystem'.[36] Since then, a large number of new ocean research programs have been initiated, and marine biologists, oceanographers, climatologists and other subject matter experts from all over the world have intensified their exchanges of data, insights and information.

The UNCLOS and the comprehensive set of agreements in force under the UN International Maritime Organization (IMO), are also among the foremost examples of successful global collaboration. IMO governs global standards for shipping's greenhouse gas emissions, safety requirements, sailing procedures, insurance schemes, certification and environmental standards, not to mention standards for maritime worker competency, training and practical skills. The Maritime Labour Convention (MLC), under the auspices of the UN International Labor Organisation (ILO), regulates the rights of seafarers with respect to topics such as minimum wage, working hours, repatriation, medical treatment and holiday leave. There are basically no other industries subject to such pervasive regulation and standardisation on a global scale as international shipping. The regulations in some areas can be flawed and inadequate, but apply to all ships, ports, seafarers and dockworkers in international shipping worldwide.

It was uplifting when the United Nations member states in June 2023, after almost twenty years of studies and negotiations, finally adopted an international binding agreement regarding the conservation and sustainable use of marine biological diversity in areas beyond national jurisdiction. The agreement encompasses two thirds of the world's ocean regions and among insiders is referred to as BBNJ, Biodiversity Beyond National Jurisdictions.[37] It covers conservation-related topics, such as marine protected areas and environmental impact assessments, and also promotes equity and capacity building of developing countries, along with fostering research and development. The agreement has, however, not yet entered into force and a number of the larger states have indicated that they will not ratify it.

Since the UNCLOS, Decade of Ocean Science, the BBNJ agreement and the IMO and ILO agreements are all under the auspices of the United Nations, they are also important in terms of upholding the international organisation's authority and reputation.

But, conversely, this also means that the effectiveness of such vitally important sets of agreements governing the ocean and maritime activities will be reinforced or weakened in accordance with fluctuations in the overall global support and legitimacy of the United Nations.

10. HOPE ON THE HORIZON?

The Permanent Five

The overall purpose of the United Nations is to prevent war and conflict and promote stability, peace and cooperation. As we have discussed, the first sentence of the organisation's charter establishes that its most important task is 'to save succeeding generations from the scourge of war, which twice in our lifetime has brought untold sorrow to mankind'. It is an estimable ambition, and today's faltering authority of the United Nations is not due first and foremost to its objectives and underlying philosophy. The most important reason lies in the outdated governance model and voting rules, which do not adequately reflect the world of today.

The UN Charter is based on the principle that all member states are sovereign and equal. Each member therefore has one vote in the UN General Assembly. This means that small states such as Vanuatu and Kiribati in principle have the same voting power as the USA and China. It is nonetheless not the General Assembly that is the organisation's most powerful and important body. The real power lies in the Security Council, the primary tasks of which are to preserve international peace and security and to hinder the outbreak of armed conflicts. All UN member states are bound by the resolutions of the Security Council. In practice, the Security Council also decides who will be the secretary-general of the organisation.

The Security Council is made up of five permanent members – the USA, China, Russia, France and Great Britain – all of whom were the predominant victors of the Second World War. The council also has ten elected members from the remaining 188 member states, each of whom serve two-year terms. The 'Permanent Five', but none of the ten elected members, have veto rights and can stop proposed resolutions. That means that none of the other 188 member states has the same power or influence as the five permanent members.

This is of course profoundly unfair and fundamentally undemocratic. The Security Council's make-up is also increasingly a historical anachronism, a reflection of yesterday's world. Today, seven of the ten most populous countries, and six of the ten largest economies, do not hold a permanent seat on the Security Council, solely due to their historical legacy at the time of the founding of the United Nations.

The veto right, which the Permanent Five have exercised several hundred times since 1945, has also in many cases paralysed discussion within the United Nations as well as its capacity to take action. The USA has systematically rejected critique of Israel's violations of Palestinians, while China and Russia have repeatedly hindered the United Nations from intervening to halt the Syrian regime's massacres of its own population. Russia has vetoed all condemnations of its attacks on Georgia and Ukraine.[38]

There is widespread recognition, also among the five permanent members of the Security Council, that the United Nations' governance structure is not adapted to the realities and requirements of today. For many years, work has therefore been underway on the revision of the organisation's voting rules and working procedures. But all attempts at meaningful reforms have foundered because they require the approval of the same states currently holding the most powerful positions. The Permanent Five have blocked almost all proposals that do not serve their own interests.

In today's world, no member state will voluntarily surrender power and influence, not even 'to save succeeding generations from the scourge of war'.

Law of the Jungle

The core ideology of the UN-led, rules-based world order is to 'replace might with right'. The United Nations is a voluntary association that is inherently asymmetric: strong states, in principle, agree to forgo opportunities that would unilaterally further their own interests at the cost of the interests of weaker states. This represents an anomaly in the history of the human race, throughout the better part of which the law of the jungle prevailed and rights were defined by might: strong countries did as they wished, the weak as they were obliged to do. Therefore, the current world order is often perceived to be of greater importance for small and medium-sized countries than for those that are large and powerful.

Despite its many inconsistencies, weaknesses and flaws, the United Nations remains a highly radical, modern and advanced construction. The organisation's sea-blue logo is a badge of honour for one of the greatest advances in the history of civilisation.

It is the strongest states that hold the key to reform, development and the enduring legitimacy and authority of this system. Due to their unique power and position, they also have a particular responsibility when it comes to acting in accordance with the UN Charter and upholding the organisation's authority and legitimacy. These are the states who in the event of any reforms must relinquish power and who, in principle, might believe that they would be better off without the encumbrance of global laws and institutions. There is little cause to nurture naive ideas about the motives and priorities of major powers. It is not altruism that leads them to accept the binding nature of their involvement, nor consideration or empathy for smaller and weaker states.

10. HOPE ON THE HORIZON?

The major world powers accept the binding obligation only because and as long as they see the benefits of a UN system and a rules-based world order which contributes to structuring and stabilising international relations. They know that a United Nations with sufficient legitimacy, authority and a clear mandate can help preserve peace, manage mutual challenges and promote economic growth and prosperity. For the major powers, self-binding can therefore be a rational choice in terms of protecting their own interests.

Without the major powers' contribution to fundamental reforms, however, there is a danger that the United Nations will be further debilitated. But still, even without reforms, the world is probably better off with a weakened and defective global organisation than without any global system whatsoever. The alternative to the UN-led, rules-based world order could easily be a world of chaos and disorder.

Such an outcome would not bode well for any of us, whether you are a citizen of a superpower or a small island state.

Group Meetings

As the United Nations is being eroded by an increasingly obsolete governance structure and growing geopolitical tensions, other bodies are acquiring greater significance in the realm of international collaboration. Regional agreements have been formed in a number of fields which are often more comprehensive, detailed and binding than those the members countries have been able to achieve in the United Nations. The foremost example is the European Union, but in Africa, South America, Central Asia and Southeast Asia more comprehensive systems of agreements are being developed for trade and economic cooperation between states.

Important and influential international organisations are also emerging across geographic regions. BRICS, which was originally made up of Brazil, Russia, India, China and South Africa, has been expanded to include Iran, Saudi Arabia, Egypt, Ethiopia and the United Arab Emirates. BRICS+ includes almost half of the world population, half of the world's oil and gas production, and one third of the global economy.[39] It is in principle a body for economic cooperation, but several of the BRICS countries have expressed explicit ambitions of challenging US power, the position of the West, and the terms of today's UN-led, rules-based world order.

We also find important alternative forums for global collaborations among the 'G countries' (G stands for Group). G77 was founded in 1964 by seventy-seven developing countries in the Global South who formed a coalition to promote their joint economic interests (G77 has since been expanded to include 134 countries). G7 was founded in 1975 as a forum for the seven wealthiest democratic countries: USA, Japan, Germany, Great Britain, Italy, France and Canada. G20 was born in 1999 as an international forum for central bank governors and finance ministers in the aftermath of the Asian financial crisis. Ten years later, after the global financial crisis in 2008, it was upgraded to a forum for prime ministers and heads of state. Today, G20 comprises the world's nineteen largest economies, the EU and the African Union. Neither G7 nor G20 has a permanent secretariat, and the hosting of the annual meetings rotates among the members.

Within the G format, countries can convene without confrontations over ideological differences, without the headaches of bureaucratic voting rules, and without the inconvenience of having to take into account the legitimate interests and concerns of the rest of the world. At these meetings, the participating countries need not pretend that they share common values, ideals and principles. If any of the countries were to object, there is little room for discussion, and even less so for joint statements, on human rights, democracy or freedom of expression. At the G20 meetings, neither Russia's attack on Ukraine, China's violations of the Uyghurs nor the USA's support of Israel's military operations in Gaza figures prominently on the agenda.

The G countries meet, discuss and form agreements based on a selection of their own interests and what they at any given time view as mutually relevant, current and important topics.

Fewer Hands Make Lighter Work

This means that the international forums for collaboration that are now emerging alongside a weakened United Nations do not espouse a holistic global perspective. They are either regional or selectively global. On the other hand, the G20 countries represent four fifths of the world economy and total greenhouse gas emissions, and two thirds of the world population.[40] Since it should in principle be simpler for the twenty-one members of G20 to reach an agreement than for the 193 member states of the United Nations, these forums can bring about more rapid policy decisions and advances in important spheres such as climate change, the ocean, nature and the environment.

Ocean stakeholders, from activists and scientists to politicians and CEOs of major maritime corporations, therefore welcomed with open arms the establishment of the Ocean20, 'O20', at the G20's summit meeting in Indonesia in the autumn of 2022. The O20 is a forum with an agenda dedicated to the ocean. In 2024, ocean-related work was further strengthened and formalised under Brazil's leadership, and I had the pleasure of co-chairing the process leading up to the G20 Leaders' Summit held in Rio de Janeiro in November 2024.

10. HOPE ON THE HORIZON?

However, in practical terms, the G format fosters fragmentation, de-ideologisation and a recalibration of global cooperation. By definition, the G7 and G20 forums represent states with actual economic, political and military power who are the puppet masters of international relations, irrespective of their historical ballast and voting rights in the United Nations. The G format can simultaneously be shaped and adapted in response to changing issues and power relations. It is, for example, easy to envision other G formats designed for different purposes: a G13 for the most environmentally ambitious countries? A G151 for the countries with a coastline or a G44 for those without?

So far, the G forums have primarily been focused on trade and economics. The shared issues of the human race are addressed only to the extent they are considered of mutual interests, but here as well we can discern seeds of hope. If the international community fails to halt global warming and protect the ocean, this will have consequences for every country, including the richest and most powerful. This recognition was the basis for an announcement made in the autumn of 2023 by the USA and China, who together are responsible for forty per cent of the world's greenhouse gas emissions, stating that they would resume their collaboration on measures to curb global warming. The former dialogue had been backburnered for several years due to the tense relationship between the two countries.[41]

It is a source of hope that a 'G2' of the two superpowers can be a driver in curbing global warming and environmental degradation, even though they are rivals and competitors in other sectors. Maybe these two countries can also find more common ground and identify more shared interests by exploiting their differences. While the USA, as we have discussed in Chapter 9, is largely self-sufficient with fresh water, food, energy and raw materials, China is strategically and economically heavily dependent on importing these resources. Unless the USA would wish to see China and its large population increasingly struggling, maybe behaving even more assertive and aggressive to secure access to these critical resources, it could be in the enlightened self-interest of the USA to invite cooperation, rather than confronting China on these issues. Framing, for example, the issue of IUU and sustainable management of wild fish stocks in terms of national security, as a common interest in ensuring international peace and stability, could also add significant weight to the collaboration needed to preserve marine resources.

With rational, informed leaders taking a long-term holistic approach, acting in the enlightened self-interests of their respective countries, it should still be possible for the world's superpowers to join forces and combine resources in finding solutions for some of the human race's largest common challenges.

The G Clef

Part of the growing appeal of G forums is their selective focus on joint interests and challenges, rather than ideologies, values and human rights. Perhaps, therefore, the G format holds the potential to adjust and calibrate the level of ambition in international talks and commitments moving forward, in much the same way that a G clef calibrates the pitch of a musical composition.

It is not difficult to imagine that, if the United Nations were to function more like a 'G-193', it would have resulted in less friction and provided more momentum in the work on common, existential challenges. This would then also bring the United Nations closer to its origins, when the principles of non-intervention and the sovereignty of member states were of central importance. There would be no United Nations if the USA and the Soviet Union had been obliged to find common ground on ideology, values and how to govern their own countries. The text of the UN Charter contains lofty formulations, not least about human rights and freedom of expression, but the interpretation of such ambitions was in practice left up to the individual member states. The members decided for themselves how they would manage their societies and govern their citizenry.

This changed when the Berlin Wall fell in 1989 and the Soviet Union collapsed. In previous chapters, we saw how the fall of the Berlin Wall was interpreted in the West as 'the end of history'. The fall of the Berlin Wall also inaugurated the period of uncontested US global hegemony and power of definition in international relations. From that moment on, liberal democracy, the principles of market economics and the West's interpretation of human rights and 'universal values' were the norm and constituted the underlying premise of the global community. But now the USA's international position has been weakened, China is growing and the world is characterised by greater ideological diversity. Without an unrivalled hegemon, neither will there be a global enforcer of 'universal' norms, the role the USA has played in a number of contexts, such as in upholding the Convention on the Law of the Sea. In such a case, we would be all the more dependent on collaboration based on genuine common interests and agreements that are respected by all parties.

The G format's approach can therefore provide a key to strengthening the United Nations, by bringing the organisation back to its origins through a sharper focus on common, existential challenges. Such an approach will, however, be painful for many of us, as it will simultaneously gloss over ideological differences and commitments to shared interpretations of human rights and 'universal values'.

Either way, all countries, regardless of size, location, power and ideology, should strengthen their dedication and commitment to the United

10. HOPE ON THE HORIZON?

Nations' ocean-related work. There is no ideology involved in recognising the significance of international collaboration for the task of protecting, preserving and restoring marine life and ecosystems. This task goes hand in hand with leveraging the ocean's resources and industrial potential in a truly sustainable manner and thereby contributing to a better life for more people on our planet. There is little that should be controversial about the idea that only a global organisation with legitimacy, authority and a strong mandate can achieve this effectively.

In the long run everyone, regardless of political leanings and ideology, will benefit from a mutually respected, joint regulatory system for the human race's largest and most important public commons.

Anything I Can Do?

Many people experience feelings of helplessness and discouragement when faced with the great challenges and powerful forces defining our times. Many shrug their shoulders in resignation, thinking that there is little they can do to make a difference when it comes to solving existential challenges such as curbing global warming or improving the health of the ocean.

But who, then, will make that difference, if not people like you and me?

First of all: nation states, organisations, companies and markets are not the outcome of natural laws. They are created, developed and run by human beings. We are 'the system'. It is individuals who fulfil functions, formulate standpoints, and make decisions affecting society and our daily lives. It is grandparents, mothers, sons, friends and lovers who lead states and fill the offices of governments, powerful corporations and international organisations. This gives each of these individuals opportunities to influence, create change and bring about improvements – and with these opportunities follows a responsibility.

Second of all: the great challenges we are facing are human-made and can only be addressed by human beings. It is the sum of our attitudes and actions that have brought us to where we are today, and that will determine our future and our destiny. It is individuals who have opinions, convictions and ideologies. It is individuals, such as you and me, who create momentum – and resistance.

Third: feelings of helplessness and discouragement can be intensified or alleviated by political discussions, the public discourse and conversations among friends. We must help one another to keep our spirits up and stay optimistic even if we should fail to meet the Paris Agreement's goal of 1.5°C, as all signs indicate will come to pass. Every one tenth of a degree counts in the context of fighting global warming: 1.6°C has much greater consequences than 1.5°C, and 1.7°C is much worse than 1.6°C.

The consequences increase exponentially and dramatically with each tiny, incremental temperature increase.

Fourth: we have the means and the tools. We already have sufficient knowledge, technology, resources and money to curb global warming, take care of the ocean, stop the destruction of nature and create a better future for more people here on Earth. We can do it if we want to, if we do it together – and if we do it now.

Fifth: each of us can contribute in our own way. Nobody can do everything, but we can all do something. We can help clean up beaches, choose green alternatives, eat less meat and throw out less food. We can recycle and sort waste, reuse that old jacket and repair our bicycle instead of buying a new one. We can exercise agency as consumers and shareholders to support companies making strides in the right direction and avoid those that are destroying the planet for our children and grandchildren. We can stop subscribing to media platforms that profit from lies and distractions. As leaders and employees at all levels we can stand up for what is right and take a stand against what is wrong. We can withdraw from trade unions that are clinging to the past and join organisations that are dedicated to accountable change. We can raise our voices and get involved to help stop the destruction of natural habitats and traditional ways of life for the coming generations. We can seek out knowledge and inform ourselves about important issues. We can play a part as elected officials, business executives, teachers, journalists, celebrities, influencers, parents, friends, colleagues and concerned citizens in generating awareness, shaping attitudes and inspiring action.

Sixth: in democratic countries we can use our vote to sort parties and politicians from the source, based on their accountability, decency and honesty. We can turn our backs on populist charlatans offering simple 'solutions' to complex challenges. Boisterous calls to 'Make Great Again!' are a recipe for misery, war and catastrophes. Global warming, sea-level rise and the loss of nature and biodiversity can only be addressed through greater collaboration and less rivalry across social, political and national divides.

And finally, whether you are a prime minister or a parking attendant, a director or a doorman, you make choices constantly. Every day we make large and small choices that can make a difference. We can choose to approve or refuse. We can choose action, passivity or indifference. We can refrain from taking a stand, but that is also a choice. We can never choose *not* to choose. It is the sum of our choices that will decide the fate of our children and the future of life on earth.

The responsibility for our beautiful and fragile blue planet will always lie with you – and me.

EPILOGUE

In my current position I work with aspects of the challenging, complex – and inspiring – issues addressed in this book. It was not long ago, however, that I still felt that many of them were quite remote, inessential and intangible. They were 'out there' and under the purview of 'others'. Global warming, the degradation of nature and loss of biodiversity were dark clouds on a distant horizon: visible and ominous, but without sufficient imminence and importance to influence the concerns and priorities of everyday life.

In just a short period of time, the clouds have drawn closer, the gravity of the situation has become more evident and the consequences more visible. It is impossible to remain unmoved by the images that reach us every day from different parts of the world of desperate human beings and animals suffering from the impact of hurricanes, heatwaves, drought, floods, landslides and forest fires. The challenges facing the planet are imposing themselves on our daily lives as sources of concern, astonishment and discussion. Climate change, environmental issues and 'sustainability' are no longer solely the province of subject matter experts, politicians and corporate leaders, nor concerned citizens of the adult world. The younger generations are taking a stand – voicing their worries and opinions, raising awareness and demanding the discussions and decisions needed today for our planet of tomorrow. I have been fortunate enough to be invited on several occasions to discuss these topics with pupils and students at primary schools, colleges and universities. The young people's questions, knowledge and energy always serve as a source of hope, inspiration and important reflections.

I do, however, feel a growing sense of disquiet. I have with time been more affected by the contrast between the young people's giddy joy about life and the recognition of the challenges we are in the process of passing on to them. The painful recognition that it is their generation who will have to shoulder the responsibility and take action where my own has failed.

It therefore made a particular impression on me when my grandson Theodor asked me one day if there will still be fish in the ocean and snow in the mountains when he and his brother Ferdinand grow up. It was a Sunday in October and I was out walking in the forest with my eldest daughter's twelve-year-old twins. They are usually as carefree, playful and cheerful as

THE OCEAN

Felix, our energetic family dog, but suddenly the existential problems of the world were close-up and personal. It was the start of a long conversation about important questions, so we sat down on the trunk of a fallen tree. I tried my best to offer compelling, comprehensible answers, communicated with a grandfather's reassuring calm and optimism.

The boys' concerns were appeased, and just a few moments later they were playing with Felix as if they hadn't a care in the world. But I felt a jab in my heart and a hollowness in my belly. It is painful when twelve-year-old boys are worried about the future. It feels wrong when children are wondering about whether they will grow up in a world they no longer recognise. And I was not truly convinced by my own words. It is becoming ever more challenging to be professionally honest and personally optimistic at the same time.

But I also know that it is fully possible to take better care of our beautiful blue planet and to ensure good, safe and prosperous lives for coming generations. We already have the knowledge, technology and resources we need. In key areas we are on the right track, although the development is moving slowly.

On a grey, blustery evening a few days later I was strolling on the beach. The ever-playful Felix ran hopefully towards me carrying sticks to be thrown and fetched. But I was immersed in my own thoughts. I was thinking about my daughters and the conversation with my grandsons. About their concerns, dreams, hopes, aspirations and expectations for what life and the future may bring.

The cloud cover broke apart, and the flat, dark surface of the ocean reflected the image of a shiny, harvest full moon. I could feel tears welling up.

Alan's tears.

ACKNOWLEDGEMENTS

This book would not have seen the day of light without the inspiration, support and feedback from prominent international experts, great colleagues, friends and family. I am deeply grateful for how they have all generously and confidently shared their perspectives, views and insights on various topics.

Alf Håkon Hoel, Professor, UiT The Arctic University of Norway, has gone through the entire manuscript in different versions and provided invaluable reflections and factual feedback. Ambassador Svein Ole Sæther, Norway's ambassador to China 2008 to 2017, has shared his views on a first version of the manuscript. Carl-Johan Hagman, President and CEO of the NYK Group Europe, has provided detailed suggestions and factual input to major sections of the book. Storm Drage Giercksky Kleiva, master student at the London School of Economics, has contributed refreshing feedback on an early version of the manuscript.

In writing the chapters of the book addressing the ocean as a biosphere and the ocean-climate nexus, I am heavily indebted to Vladimir Ryabinin, Executive Secretary of the Intergovernmental Oceanographic Commission of UNESCO and Assistant Director-General of UNESCO 2015–2024. I have also received invaluable feedback from dr. Kilparti Ramakrishna, Director of Marine Policy Center and Senior Advisor to the President on Ocean and Climate Policy, Woods Hole Oceanographic Institution; Robert Blasiak, Associate Professor, Stockholm Resilience Center; Henrik Österblom, Science Director at the Stockholm Resilience Center and Professor at Stockholm University and the Royal Swedish Academy of Sciences; Jan-Gunnar Winther, Pro-rector Research and Development, UiT The Arctic University of Norway, and Director, Norwegian Polar Institute 2005–2017; and Fredrik Myhre, Ocean Team Leader, World Wildlife Fund Norway.

For the parts of the book discussing historical developments, geopolitics and global governance, Ulf Sverdrup, Professor, Norwegian Business School (BI), and Director, Norwegian Institute of International Affairs (NUPI) 2011-2023, has generously shared his vast insights about these complex topics. For these parts of the book, I have also received invaluable feedback from Hiroko Muraki Gottlieb, Representative for the Ocean, International Council of Environmental Law and Associate, Department of Organismic and Evolutionary Biology, Harvard University; Maria Damanaki, Global Managing Director, Oceans, The Nature Conservancy, and European Commissioner for Maritime Affairs and Fisheries 2010–2014; Whitley Saumweber, Director, Stephenson Ocean Security Project, Center for Strategic and International Studies (CSIS), and President Barack Obama's Associate Director for Ocean and Coastal Policy in the White House Council on Environmental Quality; Halvard Leira, Research Director, NUPI; Bonnie Glaser, Managing Director of the Indo-Pacific Program, German Marshall Fund; Admiral Haakon Bruun-Hanssen, Chief of Defense Norway 2013–2020; Vice-Admiral Nancy Hann, Director, US National Oceanographic and Atmospheric Administration (NOAA); Elana Wilson Rowe, Research Professor, NUPI, and Associate Professor at the Norwegian University of the Life Sciences; Bård Vegar Solhjell, Director, Norwegian Agency for Development Cooperation (NORAD); Ole Jacob Sending, Research Professor, Head of Center for Geopolitics, NUPI; Johan Bergenas, Senior Vice President, Oceans, World Wildlife Fund USA; Ambassador Wegger Strømmen, Secretary General, Norwegian Ministry of Foreign Affairs, and Deputy Foreign Minister 1998–2000; Hans Tino Hansen, CEO, Risk Intelligence; and Dame Sara Thornton, Professor of Practice in Modern Slavery Policy, University of Nottingham, and Independent Anti-Slavery Commissioner for the United Kingdom 2019–2022.

The chapters on the opportunities, dilemmas and challenges offered by the 'blue economy' and current and emerging ocean industries, business perspectives, regulatory aspects and the policy options going forward, have been commented by Guy Platten, Secretary General, International Chamber of Shipping (ICS); Stephen Cotton, General Secretary, International Transport Workers' Federation (ITF); Gerardo A. Borromeo, CEO, PTC Holdings Corporation; Thomas Thune Andersen, Chair of Lloyd's Register Group (LR), and Chair of Orsted 2014–2024; Esben Poulsson, Immediate past Chairman, ICS; Lynn Loo, Professor, Princeton University, USA, and CEO, Global Center for Maritime Decarbonization, Singapore; Svein Ringbakken, CEO, The Norwegian Shipowners' Mutual War Risks Insurance Association (DNK); Thor Jørgen Guttormsen, President of the Norwegian Shipowners's Association 2010–2012 and former CEO of Høegh Autoliners and Høegh LNG; Rolf Thore Roppestad, CEO, Gard; Andreas Nordseth, Director-General, Danish Maritime Authority, and Chair, European Maritime Safety Agency (EMSA) 2017–2023; Walter Qvam, Chair SINTEF and PGS, and President and CEO, Kongsberg Gruppen 2008–2016; Wenche Grønbrek, Director of Strategy at SeaBOS, and former Global Head of Sustainable Development at Cermaq (Mitsubishi Corporation); Simen Bræin, Senior

ACKNOWLEDGEMENTS

Vice President, Statkraft; Bo Cerup-Simonsen, CEO, Mærsk Mc-Kinney Møller Center for Zero Carbon Shipping; and Harald Magnus Andreassen, Chief Economist, SpareBank 1 Markets.

General Eirik Kristoffersen, Chief of Defense Norway; John Hammersmark, Director, Security and Contingency Planning Department, Norwegian Shipowners' Association 2015–2022; and Brigadier Frode Kristoffersen, Norwegian Senior National Representative to the US Central Command, and Commander, Norwegian Special Operations Commando (FSK) 2014–2017, have gone through major sections of the book and given me the kind of candid and constructive comments that I always appreciate from these highly decorated officers who have all served in the Norwegian Special Forces.

Ignace Beguin Billecocq, Ocean and Coastal Zones Lead, Climate Champions, United Nations Framework Convention on Climate Change (UNFCCC); Kjersti Aass, Program Manager; Suzanne Johnson, Senior Advisor; and Vincent Doumeizel, Senior Advisor; all good colleagues at the UN Global Compact Ocean Stewardship Coalition, have provided detailed and constructive feedback on several parts of the book.

Lise Kingo, Independent Board Director, and CEO and Executive Director of the UN Global Compact 2015–2020, has generously shared her views on the final sections of the book.

I thank them all for their invaluable inputs and comments. Any errors, omissions and misunderstandings are, however, wholly and fully my own.

A warm thanks to Diane Oatley who has been patiently overbearing with my many revisions and last-minute additions to the text, and who has competently and meticulously translated a manuscript spanning such a wide range of subject matters, each with their own particular 'lingo' and expressions.

A warm thanks also to Christian Müller, editor at Hero Press, the Legend Times Group, who from the very beginning believed in this project, and who has guided me through the process in such a calm, professional and friendly manner.

I will forever be indebted to my good friend and long-time colleague Erik Giercksky, with whom I have had the great pleasure and privilege of leading the ocean work of the UN Global Compact since the inception in 2018. Thank you, Erik, for the inspiration and insights I have gained from our cooperation and strategic discussions, and from our daily chats about everything from domestic politics to training, arts and family life. Your energy, creativity and unwavering support have been indispensable in writing this book.

My dear Kristin, my partner in life and mother of our lovely twin-daughters, has commented on major sections of the book, contributing the wise and vital perspectives of an informed, concerned citizen who is not part of the 'blue congregation' of maritime experts. She has also generously shouldered significantly more than her fair share of the practicalities of our everyday family life to ensure that I could devote the time and attention necessary to write this book. Thank you, my dear.

And last, but not least, a few words of gratitude to my beloved daughters and grandsons, Christine, Cecilie, Silja, Mathilde, Theodor and Ferdinand, for your support and affection. Your youthful, energetic inspiration, your questions and concerns, and your dreams and hopes for what the future may hold, has instilled a special sense of mission, importance and urgency to this whole project. Therefore, this book is dedicated to each and every one of you – and to your fellow sisters and brothers of present day and coming generations.

Photo: UNGC

ABOUT THE AUTHOR

Since 1 January 2018, Sturla Henriksen has been Special Advisor, Ocean, to the United Nations Global Compact (UNGC), New York, chairing the organisation's global ocean-related projects with corporations, research institutions, international regulatory bodies and national governments. With a membership of more than 25,000 companies and civic society organisations, the UNGC is the world's largest corporate sustainability initiative and the United Nations' main body for cooperation with the international business community. The UNGC Board is chaired by the United Nations secretary-general.

Sturla Henriksen has previously been chief executive officer of the Norwegian Shipowners' Association, director at Astrup Fearnley Asset Management, and a senior executive at Accenture. He has held positions in the EU Commission Services (Brussels), the European Free Trade Association (Geneva), the Norwegian Ministry of Finance, Statistics Norway, and as officer in the Norwegian Armed Forces.

He graduated *Candidatus Oeconomiae* from the University of Oslo, and his educational background further comprises the Norwegian Army Officer School, the BI Institute of Business Administration, the Norwegian Defense College, and the Advanced Management Program, INSEAD (France).

He has been co-chair of the Oceans20 Engagement Group under the G20 presidency of Brazil, a board member of the Norwegian Institute of International Affairs (NUPI), and a member of the Advisory Board for the Executive Ocean Leadership Program, UiT The Arctic University of Norway. He has previously served on the boards of the International Chamber of Shipping (ICS) and the European Community Shipowners' Association (ECSA), and as deputy chair of the Council of Det norske Veritas (DNV). He has been a Norwegian representative to the Trilateral Commission, member of the World Economic Forum's Arctic Expert Panel, and member of the China Council Task Force on Green Maritime Operations, an advisory body to the Chinese government.

He has for many years been a licensed scuba-diver instructor.

Sturla Henriksen is the father of four daughters and has two grandsons. He lives with his family in Oslo, Norway.

Notes

PROLOGUE

1. https://oceanpanel.org/publication/ocean-solutions-to-climate-change/.

1. ALAN'S TEARS

1. https://www.space.com/20567-james-irwin-apollo-15-astronaut.html.
2. Irwin, Mary; Harris, Madalene (1978). *The Moon is Not Enough*. Zondervan Publishing House.
3. https://www.nrk.no/klima/ny-fn-rapport-om-klima_-aldri-malt-var-mere-hav-enn-i-2023-1.16908585.
4. https://www.state.gov/special-guest-remarks-at-ocean-climate-ambition-summit/.
5. https://oceanpanel.org/publication/ocean-solutions-to-climate-change/.
6. Eliot Cohen, *The Big Stick – The Limits of Soft Power and the Necessity of Military Force*, Basic Books, 2016, p. 185.
7. https://www.ibtimes.com/fisherman-catches-15-dead-bodies-his-net-over-3-days-migrants-escape-poverty-3701501.
8. Jan Zalasiewicz and Mark Williams, *Ocean Worlds – The story of the seas on Earth and other planets*, Oxford University Press, 2014.

2. POWERFUL, DARK AND MYSTERIOUS

1. https://www.unf.edu/newsroom/2024/07/shark-bites.html.
2. https://www.frontiersin.org/articles/10.3389/feart.2019.00225/full.
3. https://www.frontiersin.org/articles/10.3389/feart.2019.00225/full.
4. https://earthsky.org/earth/earths-largest-waterfall.
5. Jan Zalasiewicz & Mark Williams, *Ocean Worlds – The story of the seas on Earth and other planets*, Oxford University Press, 2014, p. 18.
6. https://www.unesco.org/en/articles/seabed-2030-announces-latest-progress-world-hydrography-day.
7. https://oceanexplorer.noaa.gov/facts/explored.html.
8. https://oceanservice.noaa.gov/facts/ocean-species.html.

9   https://oceandecade.org/.
10 https://www.unicef.org/wash/water-scarcity.
11 Philip Ball, *H2O – A Biography of Water*, Phoenix Paperback, London, 2000, p. 141.
12 Philip Ball, *H2O – A Biography of Water*, Phoenix Paperback, London, 2000, p. 144.
13 https://www.dagsavisen.no/kultur/2014/05/21/oppvarming-av-luft-og-hav-og-virkningen-pa-klimaet/.
14 https://www.lse.ac.uk/granthaminstitute/explainers/what-role-do-the-oceans-play-in-regulating-the-climate-and-supporting-life-on-earth/.
15 https://www.tu.no/artikler/hvis-vi-klarer-a-hente-opp-sa-mye-som-noen-brokdeler-vil-det-vaere-nok-til-a-forsyne-hele-jordkloden-med-energi/275826.
16 https://www.goodreads.com/quotes/273936-all-of-us-have-in-our-veins-the-exact-same.
17 https://www.frontiersin.org/articles/10.3389/feart.2019.00225/full.
18 https://www.pnas.org/content/115/25/6506.
19 https://www.visualcapitalist.com/all-the-biomass-of-earth-in-one-graphic/.
20 https://www.forskning.no/havforskning-havforskningsinstituttet-partner/velkommen-til-hvalenes-forunderlige-verden/1313724.
21 http://www.bbc.com/earth/story/20150415-the-loneliest-whale-in-the-world.
22 https://hal.science/hal-00653068/document.
23 https://www.newscientist.com/article/2182882-mantis-shrimps-punch-with-the-force-of-a-bullet-and-now-we-know-how/.
24 https://www.cell.com/matter/pdf/S2590-2385(21)00359-3.pdf.
25 https://ocean.si.edu/ocean-life/invertebrates/giant-squid.
26 https://www.abcnyheter.no/helse-og-livsstil/2012/08/25/157639/dypets-smarteste-blotdyr?redirect=true.
27 https://www.nrk.no/urix/mystisk-dypvannsfisk-lever-i-hundre-ar-og-er-gravid-i-fem-1.15543093.
28 https://forskning.no/dyreverden-havet-partner/15-ting-du-kanskje-ikke-visste-om-delfiner/1360262.
29 Jonathan Balcombe, *What a Fish Knows – The Inner Lives of Our Underwater Cousins*, Farrar, Strauss & Giroux, 2017.
30 https://www.sciencedirect.com/topics/agricultural-and-biological-sciences/diatom.
31 https://www.nature.com/articles/s41561-024-01480-8.
32 https://theconversation.com/humans-will-always-have-oxygen-to-breathe-but-we-cant-say-the-same-for-ocean-life-165148.

NOTES

33 htps://www.tv2.no, 24 March 2018.
34 https://www.mn.uio.no/ibv/tjenester/kunnskap/plantefys/leksikon/s/sverdrup.html.
35 https://cp.copernicus.org/articles/14/751/2018/.
36 https://oceanservice.noaa.gov/education/tutorial_currents/04currents1.html.
37 Patrik Svensson, *The Book of Eels – Our Enduring Fascination with the Most Mysterious Creature in the Natural World*, Ecco, USA, 2020.
38 Jonathan Balcombe, *What a Fish Knows*, Oneworld Books, UK, 2017, p. 50.
39 https://framsenteret.no/nyheter/2018/04/24/10-ting-du-ikke-visste-om-havet/.
40 https://thebreakthrough.org/articles/remember-the-guano-wars.

3. POWER AT SEA

1 *Oslo Energy Forum*, 2011.
2 Erik Hobsbawn, *Ekstremismens Tidsalder – Det 20. Århundrets Historie*, Gyldendal Norsk Forlag, 1997, p. 7.
3 Francis Fukuyama, *The End of History and The Last Man*, Free Press, USA, 1992.
4 Fareed Zakaria, *The Post-American World*, 2008, p. 34. Available at: https://projects.mcrit.com/foresightlibrary/attachments/Post_american_world.pdf.
5 https://twitter.com/africanhub_/status/1787604707109749178?s=43&t=aXg_c__h22tJotm5tF1deA.
6 Sathnam Sanghera, *Empireland – How Imperialism has Shaped Modern Britain*, Penguin Random House, UK, 2021, p. 200.
7 Sathnam Sanghera, *Empireland – How Imperialism has Shaped Modern Britain*, Penguin Random House, UK, 2021, p. 201.
8 Sathnam Sanghera, *Empireland – How Imperialism has Shaped Modern Britain*, Penguin Random House, UK, 2021, p. 198.
9 https://www.britannica.com/biography/Ibn-Khaldun.
10 https://www.history.navy.mil/research/library/research-guides/z-files/zb-files/zb-files-m/mahan-alfred.html.
11 https://www.silkroadstudies.org/resources/pdf/Monographs/1006Rethinking-4.pdf.
12 Samuel P. Huntington, *The Clash of Civilizations and the Remaking of World Order*, The Free Press, 2002, UK, p. 46-47.

13. Jared Diamond, *Guns, Germs, and Steel: A Short History of Everybody for the Last 13,000 Years*, W.W. Norton, New York, 1997, p. 409-10.
14. https://www.history.com/.amp/news/7-influential-african-empires.
15. *Foreign Affairs*, May/June 2022, p. 183.
16. Yuval Noah Harari, *Sapiens – A brief history of mankind*, Random House, 2014, p. 285.
17. Howard W. French, *Born in Blackness - Africa, Africans, and the Making of the Modern World, 1471 to the Second World War*, Liveright Publishing Corporation, US, 2021, p. 125.
18. Sven-Eric Liedman, *Den moderne verdens idéhistorie – I skyggen av fremtiden*, Dreyers Forlag, Oslo, 2016, p. 430.
19. https://www.thecollector.com/zheng-he-seven-voyages/.
20. Fareed Zakaria, *The Post-American World*, W. W. Norton and Company, 2008.
21. Peter Frankopan, *The Silk Roads: A New History of the World*. Vintage Books, New York, 2017, p.144-145.
22. Ibid, p. 94.
23. Howard W. French, *Born in Blackness – Africa, Africans and the making of the Modern World, 1471 to the Second World War*, Liveright Publishing Corporation, UK, 2021, p. 39.
24. https://www.nytimes.com/2003/12/07/books/chapters/over-the-edge-of-the-world.html.
25. http://www.surinenglish.com/lifestyle/202004/17/april-1492columbus-obtains-funding-20200417102938-v.html.
26. http://www.uschesstrust.org/the-emergence-of-two-powerful-queens-queen-isabella-of-spain-and-the-chess-queen/.
27. https://www.salon.com/2011/08/18/1493_charles_mann_excerpt/.
28. http://www.europas-historie.net/toverdenersdyrogplanter1.htm#planter.
29. https://www.salon.com/2011/08/18/1493_charles_mann_excerpt/.
30. Howard W. French, *Born in Blackness – Africa, Africans and the making of the Modern World, 1471 to the Second World War*, Liveright Publishing Corporation, UK, 2021, p. 175.
31. Howard W. French, *Born in Blackness – Africa, Africans and the making of the Modern World, 1471 to the Second World War*, Liveright Publishing Corporation, UK, 2021, p. 159.
32. https://thewire.in/history/transatlantic-slave-trade-was-not-entirely-triangular.
33. https://www.britannica.com/summary/Transatlantic-Slave-Trade-Key-Facts.

NOTES

34 Howard W. French, *Born in Blackness*, p. 323.
35 Ibid, p. 321.
36 https://www.history.com/this-day-in-history/zong-slave-ship-trial.
37 Joseph Henrich, *The Weirdest People in the World – How the West Became Psychologically Peculiar and Particularly Prosperous*, Penguin Books, UK, 2021, p. 9.
38 https://www.loc.gov/item/today-in-history/may-04/.
39 https://www.thegreatcoursesdaily.com/the-rise-of-the-east-india-company/.
40 https://www.britannica.com/topic/Western-colonialism/The-French.
41 Data From Ann Arbor, Michigan: Inter-university Consortium for Political and Social Research (ICPS).
42 https://timesofindia.indiatimes.com/india/when-bombay-went-to-east-india-company-for-10-rent/articleshow/63473137.cms.
43 https://www.rbth.com/history/327805-leaders-who-abused-alcohol.
44 Ian Grey, 'Peter the Great in England', in *History Today*, Volume 6, Issue 4. p. 229. April 1956.
45 Hans-Wilhelm Steinfeldt, *Putin*, Cappelen Damm, 2020, p. 146.
46 *The Lost Pirate Kingdom*, Netflix, 2021.
47 *The Lost Pirate Kingdom*, Netflix, 2021.
48 https://forskning.no/historie-arkeologi/piratenes-gallionsfigur/1020000.
49 Terje Tvedt, *Historiens Hjul og Vannets Makt*, Dreyers Forlag, Oslo, 2023, p. 113.
50 George Trevelyan, *George the Third and Charles Fox: The Concluding Part of the American Revolution*, Longmans, Green & Co, New York, 1912, p. 5.
51 https://www.history.com/news/7-things-you-may-not-know-about-the-u-s-navy.
52 https://www.history.navy.mil/content/history/museums/nmusn/explore/photography/american-revolution/continental-navy-ships/continental-schooner-hannah.html.
53 Howard W. French, *Born in Blackness – Africa, Africans, and the Making of the Modern World, 1471 to the Second World War*, Liveright Publishing Corporation, US, 2021, p. 334.

4. URBANISATION, COASTALISATION AND GLOBALISATION

1 https://www.thomasnet.com/insights/iphone-supply-chain/.
2 https://www.wired.com/2016/04/iphones-500000-mile-journey-pocket/.
3 Ibid, p. 49.

THE OCEAN

4 https://porteconomicsmanagement.org/pemp/contents/part7/port-city-relationships/worlds-largest-coastal-cities/.
5 Parag Khanna, *Connectography – Mapping the Global Network Revolution*, Penguin Random House, US, 2016, p. 20.
6 https://www.history.com/.amp/news/9-fascinating-facts-about-the-suez-canal.
7 https://www.mn.uio.no/astro/tjenester/publikum/almanakken/innhold/tema2004.html.
8 https://www.mn.uio.no/astro/tjenester/publikum/almanakken/innhold/tema2004.html.
9 https://www.thoughtco.com/why-we-have-time-zones-1773953.
10 https://historyhouse.co.uk/articles/speed_of_news.html.
11 Kenneth Silverman, *Lightning Man: The Accursed Life of Samuel F.B. Morse*, Da Capo Press, UK, 2004.
12 https://atlantic-cable.com/Books/Whitehouse/DDC/index.htm.
13 https://search.app/dwnNjbedTJifVe4MA.
14 https://www.nrk.no/kultur/xl/_lars-faen_-var-verdens-beste-hvalskytter-1.14585114.
15 https://snl.no/Svend_Foyn.
16 https://www.nature.com/articles/519140a.
17 James Bradley, *The China Mirage: The Hidden History of American Disaster in Asia*, Little, Brown & Company, USA, 2016.
18 https://www.globalsecurity.org/military/world/china/third-front.htm.
19 http://www.virtualmuseum.ca/edu/ViewLoitLo.do?method=preview&lang=EN&id=12941.
20 Sven-Eric Liedman, *Den moderne Verdens Idéhistorie – I Skyggen av Fremtiden*, Dreyers Forlag, Oslo, 2016, p. 453.
21 https://www.tandfonline.com/doi/abs/10.1080/01495930490274490.
22 https://en.wikipedia.org/wiki/History_of_immigration_to_the_United_States.
23 https://www.norgeshistorie.no/industrialisering-og-demokrati/1537-utvandring-fra-norge.html.
24 http://www.nativepartnership.org/site/PageServer?pagename=naa_livingconditions.
25 https://www.ipr.northwestern.edu/news/2020/redbird-what-drives-native-american-poverty.html.
26 https://escholarship.org/content/qt33p0786x/qt33p0786x.pdf.
27 https://tennesseelookout.com/2021/05/17/commentary-tennessee-must-reject-santorum-slights-on-native-americans/.
28 *Aftenposten*, 21 July 2021, pp. 4-6.

NOTES

29 https://www.theguardian.com/world/ng-interactive/2021/sep/06/canada-residential-schools-indigenous-children-cultural-genocide-map.

30 http://c250.columbia.edu/c250_celebrates/remarkable_columbians/alfred_thayer_mahan.html.

31 Lawrence Sondhaus, *Naval Warfare, 1815-1914,* p 161, Routledge, New York, 2001.

32 https://guides.loc.gov/chronicling-america-roosevelt-great-white-fleet.

33 *Dagens Næringsliv,* 24 March 2021.

34 https://www.britannica.com/technology/submarine-naval-vessel/Toward-diesel-electric-power.

35 https://www.iwm.org.uk/history/how-radar-changed-the-second-world-war.

36 https://www.govinfo.gov/content/pkg/GPO-CDOC-106sdoc21/pdf/GPO-CDOC-106sdoc21.pdf.

37 Robert Kagan, *The Ghost at the Feast – America and the collapse of the World Order 1900-1941*, Alfred A. Knopf, New York, 2023, p. 35.

38 John J. Mearsheimer, *The Tragedy of Great Power Politics*, W.W. Norton & Company, New York, 2001, p. 84.

39 Robert Kagan, *The Ghost at the Feast – America and the collapse of the World Order 1900-1941*, Alfred A. Knopf, New York, 2023, p. 120-122.

40 https://www.globalsecurity.org/military/systems/ship/scn-1913-wilson.htm.

41 https://spanishfluvictoriabc.com/spanish-flu-origin-spread-character/how-did-the-patient-zero-story-begin/.

42 https://origins.osu.edu/milestones/pandemic-flu-spanish-flu-1918-H1N1-WW1-vaccine.

43 https://www.nobelprize.org/prizes/peace/1919/wilson/acceptance-speech/.

44 https://history.state.gov/milestones/1921-1936/naval-conference.

45 Tim Marshall, *The Power of Geography*, Elliot & Thompson, 2021.

46 https://www.theatlantic.com/international/archive/2013/12/understanding-syria-from-pre-civil-war-to-post-assad/281989/.

47 https://www.csun.edu/~vcmth00m/iraqkuwait.html.

48 https://www.iranicaonline.org/articles/oil-agreements-in-iran.

49 https://www.cbsnews.com/news/bp-and-iran-the-forgotten-history/.

50 https://www.newworldencyclopedia.org/entry/Anglo-Iranian_Oil_Company.

51 https://www.cbsnews.com/news/bp-and-iran-the-forgotten-history/.

52 https://www.iranicaonline.org/articles/oil-agreements-in-iran.

53 *The Queen and the Coup*, US documentary, NRK2, 26 May 2021.
54 https://history.state.gov/milestones/1921-1936/red-linea.
55 https://history.state.gov/milestones/1921-1936/red-line.
56 https://www.census.gov/history/pdf/pearl-harbor-fact-sheet-1.pdf.
57 https://www.ushistory.org/us/51e.asp.
58 https://www.bowdoin.edu/news/2021/03/on-the-prow-of-liberty.html.
59 https://shippingtandy.com/features/a-salute-to-the-brave-and-valiant-norwegians/.
60 https://kultur.forsvaret.no/forsvarets-musikk/kongelige-norske-marines-musikkorps/nyheter/krigsseilerne.
61 https://arkivet.no/krigsseilerhistorie.
62 https://www.history.navy.mil/browse-by-topic/people/presidents.html.

5. WAVES OF GROWTH AND PROSPERITY

1 https://www.un.org/en/about-us/history-of-the-un/san-francisco-conference#:~:text=Delegates of fifty nations met in.
2 https://www.un.org/en/about-us/un-charter/full-text.
3 https://courses.lumenlearning.com/atd-herkimer-westerncivilization/chapter/the-peace-of-westphalia-and-sovereignty/.
4 https://www.thejc.com/judaism/all/the-search-for-an-alternative-zion-4W9i8fkMeTKTthnfC2U843.
5 https://miff.no/arabiske-land/2009/04/08joedenesfluktfraarabiskeland.htm.
6 https://www.un.org/depts/los/convention_agreements/convention_overview_convention.htm.
7 https://www.brettonwoodsproject.org/2019/01/art-320747/.
8 https://porteconomicsmanagement.org/pemp/contents/part1/maritime-shipping-and-international-trade/.
9 Lincoln Paine, *The Sea and Civilization – A Maritime History of the World*, Atlantic Book, UK, 2014, p. 585.
10 https://css.ethz.ch/en/services/digital-library/articles/article.html/1097618a-96c9-45b2-89f7-092198f84a7c.
11 https://www.snopes.com/fact-check/einstein-world-war-iv-sticks-stones/.
12 https://www.history.navy.mil/browse-by-topic/ships/submarines/triton-ssrn-586.html.
13 https://www.nationalgeographic.com/culture/article/you-and-almost-everyone-you-know-owe-your-life-to-this-man.
14 Kenan Malik, *Not so Black and White – A History of Race from White*

NOTES

Supremacy to Identity Politics, C. Hurst & Co (Publishers) Ltd, UK, 2023, p. 155.

15 http://www.let.rug.nl/usa/outlines/history-1963/america-in-the-modern-world/speedy-end-to-colonialism-sought.php.
16 https://snl.no/dominoteori.
17 https://www.britannica.com/list/8-deadliest-wars-of-the-21st-century.
18 https://timesofmalta.com/articles/view/We-re-here-because-you-were-there.478619.
19 https://datatopics.worldbank.org/world-development-indicators/stories/where-do-the-poor-live.html.
20 https://unohrlls.org/about-ldcs/.
21 https://www.africa.undp.org/content/rba/en/home/library/reports/graduation-of-african-least-developed-countries--ldcs----emergin.html.
22 https://www.statista.com/statistics/1072803/child-mortality-rate-africa-historical/.
23 https://www.un.org/ldcportal/content/rwanda-graduation-status.
24 Walter Rodney, *How Europe Underdeveloped Africa*, Verso, UK, 2018, p. 250-251.
25 Howard W. French, *Born in Blackness*, Liveright Publishing Corporation, USA, 2021, p. 314.
26 https://www.tandfonline.com/doi/abs/10.1080/1464988031000166 0201.
27 https://www.inonafrica.com/2017/08/24/africas-landlocked-countries-perpetual-disadvantage/.
28 Howard W. French, '*Born in Blackness – Africa, Africans and the making of the Modern World, 1471 to the Second World War*', Liveright Publishing Corporation, UK, 2021, p. 314.
29 Ibid.
30 Ibid.
31 https://www.nspirement.com/2022/09/29/where-did-chinas-gold-go.html.
32 https://www.familymoney.co.uk/uk-tax/uk-tax-essentials/history-taxation-united-kingdom/.
33 https://bradfordtaxinstitute.com/Free_Resources/Federal-Income-Tax-Rates.aspx.
34 http://piketty.pse.ens.fr/files/Historical%20evolution%20of%20income%20tax%20(Arthur%20Jatteau).pdf.
35 https://www.oecd.org/ctp/tax-policy/revenue-statistics-ratio-change-all-years.htm.
36 https://www.opec.org/opec_web/en/data_graphs/330.htm.
37 https://energyeducation.ca/encyclopedia/Oil_crisis_of_the_1970s.
38 https://energyeducation.ca/encyclopedia/Oil_crisis_of_the_1970s.

39 https://www.federalreservehistory.org/essays/latin-american-debt-crisis.
40 https://www.reaganlibrary.gov/archives/speech/presidents-news-conference-23.
41 Timothy Snyder, *The Road to Unfreedom*, Tim Duggan Books, USA, 2018, p. 260-261.
42 http://en.euabc.com/word/12.
43 https://www.wto.org/english/thewto_e/whatis_e/inbrief_e/inbr_e.htm.
44 https://www.chinabusinessreview.com/40-years-of-us-china-commercial-relations/.
45 http://www.chinadaily.com.cn/a/201808/02/WS5b728ae4a310add14f385b4a.html.
46 https://www.newyorker.com/news/evan-osnos/to-get-rich-is-glorious.
47 https://www.ait.org.tw/taiwan-relations-act-public-law-96-8-22-u-s-c-3301-et-seq/.
48 *The Economist*, 19 March 2022, p. 62.
49 https://www.undp.org/sites/g/files/zskgke326/files/migration/cn/UNDP-CH-PR.-Publications-Measuring-Poverty-with-Big-Data-in-China-reduced.pdf.
50 https://www.imf.org/en/Publications/fandd/issues/2023/12/China-bumpy-path-Eswar-Prasad.
51 https://statisticstimes.com/economy/united-states-vs-china-economy.php.

6. A NEW GLOBAL TRIANGLE

1 https://www.ecb.europa.eu/press/economic-bulletin/articles/2019/html/ecb.ebart201903_01~e589a502e5.en.html.
2 Ruchir Sharma, *The Rise and Fall of Nations*, Penguin Random House UK, 2019, p. 171.
3 *The Economist*, 26 March 2022, p. 10.
4 Ruchir Sharma, *The Rise and Fall of Nations*, Penguin Random House, UK, 2016, p. 2.
5 https://www.bbc.com/news/business-66840367.
6 https://www.washingtonpost.com/news/wonk/wp/2015/03/24/how-china-used-more-cement-in-3-years-than-the-u-s-did-in-the-entire-20th-century/.
7 https://www.usnews.com/education/best-global-universities/engineering.
8 https://www.economist.com/science-and-technology/2024/06/12/china-has-become-a-scientific-superpower.
9 https://www.wilsoncenter.org/blog-post/china-top-trading-partner-more-120-countries.

NOTES

10. Mike Martin, *The Return of Inequality – Social Change and the Weight of the Past*, Harvard University Press, USA, 2021, p. 139.
11. https://hdr.undp.org/system/files/documents/global-report-document/hdr2023-24snapshoten.pdf.
12. https://www.weforum.org/agenda/2022/07/young-adults-us-live-multigenerational-household/.
13. https://www.statnews.com/2023/11/13/life-expectancy-men-women/.
14. https://www.regjeringen.no/contentassets/64bb4d38fdfb4d3c99e7a35b234f2fa2/no/pdfs/nou202220220012000dddpdfs.pdf, p. 127.
15. https://www.oxfam.org/en/press-releases/wealth-five-richest-men-doubles-2020-five-billion-people-made-poorer-decade-division.
16. https://www.researchgate.net/publication/378764918_Inequality_in_history_A_long-run_view.
17. https://www.oxfam.org/en/press-releases/pandemic-creates-new-billionaire-every-30-hours-now-million-people-could-fall.
18. https://www.economist.com/briefing/2024/02/08/chinas-well-to-do-are-under-assault-from-every-side.
19. https://www.bbc.com/news/business-56671638.
20. Daniel Markovits, *The Meritocracy Trap*, Penguin Random House, UK, 2019, p. 47-49.
21. Speech at Oslo Energy Forum 2017.
22. https://www.ohchr.org/sites/default/files/Documents/Issues/Business/maritime-risks-and-hrdd.pdf.
23. https://ourworldindata.org/covid-vaccinations?country=OWID_WRL.
24. *The Economist*, November 4–10, 2023, p. 11.
25. https://thediplomat.com/2022/03/how-did-asian-countries-vote-on-the-uns-ukraine-resolution/.
26. https://www.theguardian.com/us-news/2018/jan/12/trump-shithole-countries-lost-in-translation.
27. https://www.nrk.no/urix/torke-og-flom-sees-fra-verdensrommet-1.16081988.

7. THE OCEAN STRIKES BACK

1. https://www.oxfam.org/en/5-natural-disasters-beg-climate-action.
2. https://e360.yale.edu/digest/extreme-weather-events-have-increased-significantly-in-the-last-20-years.
3. https://www.weforum.org/reports/new-nature-economy-report-series/future-of-nature-and-business#report-nav.
4. https://science.nasa.gov/climate-change/evidence/.

5   https://klimastiftelsen.no/publikasjoner/hvert-tonn-teller/.
6   https://www.iucn.org/resources/issues-brief/ocean-warming.
7   https://drive.google.com/file/d/1GX1Uf8Fvl3NivwOVy-HT4Gu74lhE-BQ6p/view?pli=1.
8   https://bjerknes.uib.no/artikler/nyheter/varmere-og-vatere-klima-positivt-kraftbransjen.
9   https://www.google.com/search?client=safari&rls=en&q=is+water+vapour+a+greenhouse+gas&ie=UTF-8&oe=UTF-8.
10 https://www.aftenposten.no/meninger/debatt/i/eEaXRQ/vanndampens-rolle-i-klimaendringene-er-godt-forstaatt.
11 https://www.ft.com/content/f9dd342e-09a5-4e8f-9a57-f59ff26133c7.
12 https://www.nature.com/articles/s41586-018-0158-3.
13 https://snl.no/El_Niño.
14 https://earthobservatory.nasa.gov/images/150192/tracking-30-years-of-sea-level-rise.
15 https://nap.nationalacademies.org/read/13389/chapter/5.
16 https://earthobservatory.nasa.gov/images/150192/tracking-30-years-of-sea-level-rise.
17 https://www.nrdc.org/stories/sea-level-rise-101#what-is.
18 https://no.flixable.com/title/melting-point/.
19 Nature report, summer 2020.
20 https://www.eli.org/vibrant-environment-blog/sea-level-rise-and-climate-migration-story-kiribati.
21 https://academic.oup.com/bioscience/advance-article/doi/10.1093/biosci/biab079/6325731.
22 https://www.inkl.com/reading-list/news/greenland-s-ice-sheet-lost-over-a-billion-tons-of-ice-per-day-in-2019.
23 https://www.nsf.gov/geo/opp/antarct/science/icesheet.jsp.
24 https://www.carbonbrief.org/studies-shed-new-light-on-antarcticas-future-contribution-to-sea-level-rise/.
25 https://www.inkl.com/glance/news/antarctic-doomsday-glacier-may-be-melting-faster-than-was-thought?share=LwnVvQsLyxy&section=lead-stories.
26 https://bellona.no/nyheter/klima/2021-07-metanlekkasje-den-glemte-klimatrusselen.
27 https://agu.confex.com/agu/fm21/meetingapp.cgi/Paper/898204.
28 https://bulletinofcas.researchcommons.org/cgi/viewcontent.cgi?article=1040&context=journal.
29 Jan Zalasiewicz & Mark Williams, *Ocean Worlds – The story of seas on Earth and other planets*, Oxford University Press, 2014, p. 187.

NOTES

30 *The Economist*, 04 June 2022, p. 12.
31 https://oceanservice.noaa.gov/facts/coralreef-climate.html.
32 https://www.unep.org/news-and-stories/story/why-are-coral-reefs-dying.
33 https://www.uib.no/matnat/136725/hva-vi-vet-og-ikke-vet-om-global-oppvarming.
34 https://edition.cnn.com/2024/05/04/australia/australia-great-barrier-reef-wildfires-climate-intl-hnk-dst/index.html.
35 https://www.nrk.no/urix/klimakrisen-gir-underskudd-av-gutteskilpadder-1.16058069.
36 https://www.weforum.org/agenda/2019/03/oceans-absorb-karbondioksid-challenges-emerge/.
37 https://public.wmo.int/en/media/press-release/four-key-climate-change-indicators-break-records-2021.
38 https://snl.no/havforsuring.
39 https://energiogklima.no/to-grader/innsikt/slik-gjor-karbondioksid-havet-surere/.
40 https://portals.iucn.org/library/sites/library/files/documents/01 DEOX.pdf.
41 https://oceanservice.noaa.gov/facts/eutrophication.html.
42 https://www.nmbu.no/nyheter/hvorfor-dor-oslofjorden.
43 https://theoceancleanup.com/ocean-plastic-pollution-explained/#how-much-plastic-enters-the-ocean.
44 https://www.unep.org/interactive/beat-plastic-pollution/.
45 https://www.nrk.no/nordland/det-havner-mye-mer-plast-i-havet-enn-vi-har-trodd.-forskere-og-fagfolk-er-bekymret-1.15926888.
46 https://www.wwf.no/dyr-og-natur/hav-og-fiske/plast-i-havet.
47 https://sustainabledevelopment.un.org/content/documents/Ocean_Factsheet_Pollution.pdf.
48 https://www.weforum.org/agenda/2018/11/chart-of-the-day-this-is-how-long-everyday-plastic-items-last-in-the-ocean/.
49 https://www.wwf.no/assets/attachments/PLASTIC-IGESTION-WEB-SPRDS.pdf.
50 https://www.ncbi.nlm.nih.gov/pmc/articles/PMC9269371/.
51 https://www.hi.no/hi/nyheter/2020/januar/de-minste-plastpartiklene-er-mest-skadelige.
52 https://dosits.org/science/sounds-in-the-sea/how-does-shipping-affect-ocean-sound-levels/.
53 https://savedolphins.eii.org/news/ocean-noise-drowning-out-marine-life.
54 https://www.oceancare.org/en/our-work/ocean-conservation/underwater-noise/underwater-noise-consequences/.

55 Ibid.
56 https://www.sdir.no/sjofart/fartoy/miljo/utslipp-fra-skip/utslipp-til-sjo2/ballastvann/.
57 https://www.ledernytt.no/den-nye-verdensutviklingen-vil-bedriften-din-overleve-den.6166053-112537.html.
58 https://www.unep.org/news-and-stories/press-release/unep-marine-sand-watch-reveals-massive-extraction-worlds-oceans.
59 https://mpatlas.org/.
60 https://www.cbd.int/doc/decisions/cop-15/cop-15-dec-04-en.pdf.
61 https://edition.cnn.com/2024/03/27/climate/timekeeping-polar-ice-melt-earth-rotation/index.html.

8. BLUE GROWTH FOR A GREEN FUTURE

1 https://time.com/5669038/women-climate-change-leaders/.
2 https://www.brookings.edu/articles/figure-of-the-week-the-shrinking-lake-chad/.
3 https://www.fao.org/interactive/state-of-food-agriculture/2020/en/.
4 https://www.un.org/sustainabledevelopment/blog/2024/03/un-world-water-development-report/.
5 https://www.wri.org/blog/2019/08/17-countries-home-one-quarter-world-population-face-extremely-high-water-stress.
6 *The Economist*, July 6th 2024, p. 46.
7 https://tappwater.co/blogs/blog/day-zero-6-cities-running-out-of-water.
8 https://turningthetide.watercommission.org/.
9 https://public.wmo.int/en/media/press-release/wake-looming-water-crisis-report-warns.
10 https://www.destatis.de/EN/Themes/Countries-Regions/International-Statistics/Data-Topic/Population-Labour-Social-Issues/DemographyMigration/UrbanPopulation.html.
11 https://edition.cnn.com/2024/04/18/climate/china-sinking-cities/index.html.
12 https://www.voanews.com/east-asia-pacific/indonesia-moving-capital-out-jakarta.
13 https://news.un.org/en/story/2020/11/1078592.
14 https://news.un.org/en/story/2020/11/1078592.
15 https://www.un.org/sustainabledevelopment/blog/2024/03/un-world-water-development-report/.
16 https://unesdoc.unesco.org/in/documentViewer.xhtml?v=2.1.196&id=p::usmarcdef_0000388948&file=/in/rest/annotationSVC/DownloadWatermarkedAttachment/

NOTES

attach_import_3fa8b180-d73f-48ba-a777-c3748a93313a?_=388948eng.pdf&locale=en&multi=true&ark=/ark:/48223/pf0000388948/PDF/388948eng.pdf#WWDR 2024 EN report v01.indd:.538601:5534.

17. https://asiasociety.org/files/pdf/WaterSecurityReport.pdf.
18. https://www.csis.org/analysis/irans-cloudy-accusations.
19. https://www.geopoliticalmonitor.com/cloud-seeding-and-the-water-wars-of-tomorrow/.
20. https://smartwatermagazine.com/q-a/what-percentage-water-desalinated.
21. https://idadesal.org/wp-content/uploads/2019/04/The-state-of-desalination-2019.pdf.
22. https://idadesal.org/wp-content/uploads/2019/04/The-state-of-desalination-2019.pdf.
23. https://blogs.ei.columbia.edu/2019/06/20/undersea-freshwater-aquifer-northeast/.
24. https://op.europa.eu/en/publication-detail/-/publication/0e91f9db-f4f2-11e7-be11-01aa75ed71a1.
25. https://news.un.org/en/story/2020/11/1078592.
26. https://news.un.org/en/story/2020/11/1078592.
27. https://www.theguardian.com/news/datablog/2013/jan/10/how-much-water-food-production-waste.
28. https://www.fao.org/3/cb6033en/cb6033en.pdf.
29. https://www.ipcc.ch/report/ar6/wg3/chapter/chapter-7/.
30. Aftenposten Innsikt, 08/2024, p. 60.
31. https://podcasts.apple.com/no/podcast/to-a-lesser-degree-from-the-economist/id1586004652?i=1000537507126.
32. https://www.thelancet.com/journals/lancet/article/PIIS0140-6736(23)02750-2/fulltext.
33. https://www.who.int/news-room/fact-sheets/detail/obesity-and-overweight.
34. https://www.bcg.com/publications/2018/tackling-1.6-billion-ton-food-loss-and-waste-crisis.
35. https://openknowledge.fao.org/server/api/core/bitstreams/7493258e-e420-4840-a95d-cfec8833219d/content, page vii.
36. https://sustainablefisheries-uw.org/how-many-fisheries-are-overfished/.
37. https://oceanpanel.org/wp-content/uploads/2022/05/Illegal-Unreported-and-Unregulated-Fishing-and-Associated-Drivers.pdf.
38. https://www.fao.org/iuu-fishing/en/.
39. https://foreignpolicy.com/2023/09/19/global-food-crisis-fishing-blue-foods-conflict-water-ocean-climate-resources/.

40 https://openknowledge.fao.org/server/api/core/bitstreams/9df19f53-b931-4d04-acd3-58a71c6b1a5b/content/sofia/2022/world-fisheries-aquaculture-production.html.
41 https://www.seafish.org/insight-and-research/aquaculture-data-and-insight/aquaculture-production-scales/.
42 https://www.hi.no/hi/nyheter/2023/februar/fortsatt-hoy-dodelighet-hos-oppdrettslaks.
43 https://ig.ft.com/supermarket-salmon/?segmentid=dcee0941-6e02-a9de-5643-b340f3ef2e3a.
44 https://news.mongabay.com/2024/02/norwegian-salmon-farms-gobble-up-fish-that-could-feed-millions-in-africa-report/.
45 https://ig.ft.com/supermarket-salmon/?segmentid=dcee0941-6e02-a9de-5643-b340f3ef2e3a.
46 https://seabos.org/west-coast-africa-keystone-project/.
47 https://www.barentswatch.no/artikler/alger/.
48 https://www.barentswatch.no/artikler/alger/.
49 https://www.iea.org/energy-system/renewables/hydroelectricity.
50 https://tethys.pnnl.gov/sites/default/files/publications/Thennakoon_et_al_2023.pdf.
51 Eugene Linden, *Fire and Flood – A People's History of Climate Change from 1979 to the Present*, Allen Lane, UK, 2022, p. xxvi
52 https://ember-climate.org/insights/research/global-electricity-review-2024/.
53 https://www.statista.com/statistics/476327/global-capacity-of-offshore-wind-energy/.
54 IEA report 2020.
55 https://regenpower.com/articles/can-we-power-a-big-city-with-solar-24-7/.
56 https://www.weforum.org/agenda/2022/08/heatwaves-can-hamper-solar-panels/.
57 https://news.liverpool.ac.uk/2014/06/25/watch-tofu-ingredient-could-revolutionise-solar-panel-manufacture/.
58 https://scitechdaily.com/game-changing-discovery-chinese-scientists-have-discovered-a-secret-hidden-structure-in-perovskite-solar-cells/.
59 https://www.greenbiz.com/article/future-freight-more-shipping-less-emissions.
60 https://projects.research-and-innovation.ec.europa.eu/en/horizon-magazine/emissions-free-sailing-full-steam-ahead-ocean-going-shipping.
61 https://unctad.org/news/ocean-economy-offers-25-trillion-export-opportunity-unctad-report.

NOTES

62 https://cruiseindustrynews.com/cruise-news/2022/05/cruise-industry-news-2022-market-report-fastest-growing-market-segments/.
63 https://www.cruise.no/royal-caribbean/icon-of-the-seas/.
64 https://www.dn.no/reiseliv/cruiseskip/cruise/co2/cruiseutslipp-tredoblet-pa-ti-ar-dette-er-flytende-klimagassfabrikker/2-1-1620118.
65 https://www.wcel.org/sites/default/files/publications/regulating_the_west_coast_cruise_industry_final.pdf.
66 https://www.americangeosciences.org/critical-issues/critical-mineral-basics.
67 https://www.iea.org/reports/the-role-of-critical-minerals-in-clean-energy-transitions/executive-summary.
68 https://crsreports.congress.gov/product/pdf/R/R47227.
69 https://www.iea.org/reports/the-role-of-critical-minerals-in-clean-energy-transitions/executive-summary.
70 The World Bank, 'Climate-Smart Mining: Minerals for Climate Action. Infographic', 26 February 2019, www.worldbank.org.
71 https://www.nobelprize.org/prizes/peace/2018/mukwege/lecture/.
72 https://iea.blob.core.windows.net/assets/ffd2a83b-8c30-4e9d-980a-52b6d9a86fdc/TheRoleofCriticalMineralsinCleanEnergyTransitions.pdf.
73 World Bank, *Implementing Precaution in Benefit-Cost Analysis – The Case of Deep Seabed Mining*, Policy Research Working Paper 9307, June 2020.
74 World Bank, 'Climate-Smart Mining: Minerals for Climate Action. Infographic', 26 February 2019, www.worldbank.org.
75 UN Global Compact, *Global Goals, Ocean Opportunities*, June 2019, p. 34.
76 https://www.science.org/content/article/mountains-hidden-deep-sea-are-biological-hot-spots-will-mining-ruin-them.
77 https://www.gard.no/web/articles?documentId=35617606.

9. THE NEW BATTLE OVER THE OCEAN

1 https://www.navysite.de/ships/lha4.htm.
2 Robert Kaplan, *Asia's Cauldron*, Random House, 2014.
3 https://www.un.org/en/desa/exploring-potential-blue-economy.
4 https://www.foreignaffairs.com/oceans-seas/fish-wars.
5 https://www.iea.org/reports/global-critical-minerals-outlook-2024/executive-summary.
6 https://www.state.gov/announcement-of-u-s-extended-continental-shelf-outer-limits/.

7 https://www.ft.com/content/0e550c72-c6b1-42ac-876b-22aa49c2057b?shareType=nongift.
8 https://www.researchgate.net/publication/250171761_The_Putin_Thesis_and_Russian_Energy_Policy.
9 https://crsreports.congress.gov/product/pdf/R/R46808, p. 24.
10 https://brill.com/display/book/edcoll/9789004447899/BP000019.xml.
11 https://www.thearcticinstitute.org/legal-implications-2022-canada-denmark-greenland-agreement-hans-island-tartupaluk-inuit-peoples-greenland-nunavut/.
12 https://www.un.org/depts/los/convention_agreements/texts/unclos/unclos_e.pdf.
13 https://pcacases.com/web/sendAttach/2086.
14 https://www.brookings.edu/articles/rethinking-us-china-competition-next-generation-perspectives/.
15 https://www.sipri.org/sites/default/files/2024-04/2404_fs_milex_2023.pdf.
16 https://www.linkedin.com/pulse/which-country-most-economically-self-sufficient-ramya-emandi.
17 https://sgp.fas.org/crs/misc/RS21729.pdf.
18 https://www.visualcapitalist.com/visualized-aircraft-carriers-by-country/.
19 Peter Frankopan, *The New Silk Roads – The Present and Future of the World*, Bloomsbury UK, 2018, p. 148.
20 https://www.cna.org/archive/CNA_Files/pdf/china-far-seas-navy.pdf.
21 https://crsreports.congress.gov/product/pdf/R/R46808, p. 24.
22 https://www.iiss.org/blogs/military-balance/2018/05/china-naval-shipbuilding.
23 https://www.statista.com/statistics/1256561/global-shipbuilding-capacity-by-country/.
24 https://www.carbonbrief.org/qa-how-is-climate-change-affecting-chinas-cropland/.
25 https://dialogue.earth/en/climate/10583-china-is-heading-towards-a-water-crisis-will-government-changes-help/.
26 https://data.worldbank.org/indicator/TG.VAL.TOTL.GD.ZS?locations=CN.
27 https://www.ship-technology.com/features/the-top-10-busiest-container-ports-in-the-world/?cf-view.
28 https://nationalinterest.org/blog/reboot/great-wall-reverse-take-down-china-look-pacifics-many-islands-166628.
29 https://www.realcleardefense.com/articles/2019/02/26/china_sets_a_course_for_the_uss_pacific_domain_114216.html.

NOTES

30. The Economist, 25 May 2024, p. 49.
31. https://www.forbes.com/sites/miltonezrati/2024/06/26/taiwan-rapidly-moving-away-from-china/.
32. https://www.europarl.europa.eu/meetdocs/2004_2009/documents/fd/d-cn20050426o1/d-cn20050426o1en.pdf.
33. https://time.com/6105182/us-troops-taiwan-china/.
34. https://www.taipeitimes.com/News/editorials/archives/2016/11/14/2003659209.
35. https://foreignpolicy.com/2021/11/17/biden-taiwan-china-misspoke-policy-mistake/.
36. https://thediplomat.com/2016/07/china-and-unclos-an-inconvenient-history/.
37. https://www.globaltimes.cn/content/1185568.shtml.
38. https://law.upd.edu.ph/wp-content/uploads/2020/12/PYIL-Vol.17-v0-29Dec20-The-South-China-Sea-Arbitration.pdf.
39. https://storymaps.arcgis.com/stories/6d8a684e762f4f4ca90c1a19121fcf8c.
40. https://www.reuters.com/article/world/xi-tells-mattis-china-wont-give-up-even-one-inch-of-territory-idUSKBN1JN06N/.
41. https://www.armed-services.senate.gov/imo/media/doc/Davidson_APQs_04-17-18.pdf.
42. Foreign Affairs, Volume 103, Number 5, September/October 2024, p. 35.
43. https://www.theguardian.com/world/2016/may/16/little-blue-men-the-maritime-militias-pushing-chinas-claims-in-south-china-sea.
44. https://www.business-humanrights.org/en/latest-news/cambodiachina-comprehensive-investment-and-development-pilot-zone-dara-sakor-seashore-resort/.
45. https://www.scmp.com/news/china/article/3101694/us-sanctions-chinese-company-over-dara-sakor-project-cambodia?fbclid=IwAR2X286W0aW024Kls-QfFkW-e3ZJUUDzl1mRSM-mvIfVqd1NRRsvSAwL54PE.
46. https://warsawinstitute.org/china-malacca-dilemma/.
47. https://www.theatlantic.com/international/archive/2017/08/strait-of-malacca-uss-john-mccain/537471/.
48. https://www.e-ir.info/2012/09/07/the-importance-of-the-straits-of-malacca/.
49. https://www.un.org/development/desa/dpad/publication/un-desa-policy-brief-no-153-india-overtakes-china-as-the-worlds-most-populous-country/.

50 https://diplomatist.com/2020/07/07/the-malacca-dilemma-and-chinese-ambitions-two-sides-of-a-coin/.
51 https://www.amnesty.org/en/latest/news/2021/06/china-draconian-repression-of-muslims-in-xinjiang-amounts-to-crimes-against-humanity/.
52 https://news.mongabay.com/2010/01/yurts-in-western-china/.
53 http://www.scio.gov.cn/zfbps/zfbps_2279/202310/t20231010_773734.html.
54 https://eng.yidaiyilu.gov.cn/p/0T2AURQE.html.
55 https://www.oecd.org/finance/Chinas-Belt-and-Road-Initiative-in-the-global-trade-investment-and-finance-landscape.pdf.
56 https://www.beltroad-initiative.com/belt-and-road/.
57 https://www.globaltimes.cn/page/202307/1295320.shtml.
58 https://www.foreignaffairs.com/china/belt-road-initiative-xi-imf.
59 Jonathan E. Hillman, *The Emperor's New Road*, Yale University Press, USA, 2020.
60 https://sdgs.un.org/partnerships/global-development-initiative-building-2030-sdgs-stronger-greener-and-healthier-global.
61 https://time.com/6319264/china-belt-and-road-ten-years/.
62 https://thekootneeti.in/2020/09/02/sri-lankas-geopolitics-in-the-indian-ocean/.
63 https://www.indiatimes.com/news/india/here-is-all-you-should-know-about-string-of-pearls-china-s-policy-to-encircle-india-324315.html.
64 https://edition.cnn.com/2019/05/20/china/china-kenya-sgr-rail-africa-intl/index.html.
65 https://www.chathamhouse.org/2023/05/kenyas-debt-struggles-go-far-deeper-chinese-loans.
66 https://www.trtworld.com/magazine/why-is-the-us-concerned-about-china-s-naval-base-in-djibouti-46099.
67 https://theintercept.com/2020/02/27/africa-us-military-bases-africom/.
68 http://www.xinhuanet.com/english/2020-12/12/c_139584669.htm.
69 https://www.brainyquote.com/lists/authors/top-10-ehud-barak-quotes.
70 https://www.worldometers.info/oil/oil-production-by-country/.
71 https://iea.blob.core.windows.net/assets/203eb8eb-2147-4c99-af07-2d3804b8db3f/StraitofHormuzFactsheet.pdf.
72 https://www.businessinsider.com/boat-stuck-suez-canal-costing-estimated-400-million-per-hour-2021-3?r=US&IR=T.
73 https://www.aftenposten.no/verden/i/zgP8P4/grunnstoett-skip-skaper-knipe-for-verdenshandelen-slik-skal-de-faa-det.
74 https://climate-diplomacy.org/case-studies/security-implications-growing-water-scarcity-egypt.

NOTES

75 https://www.bbc.co.uk/news/world-africa-22850124.
76 https://www.dn.no/innlegg/internasjonal-politikk/fns-sikkerhetsrad/vannkraft/innlegg-en-eksistensiell-trussel-eller-en-eksistensiell-nodvendighet/2-1-1042522.
77 https://www.intellinews.com/comment-why-has-china-s-foreign-minister-spent-three-days-in-damascus-probably-not-having-trade-discussions-216436/.
78 https://thecradle.co/Article/investigations/2350.
79 https://globalvoices.org/2021/09/24/turkeys-uyghur-dilemma-in-the-context-of-chinas-belt-and-road-initiative/.
80 https://eurasianet.org/turkeys-crazy-canal-would-impact-eurasian-trade-geopolitics.
81 https://bianet.org/english/politics/241874-retired-admirals-midnight-declaration-on-montreux-convention-triggers-coup-debate.
82 https://www.globaltimes.cn/page/202109/1233806.shtml.
83 https://carnegieendowment.org/russia-eurasia/politika/2024/05/china-russia-yuan?lang=en&center=russia-eurasia.
84 https://www.uscc.gov/sites/default/files/Research/China-and-the-Arctic_Apr2012.pdf.
85 https://www.defensenews.com/opinion/commentary/2020/05/11/chinas-strategic-interest-in-the-arctic-goes-beyond-economics/.
86 https://www.eia.gov/todayinenergy/detail.php?id=4650.
87 https://www.egmontinstitute.be/the-new-us-arctic-strategy-welcome-back-america/.
88 https://ourworldindata.org/energy/country/russia.
89 Timothy Snyder, *The Road to Unfreedom*, Tim Duggan Books, US, 2018, p. 249.
90 https://tass.com/economy/1051080.
91 https://www.abc.net.au/news/2007-08-03/russian-sub-plants-flag-under-north-pole/2520812.
92 https://www.reuters.com/article/idINIndia-28784420070802.
93 https://www.navyrecognition.com/index.php/naval-news/naval-news-archive/2022/october/12292-russian-nuclear-powered-submarine-belgorod-reportedly-deployed.html.
94 https://www.navalnews.com/naval-news/2022/10/new-images-reveal-russias-missing-submarine-belgorod-in-arctic/.
95 https://arcticreview.no/index.php/arctic/article/view/5197/8235.
96 Admiral Haakon Brun-Hansen, NRK Dagsrevyen, 14 August 2019.
97 https://committees.parliament.uk/publications/8205/documents/85026/default/.

98 https://www.nrk.no/vestland/advarer-norge-mot-styrtrik-kineser-1.11705916.
99 https://www.scmp.com/news/china-insider/article/1500823/china-gets-chance-buy-first-foothold-arctic.
100 *The Economist*, July 6th 2024, p. 39.
101 https://dialogue.earth/en/business/382806-perus-megaport-reshapes-pacific-trade-and-the-town-next-door/.
102 https://www.vox.com/world/2018/11/1/18052338/bolton-cuba-venezuela-nicaragua-speech-troika-tyranny.
103 https://www.vox.com/2019/3/27/18283807/venezuela-russia-troops-trump-maduro-guaido.
104 https://www.economist.com/the-americas/2024/01/09/the-dwindling-of-the-panama-canal-boosts-rival-trade-routes.
105 https://energyindustryreview.com/analysis/submarine-cables-risks-and-security-threats/.
106 https://www.datacenterdynamics.com/en/news/worlds-longest-subsea-cable-lands-in-djibouti-east-africa/.
107 https://www.datacenterdynamics.com/en/analysis/what-is-a-submarine-cable-subsea-fiber-explained/.
108 https://www.theguardian.com/uk/2013/jun/21/gchq-cables-secret-world-communications-nsa.
109 https://www.economist.com/international/2024/07/11/how-china-and-russia-could-hobble-the-internet.
110 https://www.hybridcoe.fi/wp-content/uploads/2023/03/NEW_web_Hybrid_CoE_Paper-16_rgb.pdf.
111 https://www.bbc.com/news/world-europe-24539417.
112 https://lloydslist.maritimeintelligence.informa.com/LL1137457/One-.ship-is-hacked-every-day-on-average.
113 https://www.vox.com/world-politics/24113745/russia-shadow-fleet-oil-tankers-environmental-risk.
114 https://www.vox.com/world-politics/24113745/russia-shadow-fleet-oil-tankers-environmental-risk.
115 https://energyandcleanair.org/the-rise-of-shadow-lng-vessels-a-new-chapter-in-russias-sanctions-evasion-strategy/.
116 https://www.bbc.com/news/world-asia-61774473.
117 https://sgp.fas.org/crs/natsec/IF11275.pdf.
118 https://www.defense.gov/News/News-Stories/Article/Article/2703096/dod-navy-confront-climate-change-challenges-in-southern-virginia/.

10. HOPE ON THE HORIZON?

1. https://press.un.org/en/2022/sgsm21352.doc.htm.
2. Rutger Bregman, *Utopia for Realists*, Bloomsbury Paperbacks, 2018, p. 4.
3. https://www.brookings.edu/research/the-unprecedented-expansion-of-the-global-middle-class-2/.
4. *The Economist*, September 21st 2024.
5. Hans Rosling, *Factfulness, Ten Reasons Why We're Wrong About the World – And Why Things are Better Than You Think*, Sceptre, UK, 2018.
6. https://ourworldindata.org/less-democratic.
7. https://www.statista.com/statistics/912012/number-of-trade-policy-interventions-worldwide-harmful-or-liberalizing/#statisticContainer.
8. https://www.theworldcounts.com/challenges/planet-earth/state-of-the-planet/overuse-of-resources-on-earth.
9. https://www.oecd.org/newsroom/growth-continuing-at-a-modest-pace-through-2025-inflation-declining-to-central-bank-targets.htm.
10. https://www.un.org/en/global-issues/population.
11. https://wir2022.wid.world/chapter-1/.
12. https://www.theguardian.com/global-development/2023/nov/16/rich-countries-hit-climate-finance-goal-two-years-late-data.
13. https://unctad.org/meeting/debate-trail-capital-flight-africa..
14. https://phys.org/news/2021-10-climate-scientists.html.
15. https://theconversation.com/carbon-budget-for-1-5-c-will-run-out-in-six-years-at-current-emissions-levels-new-research-216459.
16. https://www.bbc.com/news/world-58549373.
17. Peter Katzenstein, 'Small States in World Markets: Industrial Policy in Europe', Cornell University Press, USA, 1985.
18. https://unglobalcompact.org/take-action/think-labs/just-transition/about.
19. https://www.ics-shipping.org/press-release/new-proposal-urges-governments-to-bite-the-bullet-to-meet-net-zero-2050-targets/?_gl=1*eo5mq6*_up*MQ..*_ga*MTkyNDk3NjcoMC4xNzI2MTEo NjI1*_ga_KEQ52XLWBK*MTcyNjExNDYyNS4xLjAuMTcyNjEx NDYyNS4wLjAuNjM5NjY4NTUy.
20. https://www.dn.no/kommentar/klimaendringer/tyskland/robert-habeck/ubehagelige-gronne-nodvendigheter/2-1-1144653.
21. https://foreignpolicy.com/2022/01/07/climate-change-democracy/.
22. https://www.presidency.ucsb.edu/documents/campaign-address-

progressive-government-the-commonwealth-club-san-francisco-california.
23. https://www.nber.org/system/files/working_papers/w32450/w32450.pdf?utm_campaign=PANTHEON_STRIPPED&utm_medium=PANTHEON_STRIPPED&utm_source=PANTHEON_STRIPPED.
24. https://www.dropbox.com/scl/fi/lz4n0073sie1wzj3tv6a1/CRSUS100105_proof.pdf?_hsenc=p2ANqtz-9DYTMlc8Pqrvr-0ToRzITU2DfJrQ6sDndXSb6CgqUjcXM8jX742z607nCKtUk-M9R70g-__0t1etG-7ZdnMq7K4ybH4KQKVOS0WM0m1xNvb-fMH53b-s&_hsmi=308505593&rlkey=5heavxiaxtpydxpiyrg8l3071&e=3&st=076i38wi&utm_campaign=Press+Package&utm_content=308505593&utm_medium=email&utm_source=hs_email&dl=0.
25. https://infinitummovement.no/panteordningen-sprer-seg-til-andre-land/.
26. https://www.maersk.com/news/articles/2023/11/22/maersk-signs-landmark-green-methanol-offtake-agreement.
27. https://taxfoundation.org/data/all/eu/carbon-taxes-europe-2024/.
28. https://www.activesustainability.com/climate-change/100-companies-responsible-71-ghg-emissions/.
29. https://www.energymonitor.ai/industry/exclusive-how-just-25-oil-companies-are-set-to-blow-the-worlds-1-5c-carbon-budget/?cf-view.
30. https://www.ciel.org/wp-content/uploads/2023/06/CIEL_brief_Deep-Trouble-The-Risks-of-Offshore-Carbon-Capture-and-Storage_June2023.pdf.
31. https://www.unep.org/resources/report/global-climate-litigation-report-2023-status-review.
32. https://www.nrk.no/urix/eldre-kvinner-frykter-de-vil-do-i-hetebolger-_-fikk-historisk-seier-i-retten-1.16836216.
33. https://www.desmog.com/2023/09/21/why-california-is-taking-big-oil-to-court-and-why-this-fossil-fuel-lawsuit-matters/.
34. https://bulloakcapital.com/blog/if-california-were-a-country/.
35. https://www.aljazeera.com/news/2024/5/22/island-states-win-historic-climate-case-in-world-oceans-court#.
36. https://www.unesco.org/en/decades/ocean-decade.
37. https://treaties.un.org/pages/ViewDetails.aspx?src=TREATY&mtdsg_no=XXI-10&chapter=21&clang=_en.
38. https://www.securitycouncilreport.org/un-security-council-working-methods/the-veto.php.

NOTES

39 https://www.visualcapitalist.com/visualizing-the-brics-expansion-in-4-charts/.
40 https://www.oecd.org/tax/g20-economies-are-pricing-more-carbon-emissions-but-stronger-globally-more-coherent-policy-action-is-needed-to-meet-climate-goals-says-oecd.htm.
41 https://www.voanews.com/a/us-and-china-renew-dialogue-on-climate-ahead-of-cop28/7358464.html